Structural Principles

Structural Principles

I. Engel
School of Architecture
Washington University

PRENTICE HALL, Englewood Cliffs, New Jersey 07632

Library of Congress Cataloging in Publication Data

Engel, I. (Irving), (date)
  Structural principles.

  Includes bibliographical references and index.
  1. Structures, Theory of. I. Title.
TA645.E54 1984   624.1′7   83-26929
ISBN 0-13-854019-5

Editorial/production supervision:
  Barbara H. Palumbo
Cover design: Lundgren Graphics, Ltd.
Manufacturing buyer: Anthony Caruso

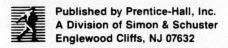
Published by Prentice-Hall, Inc.
A Division of Simon & Schuster
Englewood Cliffs, NJ 07632

Printed in the United States of America

10 9 8 7 6 5

ISBN 0-13-854019-5

Prentice-Hall International (UK) Limited, *London*
Prentice-Hall of Australia Pty. Limited, *Sydney*
Prentice-Hall Canada Inc., *Toronto*
Prentice-Hall Hispanoamericana, S.A., *Mexico*
Prentice-Hall of India Private Limited, *New Delhi*
Prentice-Hall of Japan, Inc., *Tokyo*
Simon & Schuster Asia Pte. Ltd., *Singapore*
Editora Prentice-Hall do Brasil, Ltda., *Rio de Janeiro*

To my wife, *Margaret*

Contents

Contents

Foreword

This foreword will span backward into my past, as well as forward into the textbook. Teaching structural principles to students in architecture involves the basic problems of selecting material (theory and application) and of learning how to teach it (method and an "attitude," as the author once put it). This book speaks for itself in both respects from the viewpoint so well expressed in the Preface. There remains for me to trace its roots, which presumption is perhaps excused in the Acknowledgments.

In 1909 Cornell's College of Architecture initiated an arrangement for the construction aspect of its curriculum, which evolved into a tradition. The teachers were faculty members whose background combined an architectural training with engineering courses in structures and related professional experience. It was a chain of three principal links, with myself in the middle, reaching backward to 1909 through George Young, Jr., and forward to the present through Irving Engel. I still relish Professor Young's anecdote about his mentor, Professor Church in Civil Engineering, who wrote a definitive textbook on the strength of materials: "Only Professor Church and his Maker really understood the first edition, and in the second edition it was Church alone." George Young was a major influence in the education of several generations of architects at Cornell. I was among them, nourished by his roots.

My own roots sprouted under Urquhart and O'Rourke, during basic and graduate study in Civil Engineering at Cornell, and were fed by valuable structural design experience with Elwyn E. Seelye. As a teacher, my inheritance from George Young remained consistent in "attitude," with the influence of Hardy Cross ideas clearly evident, and I see it reflected in the "third link," Professor Engel. This book and its author have a meaning for me akin to the sense of pride and pleasure one feels toward a son or grandchild.

Ludlow D. Brown
Orleans, Massachusetts

Preface

This book covers the informational aspects of the beginning studies in structures, and the principles that underlie their design, for students of architecture and for anyone else who is interested in acquiring *fundamental* knowledge of the behavior of building structures. The text does not deal with engineering structures such as bridges, dams, etc. Although basic courses in physics and calculus might be helpful, it is assumed that the student does not have this background. Consequently, unlike engineering textbooks covering similar material, calculus is not used.

Although it is difficult to enumerate the many tasks with which the architect is faced, and the variety of complex processes involved in the design of a building, one thing is certain: A physical product must eventually be created. There are many factors involved in the development of the product; consideration of the structure is, of course, a significant factor. To consider properly this aspect of architectural design, it is my opinion that the architect must understand the *principles* of structures in both mathematical and conceptual terms. Much of the decision-making process involved in the determination of the structural configuration must be based on judgment. It is hoped that this book will provide a background of experience that will nourish and color the judgment of the architectural designer.

PLAN OF THE BOOK

As previously mentioned, the book is designed for the student who has little or no background in physics or calculus. For those who have had courses in physics, the principles presented in Chapters 2 through 4 should be familiar and will be useful for review purposes.

Most of the material in this book has been classroom tested for several years, and the sequence in which the material is presented has been carefully arranged based on this experience. Each chapter requires knowledge and information from previous chapters. It is recommended that the reader do as many of the problems presented at the end of each chapter so that the principles are understood before going on to the next chapter. The Appendix contains data that will be useful in the problem-solving exercises, and reference is made in the text to the appropriate data needed to solve a given problem. Where information is given in the Appendix about properties of steel and aluminum (stresses versus strain diagrams), such information has been generalized, and the designations of various grades of these materials have been simplified for the purpose of this book, which addresses itself to students who are just beginning their studies in structures.

Roughly the first half of the book is devoted to the principles, derivation of formulas, and techniques necessary for the processes of analysis and design, which take place in the latter part of the book. Virtually no space has been given to analysis and design of structural members using empirical methods or requirements suggested by building codes or industry specifications. Whereas consideration of such requirements is vitally important in the long run, it is my opinion that they are more effectively addressed in studies that come after the principles level. Such studies, in fact, can be fruitful only after the student has developed a great deal of understanding in the principles and terminology of structural analysis and design. Consequently, many of the techniques used in the analysis and design processes presented in this book are necessarily, and as they should be, rudimentary in nature.

Some areas of the book may be suitable for self-study, but the text was not primarily intended for use in this manner. In my years of teaching this subject, I have developed the opinion that many of the principles presented must be read by the student and elaborated on by an instructor. Perhaps it is best to say that this book is intended as a supplement to classroom activities rather than vice versa. Therefore, the manner in which this book is used is best left to the individual style and character of the teacher.

I. Engel
St. Louis, Mo.

Acknowledgments

Much of the technique used for the presentation of the information in this book was influenced during my two years of graduate study at Cornell University. Specifically, I was greatly influenced by one of the finest teachers I have had the privilege of knowing during my years as a student of architecture and my years as an educator in the field of architecture. I am speaking of Professor Ludlow D. Brown (now Professor Emeritus) of Cornell University. While I was a graduate student at Cornell University, I was most fortunate to have been Professor Brown's graduate assistant. When I came to Cornell University to further my education in the field of Architectural Science, I was already a licensed architect with some years of experience in architecture and structural engineering. In terms of the principles of structures (with which this book deals), there wasn't much for me to learn. However, I was interested in the field of architectural education and, when it came to the teaching of principles, I did, indeed, have a great deal to learn. Professor Brown, with his intellect, wit, wisdom, and guidance, helped me to learn how to teach. I will forever be grateful for his guidance and encouragement. If I can, during my tenure as an educator in the field of architecture, be as good a teacher as he, then architecture and our society at large will be properly served.

In addition, this book could not have been completed without the one person who was my typist, editor, and most severe critic. I am speaking of my wife, Margaret, without whose perseverance this document would simply not exist.

These are the people I wish to acknowledge. Apart from them, this book is totally mine. Any errors or lack of clarity are mine. Any negative comments must come to me. Any sections that may bring delight to the reader may be to my credit or they may be due to the influence of Professor Brown.

1

Introduction

OPENING COMMENTS

The technology that must be considered in the design of a building of any type can be extraordinarily complex, especially by contemporary standards. In essence, a building may be thought of as a machine consisting of a variety of parts, all of which must be properly designed if they are to fulfill their intended function. The structure of a building is, in a collective sense, one of those parts. It may be thought of as one of the more critical parts since, for all practical purposes, the structure of a building is not replaceable (i.e., like an air-conditioning unit), and normally it is not easily fixed if something was miscalculated or poor judgment was used in the design process. A good structural design is based largely on a sensible and feasible configurational arrangement of elements, and this is generally inseparable from the architectual design process. In addition to the configurational aspects of a building design, the proper choice of structural materials, an engineering analysis to confirm that the choice of configuration and materials is sensible, and an engineering design of the elements that form the configuration are also primary ingredients in a technically well-designed building.* How these issues are addressed will vary according to the building type, the geographical location, and the forces to which the building structure will be subjected.

The choice of materials and the manner in which structural elements are arranged constitute, fundamentally, what is generally referred to as the "structural system." Whatever the structural system of a building may be, the arrangement of parts should be orderly, sensible, and feasible. While the word *system* is used occasionally throughout this book, we will be concentrating on learning about the behavior of various elements that will make up the final system of the structure, and on a variety of issues that should be

---

*All of these issues may be satisfied perfectly, but this does not guarantee that a building will be aesthetically or functionally successful.

TENSION

FIGURE 1–1

COMPRESSION

FIGURE 1–2

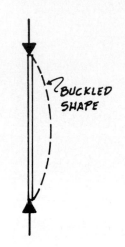

BUCKLED SHAPE

FIGURE 1–3

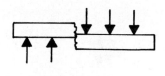

VERTICAL SHEAR

FIGURE 1–4

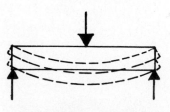

HORIZONTAL SHEAR

FIGURE 1–5

considered in the development of the structural configuration in a building design.

You are now embarking on studies in a physical science. You will be led very carefully through a maze of information, techniques, and issues to consider toward the goal of a sensible and competent structural design for a building. Sometimes you may have to find your way back through the maze to pick up something that was missed on your first trip through.

In order to start out in the proper direction, it is necessary at the outset to establish a basic vocabulary and to define terms that will be used throughout the text. The first chapter is, therefore, largely devoted to that end. It is not necessary for the student to attempt to memorize the various terms and definitions that will be presented because they are used so frequently in the text that they will, in a very short time, automatically become part of everyday structural language.

STATES OF STRESS

A *stress,* for our purposes, may be defined as a force exerted between portions of a body. Structural members are subject to a variety of stresses due to the man-made and natural forces they must resist. The primary stresses we will be concerned with are discussed below.

TENSION: This is a state of stress whereby adjacent particles of material tend to pull away from each other. Tension may be created by application of external forces, as shown in Fig. 1–1. This state of stress produces a "stretch" in the material.

COMPRESSION: This is simply the opposite of tension; that is, adjacent particles tend to push on each other. Compression may be produced by the application of external forces; as shown in Fig. 1–2. Compression produces a shortening of the member.

One of the more potentially serious consequences of compression, aside from crushing of the material, is buckling. This can occur when a long, slender member is subjected to a compressive force, as shown in Fig. 1–3.

SHEAR: This state of stress is the tendency for adjacent parts of a member to "slide" away from each other. There are two kinds of shear that may be present in a structural member: vertical shear and horizontal shear. Vertical shear is the tendency for sliding to occur along vertical planes, as shown in Fig. 1–4. Horizontal shear is the tendency for sliding to occur along horizontal planes. Figure 1–5 depicts a typical condition where a horizontal shear stress exists. This figure shows a structural member spanning between two supports and being subjected to a load, which causes the member to bend, as shown (greatly exaggerated) by the dotted lines. If the member is visualized as being made of several layers of material stacked vertically, it should be evident that these horizontal layers will slide with respect to each other when the member bends under the load. The student can verify this phenomenon by taking several strips of heavy cardboard, or balsa wood, of exactly the same length, and then placing the stack of material between two supports and applying a load to bend the member. If the ends of the member are observed, it will be seen that the individual pieces, which lined up with each other before the load was applied, no longer line up, indicating that some slippage has taken place along horizontal planes. It is the resistance to this tendency that produces the stress in an actual structural member made of one piece of material.

It must be mentioned here that in the case of the structural member of Fig. 1–5 vertical shear will also exist, as shown in Fig. 1–6. In general, it may be said that any structural member subjected to loads that cause bending will have both vertical and horizontal shearing stresses.

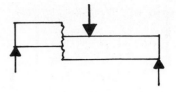

FIGURE 1–6

FLEXURE (BENDING): A structural member spanning between supports and subjected to a load, as indicated in Fig. 1–7, delivers the load to the supports through flexure. All three states of stress previously described exist simultaneously. Compression is present in the top of the member and tension in the bottom. At the same time, there is the tendency for each part of the member to slide away from adjacent parts, both vertically and horizontally, indicating shearing stresses.

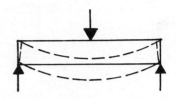

BENDING

FIGURE 1–7

TORSION: This is a ''twist'' that is introduced to a structural member, which may be present due to a variety of conditions. In architectural structures, however, torsion is generally created due to the eccentric placement of a load, as shown in Fig. 1–8. Torsion is a condition that should be avoided where possible.

It must be emphasized that the various states of stress that have been defined normally exist in combination with each other. This has been indicated in the definition of *flexure,* a most common occurrence in structures. There are, however, some unique structural configurations, which will be studied in Chapter 9, where pure states of stress (i.e., tension or compression) exist.

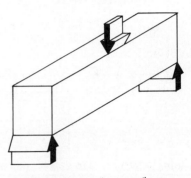

TORSION – A "TWIST" DUE TO ECCENTRIC LOADING

FIGURE 1–8

STRUCTURAL MATERIALS

Of the four most common structural materials in use today, two have been developed very recently in terms of the overall picture of building history. These two materials are steel and steel-reinforced concrete, which were developed for structural purposes in the latter part of the nineteenth century. The other materials most commonly used for structures are timber and masonry. Masonry may be a variety of materials such as stone, brick, concrete block, etc.

STEEL: Structural steel is made of iron (about 98%) and small amounts of several other elements, most notably carbon and manganese. The alloying constituents used with iron, in carefully controlled amounts, give steel the unique properties of high strength, hardness, and at the same time, *ductility,* which is the ability to undergo large deformations without fracture. Generally, it may be said that structural steel has the highest strength per unit of weight compared to other commonly used structural materials.

Structural steel members are made in a variety of standard shapes, some of which are shown in Fig. 1–9. While it is generally best, for economic reasons, for the designer to choose from the standards, it is possible to develop other shapes as required by welding steel plates together or by combining two or more of the standard shapes, as indicated in Fig. 1–10. Because of the weldability of structural steel, an almost infinite variety of shapes can be created to satisfy a variety of conditions that may be encountered in building design.

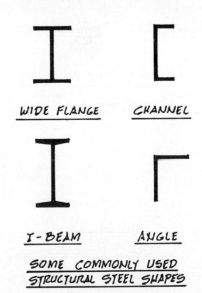

WIDE FLANGE   CHANNEL

I-BEAM   ANGLE

SOME COMMONLY USED STRUCTURAL STEEL SHAPES

FIGURE 1–9

REINFORCED CONCRETE: Concrete is a mixture of cement, aggregates, and water. The aggregates most commonly used are sand and gravel. These aggregates produce a ''normal weight'' concrete that has a unit

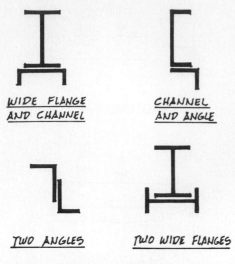

WIDE FLANGE AND CHANNEL

CHANNEL AND ANGLE

TWO ANGLES

TWO WIDE FLANGES

SOME POSSIBLE COMBINATIONS OF STANDARD SHAPES

FIGURE 1–10

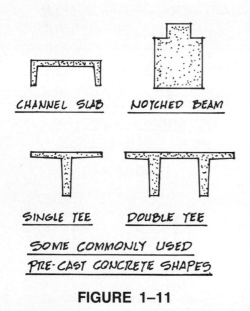

CHANNEL SLAB

NOTCHED BEAM

SINGLE TEE

DOUBLE TEE

SOME COMMONLY USED PRE-CAST CONCRETE SHAPES

FIGURE 1–11

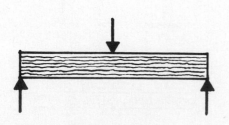

FIGURE 1–12

weight of approximately $150\#/ft^3$. Lightweight aggregates are sometimes used for structural concrete, which can reduce the unit weight to approximately $110\#/ft^3$. The use of lightweight aggregates, which are normally made of expanded clay, can be helpful where, for a variety of reasons, the weight of the building elements must be kept as low as possible.

Since concrete is essentially a brittle material, it cannot resist tension to any significant degree. Consequently, steel reinforcing rods must be placed in those areas within a concrete structural member where tension will be present. This produces *reinforced concrete*.

Concrete has the advantage of being a plastic material that can be formed into a variety of shapes. However, it should be realized that the forms used are made of materials that are not so plastic. The designer must consider the dilemmas that may arise in the development of formwork for an unusual shape.

Reinforced concrete structural members may be *cast in place* or *pre-cast*. *Cast in place* means that the concrete which forms the member is poured in its final location within a structure. This technique leads to a monolithic concrete building that has a great deal of inherent rigidity due to the integral nature of the joints between members. *Pre-cast* concrete means that the individual members are made in a location other than their final place within the structure. Generally, higher quality and greater accuracy can be achieved through pre-casting because of highly controlled conditions. Some commonly used pre-cast shapes are shown in Fig. 1–11.

TIMBER: There is often much confusion regarding the use of the terms *timber, lumber,* and *wood*. There need not be, since for our purposes the terms may be regarded as interchangeable. Technically, *timber* may be defined as wood that is prepared for use and suitable for building purposes; it is the primary term chosen for use in this discussion, although the terms *lumber* and *wood* may be equally suitable.

Timber is a natural material and is available in a variety of species. For structural purposes, the most commonly used species are the pines and the firs. Timber is a fibrous material with a high tensile strength. In general terms, however, its strength is much lower than concrete and, of course, steel. One of the disadvantages of timber is its extremely low resistance to horizontal shear. This low resistance is due to the direction of the grain in a wood member spanning between supports, as shown in Fig. 1–12. The grain, in fact, represents weak planes in the horizontal direction where resistance to shear is minimal.

Structural members made of timber may be solid-sawn or laminated. Laminated timber is made of small sections of sawn pieces glued together with high strength glues. Typical cross sections of sawn and laminated timber are shown in Fig. 1–13. Because laminated timber is made of small pieces of sawn lumber glued together, a variety of shapes and spans may be satisfied. Several commonly used structural configurations in laminated timber are shown in Fig. 1–14. The use of small pieces of wood to make a laminated timber member also allows for selectivity in the quality and placement of the pieces, thereby permitting higher strengths to be assigned to laminated timber.

MASONRY: The term *masonry* does not denote a particular material, but rather a type of construction whereby small units are put together with mortar. Masonry may be stone, brick, concrete block, etc. The most common forms of masonry used today are brick and concrete block. Brick is made of clay molded into small units and fired in a kiln in order to give it strength. Concrete block is made of a concrete mix, which may use normal or lightweight aggregates, and is molded into small units which can be handled by a mason. Masonry may be used as a veneer, where it is not part of the supporting structure, or a masonry wall may be used to carry loads. The strength of a masonry wall is controlled by the quality of the mortar used between the units. The strength of a load-carrying masonry wall may be significantly increased by reinforcing the wall with steel reinforcing rods. Reinforcing for masonry walls is recommended (if not required) where seismic activity is probable.

CLASSIFICATION OF STRUCTURES

In general, structures may be classified according to the state of stress by which they transfer forces; they may also be classified, very generally, by geometry.

Classification by State of Stress

COMPRESSION STRUCTURE: The compression structure is one that distributes its load and delivers it to the ground through, essentially, massive amounts of compressive materials such as stone or concrete. The compression structure is generally characterized by massive proportions. Architectural structures that fall into this category are largely a matter of history, such as the Pyramids, Gothic cathedrals, etc. In contemporary terms, engineering structures, such as the gravity dam, provide examples of true compression structures. Examples of compression structures are shown in Fig. 1–15.

TENSION STRUCTURE: The tension structure is one that delivers load to the supports through pure tensile stress. The tensile structure is characterized by its lightness and by the use of flexible elements such as steel cables, rope, fabric, etc. Examples of typical tension structures are shown in Fig. 1–16. It should be noted that tension structures must contain some compression elements in order to transfer the load to the ground, such as the towers on a suspension bridge.

Classification by Geometry

SURFACE STRUCTURE: The surface structure may be compressive or tensile. It is characterized by a very small thickness in relation to its other dimensions. Often these kinds of structures are referred to as ''membrane'' structures, which implies that they are made of flexible material such as canvas or vinyl. In fact, surface structures may be made of either flexible or rigid material. The canvas tent is an example of a surface structure made of a fabric that is stressed in a state of pure tension. Air-inflated (pneumatic) structures also provide examples of surface structures made of flexible material and acting in a pure state of tension. The fabric structure shown in Fig. 1–16 represents a simple configuration of a fabric surface structure. Depending on the arrangement of the supports, an infinite variety of configurations may be created.

Surface structures may, as mentioned previously, be made of a rigid material. The most common surface structures of this type are made of reinforced concrete, and they are often referred to as *thin shells*. Figure 1–17 shows some common types of reinforced concrete thin shells. Since concrete

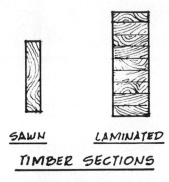

SAWN   LAMINATED

TIMBER SECTIONS

FIGURE 1–13

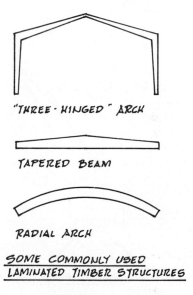

"THREE-HINGED" ARCH

TAPERED BEAM

RADIAL ARCH

SOME COMMONLY USED
LAMINATED TIMBER STRUCTURES

FIGURE 1–14

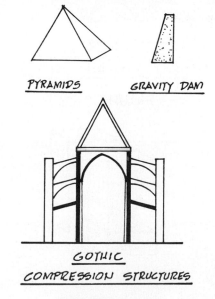

PYRAMIDS   GRAVITY DAM

GOTHIC
COMPRESSION STRUCTURES

FIGURE 1–15

SUSPENSION BRIDGE

FABRIC

TENSION STRUCTURES

FIGURE 1–16

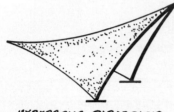

HYPERBOLIC PARABOLOID

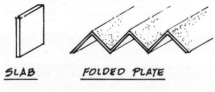

SLAB   FOLDED PLATE

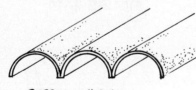

BARREL VAULT

SURFACE STRUCTURES

FIGURE 1–17

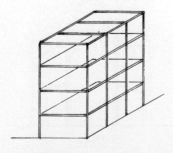

A SIMPLE SKELETAL STRUCTURE

FIGURE 1–18

is poured as a liquid, a large variety of shapes may be created, but many of these require very complex formwork.

SKELETAL STRUCTURE: The most common structural configuration today is that of the "structural frame" or skeletal structure. Such a structure is made of "stick" elements arranged in a way to form a stable frame and to serve the intended function of the building. A very simple example of a skeletal structure is shown in Fig. 1–18. The "sticks" that form the skeletal structure may be steel, wood, or reinforced concrete (cast in place or pre-cast). The typical skeletal structure is composed of flexural elements (the horizontal "sticks") and compression elements (the vertical "sticks"). When dealing with a structure made of "sticks," one of our concerns, is that such configurations will tend to rack when subjected to lateral forces or eccentric placement of vertical loads. Consequently, "stick" structures must be braced, or somehow stiffened, in order to be stable. The issue of lateral stability and bracing will be discussed in some detail in Chapter 5.

The skeletal structure is such an important part of today's architecture that further discussion, along with detailed definitions, will be presented in the next section of this chapter, entitled "Structural Nomenclature."

It must be emphasized that the preceding discussion on classification of structures must be regarded as a series of generalizations. Combinations of the various classifications are quite normal and are inherent in the architectural structures of today.

STRUCTURAL NOMENCLATURE

Up to this point, a variety of expressions and definitions have been presented that are basic to the general language of structures. In the interest of expanding structural vocabulary as a prerequisite for the remainder of the text, we will now deal with some specifics. To do this, we will proceed with a survey of building parts that will be presented in much the same manner as a building is constructed—that is, starting at the lowest part and proceeding upward.

SOILS: All buildings, regardless of their configuration, must ultimately derive their support from the ground. The quality of the soil or rock that underlies every building can have an influence on the building configuration, the choice of building material where weight must be kept to a minimum, and the placement of the building on the site.

The quality of material that underlies a building is normally determined by tests carried out by a soil-testing laboratory. Since the quality of the material can influence the design of a building, the architect should retain the appropriate consultants to carry out the testing program at the beginning of the preliminary design stage.

Since soil is a natural material that has been deposited over a long period of time and in a variety of ways, the quality varies greatly. Soils are made up of a variety of materials, such as clay, sand, silt, etc., which normally exist in combination with each other in a variety of proportions. For this reason, the capacity of soil to carry the weight of a building safely will vary greatly. Of course, if one encounters a solid strata of bedrock on which the building may be placed, there will be little concern about load-carrying capacity. Often, however, bedrock is much too deep and, except in the case of very large buildings, it may not be feasible to carry the loads down to that level.

Perhaps the most common soils on which buildings must rest are the clayey soils, which are mostly clay with some other materials. Clay soils are subject to a variety of phenomena and must be carefully evaluated for tendencies to swell when water is introduced or to shrink excessively when allowed to dry, and for consolidation. Consolidation is a process whereby water is squeezed out due to a load being applied, much like water being squeezed out of a sponge. When this happens, the volume of the soil decreases and the building resting on it will settle. Unlike the water in a sponge, which flows freely, allowing instantaneous dimensional change, the water held in a clay soil is confined, to a large degree, and requires time to be squeezed out. Consolidation of soil is, therefore, a time-dependent phenomenon and a building may settle for a long period of time. The amount of settlement that will take place can be predicted with some accuracy by laboratory testing of soil samples. Small amounts of settlement are normally anticipated and are tolerable if the building settles uniformly. Differential settlement (where one part of a building settles more than other parts) can often cause damage to architectural finishes. Where differential settlement is excessive, the structural frame may be subjected to levels of stress, due to warping and twisting, that were not anticipated in the design.

FOUNDATIONS: The foundation is that part of a building between the soil, or rock, and the structure. The purpose of the foundation is to receive the building loads and distribute them over a wide area, thereby reducing the pressure on the soil to a magnitude that can safely be resisted.

In the broadest and most general sense there are two classifications of foundations: shallow foundations and deep foundations. There are a number of variations of each of these types that may be designed for varying soil conditions. In fact, where the quality of soil varies greatly under a building, a combination of shallow and deep foundations may be used.

The most common form of shallow foundation is the *spread footing*, which is essentially an enlargement of the base of a member delivering load to the ground. The purpose, of course, is to spread the load out over a large area of soil. A typical spread footing is shown in Fig. 1–19. Spread footings carry a single building load; consequently, there will normally be many such footings under a single building. The spread footing size and thickness are determined by the intensity of the load coming down and the load-carrying capacity of the soil. The depth at which a spread footing is placed depends upon the distance below the ground surface where suitable soil is located. The depth of a spread footing, however, is somewhat limited by the size of the excavation required. The minimum depth below the ground surface at which a footing may be placed is governed by the depth at which water in the soil may freeze. If freezing takes place below the bottom of a footing, with consequent expansion of the soil, then the footing will be pushed upward, causing damage to the building. The depth at which freezing may occur varies regionally and is known as the *frost line*. The base of a footing should be located below the level of the frost line. The minimum depth required is normally given in local building codes.

A variation of the spread footing occurs where load-carrying elements are so closely spaced that it may be more feasible, or necessary, to pour a single footing that is designed to carry two or more building loads. Such a footing is called a *combined footing*, and a typical example is shown in Fig. 1–20. As with the spread footing, the size and thickness of a combined footing is governed by the quality of the soil on which it rests and the magnitude of the building loads.

A combined footing, as mentioned, is designed to carry two or more building loads. However, when *all* of the building loads are carried on a single solid footing, we refer to this as a *mat foundation*, shown in Fig.

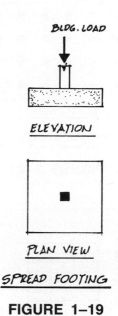

FIGURE 1–19

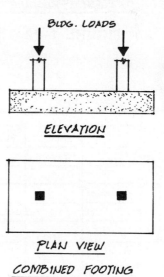

FIGURE 1–20

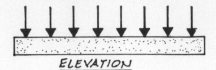

ELEVATION

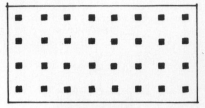

PLAN VIEW

MAT FOUNDATION

FIGURE 1–21

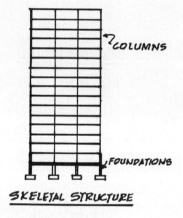

SKELETAL STRUCTURE

FIGURE 1–22

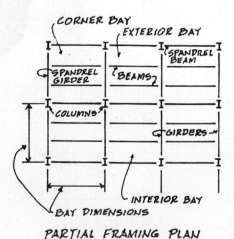

PARTIAL FRAMING PLAN
SKELETAL STRUCTURE

FIGURE 1–23

1–21. A mat foundation may be necessary where the quality of soil is poor and the area over which the building loads must be distributed is very large.

Deep foundations are generally pile foundations. A pile, in essence, is a "stick" that is driven into the ground until it offers resistance. This resistance signifies a significant load-carrying capacity. The necessary resistance may be achieved in two ways: by friction or bearing. Friction piles derive their load-carrying capacity through the friction between the pile and the soil. Bearing piles derive their strength by direct bearing on, usually, a sound strata of bedrock. The "sticks" from which piles are made are of timber, concrete, or steel. Timber piles (essentially, telephone poles) are normally used as friction piles. Concrete and steel piles are normally used as bearing piles. Often, where bearing piles must be driven to great depths, a combination of steel and concrete is used. In a case such as this, hollow steel shells are driven into the ground to the point of bearing, and then they are filled with concrete.

As mentioned at the beginning of this discussion, there are many variations of foundation types. There are also a number of variables involved in the choice of a foundation type and its design. Testing of soil, interpretation of data, foundation choice and design are complex matters that are well beyond the scope of this discussion, which has been presented for the sake of terminology.

SKELETAL STRUCTURES: Skeletal structures may be of any scale, from one story in height to high-rise buildings. Fig. 1–22 shows an elevation of a typical skeletal structure; Fig. 1–23 shows a typical framing plan. A framing plan is a structural drawing that shows only the structural elements supporting the floor. Reference should be made to these two figures for the following definitions:

Bay: A bay, in a skeletal building, is a rectangular portion of the building defined by dimensions between columns.

Beam: A horizontal member that supports a floor.

Girder: A beam that supports other beams.

Spandrel beam: A beam at the perimeter of the building.

Spandrel girder: A girder at the perimeter of the building.

Column: The vertical member that supports beams and girders and delivers loads to the foundation of the building.

Normally, bays in a skeletal structure are rectangular, and all bays within a building are of the same size. This is important for the sake of economy. However, it should be realized that bay sizes may vary within a given building, and they may not be rectangular. In fact, it is possible to space columns at random, in which case the bay may not be a definable unit of space. Random arrangements of structural elements are normally avoided, however, because each member must be designed and fabricated differently, thereby increasing costs significantly.

Bearing wall structures: A bearing wall structure is one, as the name implies, in which walls are used in lieu of columns to receive floor or roof loads and deliver them to the foundation. A simple example of a bearing wall structure is shown in Fig. 1–24.

Bearing walls are often made of masonry or poured reinforced concrete. However, one of the most common types of bearing wall structures used today is wood frame construction, as found in residential architecture. While the wood frame is made of individual "sticks," they are very closely spaced; this, in essence, forms a linear support, as opposed to a point support (column), as found in a skeletal structure.

In the construction practices of today the masonry, or poured concrete, bearing wall is not a particularly common form of building except in the case of one- or two-story buildings. Bearing wall construction, however, can be used for buildings of many stories and in the past this was not uncommon.

FORCES ACTING ON ARCHITECTURAL STRUCTURES

A *force* may be defined as an action upon a body which changes, or tends to change, the state of motion of that body. The function of a structure is to resist the various forces to which a building is subjected. The most prevalent of the forces acting on a building are the forces due to gravity. In building design, the forces due to gravity are broken into two categories:

1. *Dead load*. This includes the weight of all permanent items, such as the building structure itself, permanent partitions, flooring material, ceilings, mechanical ductwork, lighting fixtures, etc.
2. *Live load*. This includes the weight of all items that are not necessarily permanent, such as people, furniture, filing cabinets, etc.

The dead load of a building can be calculated precisely by knowing the weights of the materials used and the permanent items to be installed.

The live load cannot be calculated precisely, since the presence or absence of nonpermanent items is unpredictable. In the design of a structure, the live load can only be estimated based on the function of the space. Recommendations for the live load to be used in the design of a building are given in building codes and vary with the function of the space. Some typical live load recommendations are given on Data Sheets D–14 and D–15 in the Appendix. Both dead load and live load are computed on a pounds-per-square-foot basis; that is, a certain number of pounds on every square foot of floor (or roof) surface.

In addition to gravity forces, which are always present, there are other forces that must often be considered in building design. These may be due to wind, earthquake, impact, vibration, temperature change, snow and ice, pressure due to liquids or earth, etc. The intensity or probability of some of these forces varies regionally; several maps are shown in the Appendix on Data Sheets D–16 through D–18, which show these variations.

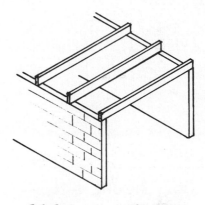

BEARING WALL STRUCTURE

FIGURE 1–24

FORCE CHARACTERISTICS

In order to deal with forces acting on buildings, we must be able to determine the following characteristics of the forces:

1. *Magnitude*. This is force expressed as weight, such as pounds, tons, kilograms, etc. One of the most common terms used in structural design is *kips,* the abbreviated term for kilopounds, which means 1000 pounds. In many cases the magnitude of a force must be based on recommendations that are derived from experience, or probability. This means that the magnitudes of many forces used in building design are largely a matter of sound judgment.
2. *Direction*. This is a statement of the direction of a force, such as an angle, compass reading, etc. Gravity forces act vertically, but other forces may act on buildings in a variety of ways. In order to design a

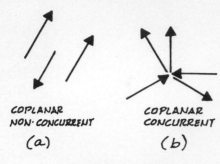

COPLANAR
NON·CONCURRENT

COPLANAR
CONCURRENT

(a)   (b)

FIGURE 1–25

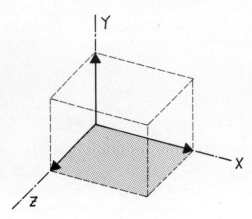

NON-COPLANAR, CONCURRENT

FIGURE 1–26

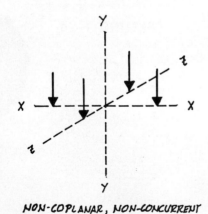

NON-COPLANAR, NON-CONCURRENT

FIGURE 1–27

stable building, we must be able to identify the manner in which forces will act on structural elements.

3. *Position.* This deals with identifying the location of the forces applied to the various structural elements. The position of a force is normally given as a dimension from a reference point.

ARRANGEMENT OF FORCES

The forces with which we will be dealing may be arranged in a number of ways. In order to deal with a problem it is necessary to understand these arrangements, which are:

1. *Coplanar forces.* This includes all forces that act along lines that lie in the same plane. This is a two-dimensional arrangement; examples are shown in Fig. 1–25.
2. *Non-coplanar forces.* This is a three-dimensional arrangement with no single plane in common. An example of a non-coplanar arrangement is shown in Fig. 1–26.
3. *Concurrent forces.* This is an arrangement in which the lines of action of all forces intersect at a common point. Such arrangements may be coplanar or non-coplanar, as shown in Figs. 1–25b and 1–26.
4. *Non-concurrent forces.* In this arrangement there is no point of intersection of the forces. Parallel forces constitute an example of non-concurrent forces. Such an arrangement may be coplanar or non-coplanar, as shown in Figs. 1–25a and 1–27.

It should be recognized that, in a given situation, various combinations of the above may exist. For example, a system of forces may be defined as *coplanar-concurrent, non-coplanar-concurrent,* etc. Parallel forces are a special case. They may be only coplanar or non-coplanar.

2

Concurrent and Coplanar Forces

RESOLUTION OF CONCURRENT FORCES

In structural analysis it is usually convenient, if not necessary, to deal with forces in terms of a rectangular system: that is, vertical and horizontal forces. However, a force applied to a structural element may be applied at an angle and therefore is neither vertically nor horizontally directed. For example, forces due to wind are conventionally taken to act normal to sloping surfaces, as shown in Fig. 2–1. Also, a vertical or horizontal force may be applied to a structural element that is oriented in neither a vertical nor horizontal position, but at an angle, such as a gravity force acting on a sloping surface, as shown in Fig. 2–2. Consequently, it will be important for the reader to understand the vertical and horizontal effects of a force applied at an angle, as well as the effects of vertically or horizontally directed forces applied to inclined structural elements. To this end, the following terms and their definitions will be useful:

1. *Components* of a given force: Two forces having the same effect as the given force, as shown in Fig. 2–3. The given force is shown heading upward and to the right. The same effect can be represented by two forces, vertical and horizontal, whose directions must be upward and to the right.
2. *Resultant* of two or more given forces: This is the *single* force that has the same effect as the given forces, as shown in Fig. 2–4. In this case, the two forces, one heading upward and one heading to the right, may be represented by a single resultant that must be heading up and to the right.

In order to deal with problems involving forces of this nature, it will be necessary to learn how to resolve a single force into its components, or a group of forces into a single resultant. The basic tool for the resolution of

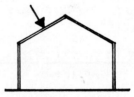

INCLINED FORCES

FIGURE 2–1

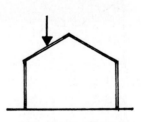

VERTICAL FORCE ON
AN INCLINED PLANE

FIGURE 2–2

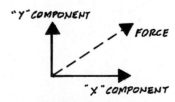

"Y" COMPONENT
FORCE
"X" COMPONENT

FIGURE 2–3

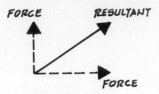

FIGURE 2-4

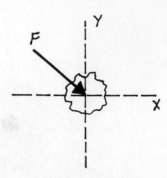

FIGURE 2-5

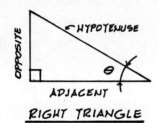

RIGHT TRIANGLE

FIGURE 2-6

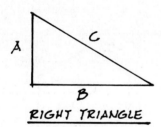

RIGHT TRIANGLE

FIGURE 2-7

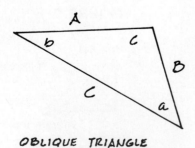

OBLIQUE TRIANGLE

FIGURE 2-8

forces is the *force diagram,* which is made up of vector quantities, using either an analytic or graphic method for evaluation.*

It should be understood at this point that the force diagram can represent only the *magnitude* and *direction* of forces, but not the *position* of forces, which is the third force characteristic defined in Chapter 1. In this chapter we will be concerned only with concurrent and coplanar forces. The position of all concurrent forces is fixed by their point of intersection and does not constitute a variable. Therefore, the force diagram is sufficient to deal with concurrent force problems, both coplanar and non-coplanar. In Chapter 3 we will deal with non-concurrent forces; then the force characteristic of position will have to be considered.

In order to demonstrate the procedures for the resolution of forces, consider a body acted upon by a coplanar force, as shown in Fig. 2–5. This force will tend to displace the body in the direction of the *line of action* of the force. The force *(F)* shown in the figure will tend to displace the body downward and to the right. For the sake of convenience it will be useful to determine the horizontal and vertical effects of the force; that is, its *components* in the *X* and *Y* directions. There are two methods available to help us achieve this goal. They are the *graphic* method and the *analytic* method.

Before we get into any specific examples, it will be useful to review the basic trigonometry that will be needed.

TRIGONOMETRY REVIEW

For dealing with right triangles the following basic trigonometric functions will be needed (refer to Fig. 2–6):

$$\sin \theta = \frac{\text{opposite}}{\text{hypotenuse}}$$

$$\cos \theta = \frac{\text{adjacent}}{\text{hypotenuse}}$$

$$\tan \theta = \frac{\text{opposite}}{\text{adjacent}}$$

While there are more trigonometric functions, these will serve our purposes where right triangles are concerned.

Another useful bit of information for dealing with right triangles is the *Pythagorean Theorem,* which states:

The square of the hypotenuse of a right triangle is equal to the sum of the squares of the two sides.

Stated mathematically (refer to Fig. 2–7):

$$A^2 + B^2 = C^2$$

Where *oblique* triangles must be dealt with, a relationship that will be useful is the "Law of Sines," which states mathematically (refer to Fig. 2–8):

$$\frac{A}{\sin a} = \frac{B}{\sin b} = \frac{C}{\sin c}$$

*A vector quantity is a physical quantity whose magnitude and direction may be represented by a vector.

PARALLELOGRAM OF FORCES

An important principle, which will be used along with the trigonometric relationships stated, is known as the "Parallelogram of Forces." It tells us that:

> When two forces act concurrently at a point, the *resultant* force can be represented as to magnitude and direction by the diagonal of a parallelogram, the sides of which are parallel to and proportional to the two forces.

For example, to find the resultant of the concurrent forces X and Y shown in Fig. 2–9, simply construct the parallelogram (which, of course, will be a rectangle when dealing with vertical and horizontal forces) as shown in the figure, and then determine the proportion of the diagonal.

While the Parallelogram of Forces principle was stated specifically as a means for determining the resultant force, it, of course, can also be applied in the reverse manner. That is, given a single inclined force *(R)*, as shown in Fig. 2–10, the components X and Y may be found by constructing the parallelogram, as shown in the figure.

Primarily, we will be dealing with parallelograms that are rectangles. On occasion, however, we will encounter a problem in which the parallelogram is nonrectangular, and the principle, as stated, will serve us equally well. We will look at a condition like this in the example problems.

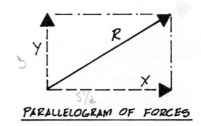

PARALLELOGRAM OF FORCES

FIGURE 2–9

LIMITS OF ACCURACY

Now that we are ready to start dealing with numbers, it will be useful for the student to understand the basis for the degree of accuracy used in the numerical computations throughout this book. This has to do, specifically, with the basis for judgments used in rounding off numbers, and the number of decimal places used in the computational procedures.

In Chapter 1, it was mentioned that the *live load* in a building cannot be calculated precisely because the presence or absence of nonpermanent items is unpredictable. Since the live load used in the design of a building is simply an estimate, based on the function of the space, the percentage of error in the value used for the live load is, by design, usually high and on the safe side. The fact is that we don't really know what the live load in a space may be at any given time. We do know that the values used for live load should, based on sound judgment and experience, cover the worst possible loading situation (which probably will never occur). In addition to the uncertainty of live load values, there are other unknowns involved in the structural design of a building. For example, it was written, in Chapter 1, that the dead load can be calculated precisely. Theoretically this is so, but only if we know what the dimensions of the structural elements are at the outset (for example, the thickness of a concrete slab). But we don't have this information at the outset of the design process. Consequently, we must estimate a value for dead load to be used based on experience and sound engineering judgment. Normally, the value will be on the safe side.

Aside from the structural elements, there are other dead load items whose weight is difficult to determine on a pounds-per-square-foot basis. This would include items such as lighting fixtures, mechanical ductwork, sprinkler systems, etc., all of which would be supported by the structural system. In order to deal with the dead load of items such as these, whose final location would not be accurately known during the design process, we generally apply a certain number of pounds per square foot to every square

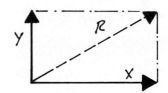

PARALLELOGRAM OF FORCES

FIGURE 2–10

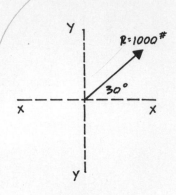

FIGURE 2–11

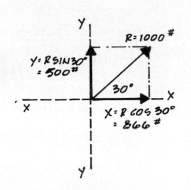

FIGURE 2–12

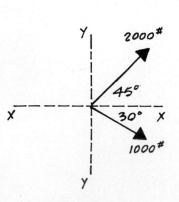

FIGURE 2–13

foot of the floor (or roof). This will cover the dead load for miscellaneous items, wherever they may be placed. The value used for this would be based on judgment, experience, and the nature of the building; it would normally be on the safe side.

In addition to the questions concerning live loads and dead loads, there are still other areas of uncertainty in the design of an architectural structure. These would include questions regarding the actual strength of structural materials, the accuracy of construction, etc. All of the potential problems are covered with the application of a variety of factors of safety that are used in the structural design process.

In light of the preceding discussion, it should be realized that it is an exercise in futility to carry numbers in the computations to three or four decimal places. In this book the numbers will generally be rounded off to, at most, two decimal places.* How numbers are rounded off is a matter of judgment based on the percentage of error that is being introduced. Large numbers may be rounded off to one decimal place or even to the whole number. For example, a number such as 100.15 may be rounded off to 100.2 or simply 100 without any discernible percentage of error. On the other hand, one must be more careful about rounding off smaller numbers. For example, if the number 1.5 is rounded off to 2, a substantial percentage of error is introduced. In other words, the size of the number has a great deal to do with the manner in which it is rounded off.

It must be emphasized that the idea of rounding off numbers, and the fact that there are various unknowns involved in the structural design of a building, should not be taken as a signal to get somewhat careless in the approach to problem solving. In the structural design process, rounded-off numbers do not cause building failures . . . *wrong* numbers and poor thinking in the approach to a problem can do so.

Example 2–1 (Figs. 2–11 and 2–12)

In Fig. 2–11 a force *(R)* of 1000# is given at an inclination of 30° from the horizontal; it is desired to know its vertical *(Y)* and horizontal *(X)* components. In order to solve this, it is only necessary to construct the parallelogram, as shown in Fig. 2–12, and use the trigonometric relationships for a right triangle. To find the *X* component,

$$\cos 30° = \frac{X}{R} \text{ and } X = R \cos 30°$$

To find the *Y* component,

$$\sin 30° = \frac{Y}{R} \text{ and } Y = R \sin 30°$$

Since the forces involved are directly proportional to the lengths of the vectors, we may deal with the forces directly.

$$\therefore X = R \cos 30° = (1000\#)(.866) = 866\#$$

and

$$Y = R \sin 30° = (1000\#)(.500) = 500\#$$

Example 2–2 (Figs. 2–13 and 2–14)

Example 2–1 was solved by purely analytical means. In this example we shall consider a graphic solution. It is desired here to determine the resultant effect of the two

---

*The only exception to this is in the first part of Chapter 7, where the subject matter deals with very small numbers that must be taken to at least three decimal places.

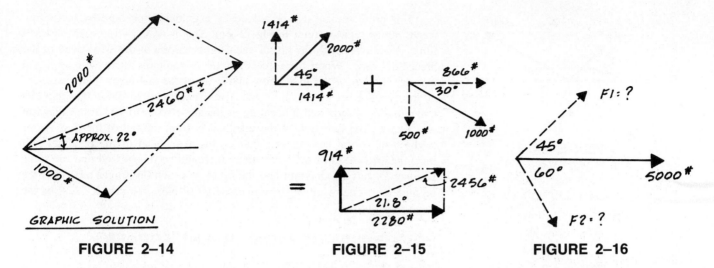

GRAPHIC SOLUTION

FIGURE 2–14   **FIGURE 2–15**   **FIGURE 2–16**

forces shown in Fig. 2–13. This can be determined graphically, using the Parallelogram of Forces principle. In order to do this, lay out the forces as vectors, as shown in Fig. 2–14. In this case they have been drawn in the appropriate direction at a scale of 1 in. = 1000#. If the parallelogram is completed and the diagonal is drawn, it represents the resultant effect of these forces. The magnitude is determined by measurement of the diagonal using the same scale that was used for the original layout. The direction is determined with the use of a protractor. It should be noted that *absolute* accuracy in the graphic method is dependent on the care taken with the drawing and the scale used. Absolute accuracy, however, in such problems is not of great importance and, consequently, a graphic solution is sufficient. The final solution to this problem is shown in Fig. 2–14.

Example 2–3 (Figs. 2–13 through 2–15)

The problem given in Example 2–2 would present some difficulty in solving directly by analytical means. The two triangles that make up the parallelogram shown in Fig. 2–14 are oblique triangles, and the Law of Sines would be used, which is not difficult, but two angles of the oblique triangle would have to be known. As this problem is stated, this information is not known at the outset. However, this problem *can* be solved analytically by breaking each force shown into vertical and horizontal components and adding the results algebraically. This would lead to a net vertical and horizontal effect whose resultant can be found by a right-triangle solution. This procedure is shown in Fig. 2–15.

Example 2–4 (Figs. 2–16 through 2–18)

In a reverse situation from that shown in Example 2–2, it may be desired to find components of a force in directions other than vertical or horizontal, as shown in Fig. 2–16. Analytically, this presents no problem since all information is given so that the Law of Sines may be used. The parallelogram of forces and the solution are shown in Fig. 2–17.

This problem may also be solved graphically by drawing the given 5000# force to scale and then laying out the parallelogram based on the given angles desired for the components. The graphic solution is shown in Fig. 2–18.

Example 2–5 (Figs. 2–19 through 2–21)

When dealing with three or more concurrent forces acting at a point, as shown in Fig. 2–19, the resultant can be found analytically by breaking each force into its vertical and horizontal components and then adding the results algebraically. The process is shown in Fig. 2–20. The process was started with the 283# force and then proceeded in a clockwise direction. It is arbitrary, however, where the process is started and in what order the forces are broken into components.

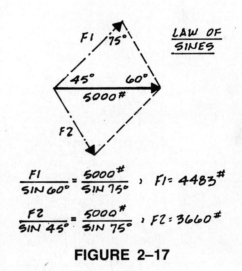

LAW OF SINES

$$\frac{F1}{SIN\,60°} = \frac{5000^{\#}}{SIN\,75°}\;,\;F1 = 4483^{\#}$$

$$\frac{F2}{SIN\,45°} = \frac{5000^{\#}}{SIN\,75°}\;,\;F2 = 3660^{\#}$$

FIGURE 2–17

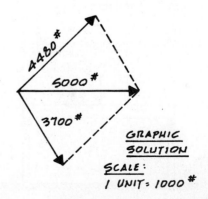

GRAPHIC SOLUTION

SCALE:
1 UNIT = 1000#

FIGURE 2–18

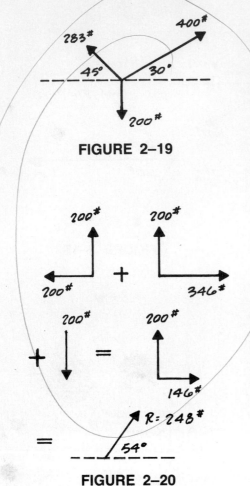

FIGURE 2–19

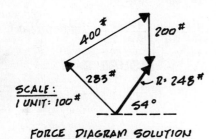

This problem can also be solved by a graphic procedure by laying the forces out as vectors (refer to Fig. 2–21). This is done by representing each force as a vector, with the tail of each force starting at the arrowhead of the preceding force. When the force diagram is complete there will be a gap between the tail of the first arrow and head of the last arrow. The resultant of the forces is represented by the arrow necessary to close this gap. Its magnitude is determined by scaling at the same scale to which the diagram was drawn. The direction of the resultant is from the tail of the first vector to the head of the last vector. That is, the arrowhead of the resultant will meet the arrowhead of the last vector. It should be recognized that the force selected to start the diagram and the order in which the forces are taken are unimportant. The resultant force necessary to "close the circuit" will be the same regardless.

EQUILIBRIUM OF CONCURRENT FORCES

In the design of a structure, one of the chief concerns is, of course, that all structural elements subjected to forces remain in a state of rest. The *action* by forces on every structural member must be resisted by a *reaction* provided by the supporting structural elements. In Fig. 2–22, the loaded beam is being supported by the columns. The columns must furnish the necessary reactions, R_1 and R_2, in order to keep the structure from collapsing. The determination of the reactions for a problem of this sort will be discussed later. For now, let us concern ourselves with determining the *equilibrating* force (sometimes called the *equilibrant* or *antiresultant*) for a system of coplanar-concurrent forces. Finding the equilibrant for coplanar-concurrent forces is a relatively simple matter. Both analytic and graphic methods may be used.

FIGURE 2–20

Analytic Method

The process necessary to determine the equilibrant by analytic methods is no different than the process used to determine the resultant of concurrent forces. Once the resultant is determined it may be said that the equilibrant will be a force equal in magnitude and opposite in direction to the resultant. Such a resultant may also be equilibrated by two forces (vertical and horizontal) that are equal in magnitude and opposite in direction to the vertical and horizontal components of the force (see Figs. 2–23 and 2–24).

In view of the preceding discussion, it may be said that the conditions necessary to produce equilibrium with only coplanar-concurrent forces acting are:

1. The algebraic sum of all vertical components of forces must be zero.
2. The algebraic sum of all horizontal components of forces must be zero.

These principles may be expressed in mathematical form:

1. $\Sigma F_y = 0$
2. $\Sigma F_x = 0$

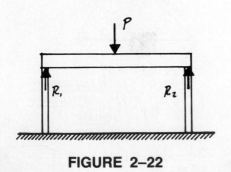

FORCE DIAGRAM SOLUTION

FIGURE 2–21

FIGURE 2–22

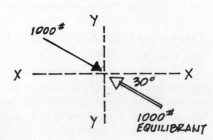

FIGURE 2–23

Graphic Method

The graphic process for determining the equilibrant of coplanar-concurrent forces is virtually identical to the process used to determine, graphically, the resultant, with one small difference. The equilibrant of the forces shown in Fig. 2–19 will now be determined using the following process (refer to Fig. 2–25):

Each force is laid out as a vector, with the tail of each force starting at the arrowhead of the preceding force. When the vector diagram is complete there will be a gap between the tail of the first vector and the arrowhead of the last vector. The equilibrant of the forces is the vector necessary to close this gap. The direction of the equilibrant is from the arrowhead of the last vector to the tail of the first vector (exactly the reverse of the resultant).

FRICTION

In later chapters, extensive application will be made of the principles studied in the preceding pages. However, it would be useful to discuss briefly at this time the phenomenon of friction which, for analysis, requires the use of the principles of concurrent forces.

Friction offers resistance to motion when two rough surfaces in contact tend to move on one another. In many cases the effect of friction is small in comparison with other forces and may normally, in the case of architectural structures, be neglected. Regardless, we will briefly concern ourselves with this phenomenon and some of the basic terminology involved.

When a body, of weight W, as shown in Fig. 2–26, is resting on a surface and a force P is applied, there will be a frictional resistance to sliding (F) and a reaction (G). The friction existing when sliding is about to start is called the *friction of impending motion*. The resultant of the two reactions, G and F, acts at an angle θ, as shown in the figure, such that:

$$\tan \theta = \frac{F}{G}$$

Research has shown that the ratio between the friction and the weight of the body $(F/G = \tan \theta)$ remains constant for all sizes and weights of bodies but varies with the materials in contact. This ratio $(F/G = \tan \theta)$ is called the *coefficient of friction* and is expressed in the form of a decimal. A table giving the coefficients of friction for a variety of contact surfaces is given on Data Sheet D–11 in the Appendix.

If a weight (W) is placed on a sloping surface, as shown in Fig. 2–27, then W can be resolved into components parallel and perpendicular to the slope, as shown by P and N. The reactions to these components are N' and F, where F is the frictional resistance to sliding. If the tangent of the angle θ (which is also the angle to which the surface is inclined) is equal to the coefficient of friction, then motion is impending. Therefore, if the surface is inclined at an angle θ which coincides with coefficient of friction, then the body is about to slide. If a body resting on a rough surface is acted upon by various forces, and the resultant of these forces makes an angle with a line perpendicular to the contact surface, which is greater than θ, slipping will occur. Conversely, if the angle is less than θ, no slipping will occur.

While the concepts surrounding the phenomenon of friction are generally of minimal importance in the design of architectural structures, there are, on occasion, questions raised about the behavioral aspects of friction between two surfaces in contact with each other (for example, a steel beam resting on a masonry wall). The brief discussion on friction has been included to provide an awareness of the behavior involved.

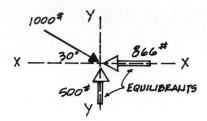

FIGURE 2–24

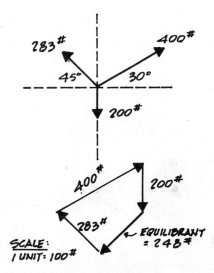

FORCE DIAGRAM

FIGURE 2–25

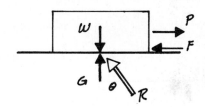

FIGURE 2–26

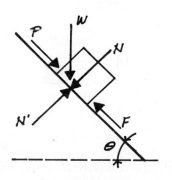

FIGURE 2–27

PROBLEM 2-1

RESOLUTION OF FORCES - COPLANAR & CONCURRENT

(1) DETERMINE THE MAGNITUDE OF THE HORIZONTAL AND VERTICAL COMPONENTS WHEN θ = 30°, 45°, 60°.

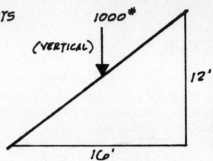

(2) DETERMINE THE MAGNITUDE OF THE COMPONENTS PARALLEL AND PERPENDICULAR TO THE SLOPE.

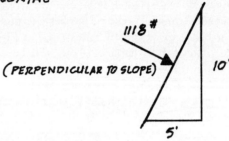

(3) DETERMINE THE MAGNITUDE OF THE HORIZONTAL AND VERTICAL COMPONENTS.

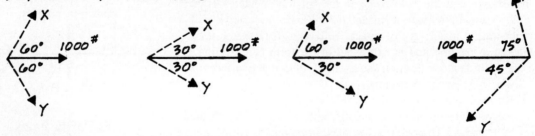

(4) DETERMINE THE MAGNITUDE OF COMPONENTS X & Y.

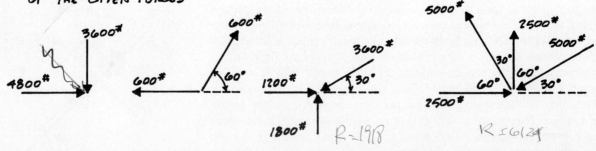

(5) DETERMINE THE MAGNITUDE AND DIRECTION OF THE RESULTANT OF THE GIVEN FORCES

PROBLEM 2–2

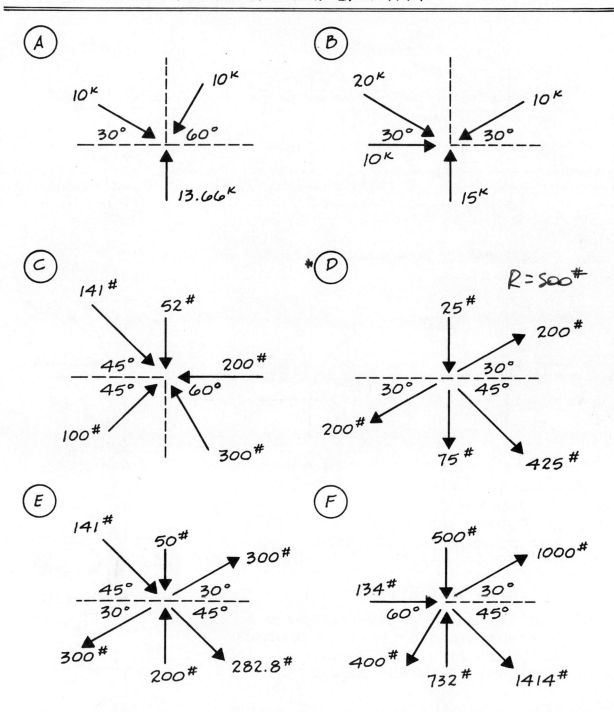

RESOLUTION OF FORCES

• DETERMINE THE MAGNITUDE AND DIRECTION OF THE RESULTANT OF THE CONCURRENT FORCES SHOWN IN EACH PART.

(A)
10ᴷ 10ᴷ
30° 60°
13.66ᴷ

(B)
20ᴷ 10ᴷ
30° 30°
10ᴷ
15ᴷ

(C)
141# 52#
45° 200#
45° 60°
100# 300#

*(D) R = 500#
25# 200#
30°
30° 45°
200#
75# 425#

(E)
141# 50# 300#
45° 30°
30° 45°
300# 200# 282.8#

(F)
500# 1000#
134# 30°
60° 45°
400# 732# 1414#

PROBLEM 2–3

EQUILIBRIUM OF CONCURRENT FORCES

(I)   DETERMINE THE EQUILIBRANT FOR EACH OF THE FOLLOWING BY
       ANALYTICAL MEANS.

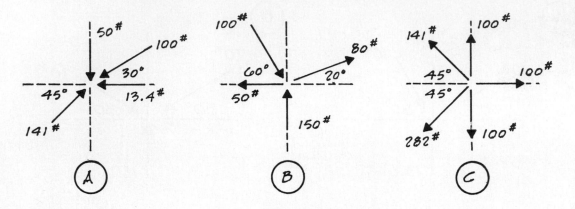

(II) DETERMINE THE EQUILIBRANT FOR EACH OF THE FOLLOWING
       GRAPHICALLY.

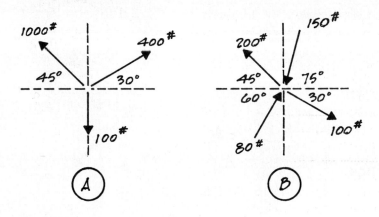

(III) WHAT FORCES ACTING ALONG LINES X & Y
       WILL EQUILIBRATE THE OTHER FORCES.

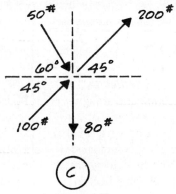

(IV) FIND THE MAGNITUDE AND DIRECTION OF A
       FORCE INCLINED AT 30° BELOW THE HORIZONTAL
       AND A HORIZONTAL FORCE, WHICH WILL
       EQUILIBRATE THE FORCES SHOWN.

3

Non-Concurrent Forces

INTRODUCTION

Up to this point, only concurrent forces have been considered; consequently, we have only been interested in their magnitudes and directions. However, when dealing with forces that are non-concurrent, we must consider the third force characteristic indicated in Chapter 1, that is, the *position* of the forces.

Bodies acted upon by a group of concurrent forces are subjected to a tendency of *translation*. That is, there is a tendency for the body to move in the direction and along the line of action of the resultant of the concurrent forces. The tendency of a body to translate, however, is not the only tendency for motion with which we will be concerned in the study of structural principles. For example, visualize a wheel turning on a fixed axle (Fig. 3–1). It should be evident that the wheel, as a whole, does not translate, but every part of the wheel is moving with respect to the fixed axle. What is taking place here is not motion of translation but motion of rotation.

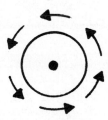

FIGURE 3–1

DETERMINING THE MAGNITUDE OF ROTATION

Consider a "seesaw" type of structure as shown in Fig. 3–2. It should be recognized that there will be no rotation when the equal forces *(F)* are equidistant from the support. The tendency for one force to cause rotation about the point of support (point *o*) will be balanced by an equal and opposite tendency due to the force on the other side. However, if the distances from the forces to the point of support are not equal, rotation will take place. In Fig. 3–3, if distance *b* is greater than distance *a,* the structure will rotate in the direction shown. If, as shown in Fig. 3–4, the distances of the loads from the support are the same and force *P* is greater than force *F,* rotation will also take place in the direction shown. It is possible to arrange the magnitudes of the forces and their distances to the point of support in such

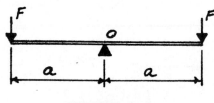

FIGURE 3–2

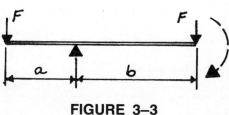

FIGURE 3–3

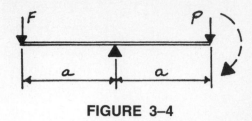

FIGURE 3–4

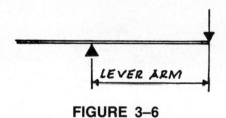

FIGURE 3–5

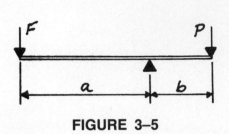

FIGURE 3–6

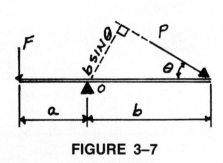

FIGURE 3–7

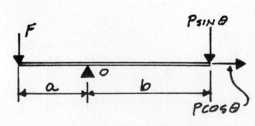

FIGURE 3–8

a manner that no rotation will take place even when dealing with unequal forces and distances.

In Fig. 3–5, no rotation will take place when the product of one force and its distance from the support is equal to the product of the other force and its distance to the support; that is, *(F) (a) = (P) (b)*. This phenomenon is known as "The Principle of the Lever," which was first defined by Archimedes (287–212 B.C.). The principle, in essence, states that:

> Two forces will balance at distances reciprocally proportional to their magnitudes.

The *perpendicular* distance from the line of action of a force to a point at which the magnitude of rotation is desired is known as the *lever arm* (see Fig. 3–6) and the product of a force and a lever arm is known as the *moment of the force,* or simply the *moment.*

DETERMINING THE LEVER ARM

Using the same "seesaw" structure, force P is now applied at an angle θ from the horizontal, as shown in Fig. 3–7. The tendency of force P to produce rotation about the support point o is less than it would be if the same force were vertical. In order to determine the moment due to force P with respect to point o, resolve P into its vertical and horizontal components as shown in Fig. 3–8. Since these components are, by definition, forces that *produce the same effect as the original force P,* the moment due to the vertical and horizontal about point o must be the same as the moment due to the force P about the same point.

The components of P are $P \sin \theta$ (vertical component) and $P \cos \theta$ (horizontal component), and the lever arms of the components are b for the vertical component and zero for the horizontal component. The zero lever arm for the horizontal component is so because it will not cause a tendency for rotation about point o. The total moment, therefore, is $P(\sin \theta)b + P(\cos \theta)0 = P(\sin \theta)b$.

Now if a perpendicular is drawn from point o to the *line of action* of P (see Fig. 3–7), its length will be $(\sin \theta)b$. If this is used as the lever arm, the moment due to force P about point o is $P(\sin \theta)b$, which agrees with the moment due to the components.

Two important principles are derived from the preceding discussion:

1. The moment of any force is found by multiplying the magnitude of the force by the *perpendicular* distance from its line of action to the point about which the moment is desired.
2. Any force whose line of action passes through a point (such as the horizontal component of P in the preceding demonstration) produces no moment about that point.

It should be recognized from the first principle above that the unit of moment is (Force) × (Distance), which may be expressed as ft lbs, ft kips, in. lbs, in. kips, etc.

EQUILIBRIUM AGAINST ROTATION

In order for a body to be in equilibrium against rotation, the summation of moments about a point must be equal to zero (expressed mathematically,

$\Sigma M = 0$). In other words, all tendencies for a body to rotate clockwise must be balanced by a moment of equal magnitude that will tend to rotate the body counterclockwise. In Fig. 3–7, for example, in order to have equilibrium against rotation, the magnitude of *(F) (a)* must be equal to $P(\sin \theta)b$. For equilibrium against translation, the support point must provide reactions of $F + P(\sin \theta)$ vertically ($\Sigma F_y = 0$), and $P(\cos \theta)$ horizontally ($\Sigma F_x = 0$).

It may then be said that the three basic equations which *must be satisfied* for a state of static equilibrium to exist are

$$\Sigma F_x = 0, \ \Sigma F_y = 0, \ \Sigma M = 0$$

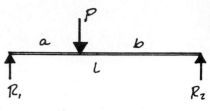

FIGURE 3–9

PARALLEL FORCES

Parallel forces constitute a special case of non-concurrent forces. Consider the beam of Fig. 3–9, subjected to a load as shown and supported at the ends. These supports must furnish the necessary equilibrating reactions. In order to determine the magnitude of the reactions R_1 and R_2, we will use the basic equations of static equilibrium.

1. $\Sigma F_x = 0$. This tells us that the algebraic summation of all forces in the horizontal direction must equal zero. That is, for every action in the horizontal direction there must be an equal and opposite reaction. In the problem under consideration, however, there are no horizontal forces applied to the structure. Consequently, the algebraic summation of forces in the horizontal direction is already zero. This equation is of no use in helping us to determine the values of R_1 and R_2.

2. $\Sigma F_y = 0$. This states that the algebraic summation of all forces in the vertical direction must equal zero. That is: $R_1 + R_2 = P$. Unfortunately, this equation involves two unknowns. The third equation for static equilibrium will solve this problem.

3. $\Sigma M = 0$. This equation tells us that this structure is not rotating with respect to any point. That is, for all tendencies to rotate in one direction there must be an equal and opposite tendency. In the problem being considered there will be a *tendency* for rotation to exist since the position of the forces is such that their lines of action do not pass through a common point (non-concurrent). For example, visualize that one of the supports is removed (refer to Fig. 3–10). It should be clear that the load will cause rotation about the remaining support. The magnitude of this tendency to rotate is equal to the product of the force and its perpendicular distance to the support or *(P)(b)*. The reaction R_1 must furnish an equal but opposite tendency to rotate with respect to the same point (see Fig. 3–11). The magnitude of this rotational tendency is equal to the product of the force and its perpendicular distance to the reference point or *(R₁)(L)*. Since these rotational tendencies must be equal, then

$$(R_1)(L) = (P)(b)$$

and

$$R_1 = \frac{(P)(b)}{L}$$

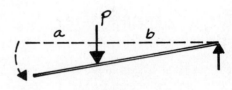

FIGURE 3–10

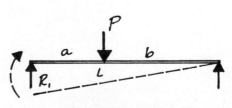

FIGURE 3–11

By a similar process, the reaction R_2 is found to be

$$R_2 = \frac{(P)(a)}{L}$$

Let's look at several numerical examples and apply the ideas discussed for finding the reactions.

Example 3–1 (Fig. 3–12)

In order to determine the reactions R_1 and R_2 that will satisfy the conditions for static equilibrium, we must satisfy the three basic equations.

1. $\Sigma F_x = 0$. Since there are no horizontal forces involved, this equation is already satisfied and is of no use to us in the determination of the vertical reactions.
2. $\Sigma F_y = 0$. This tells us that $R_1 + R_2 = 20^K$. This involves two unknowns; therefore, we must have another equation.
3. $\Sigma M = 0$. This expression will provide a solution for us. If we take moments about the point on the beam through which R_2 passes, we can solve for R_1.

$$(R_1)(24') = (20^K)(16')$$

and

$$R_1 = \frac{(20^K)(16')}{24'} = 13.33^K$$

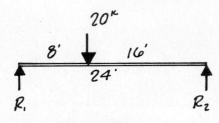

FIGURE 3–12

We could have taken moments about the point on the beam through which R_1 passes, which would provide a solution for R_2. The choice of taking moments about R_1 or R_2 is arbitrary. What is important is that a center of moments be selected through which one of the unknowns passes which, of course, will eliminate that unknown from the moment equation.

Now that we have a value for R_1, the simplest way to solve for R_2 is to go back to the equation $\Sigma F_y = 0$, where it was established that

$$R_1 + R_2 = 20^K$$

Using the value determined for R_1 and solving for R_2,

$$R_2 = 20^K - 13.33^K = 6.67^K$$

The solution for external reactions is now complete.

Example 3–2 (Fig. 3–13)

Let's look at a case with more than one load applied, as shown in Fig. 3–13. It has already been established that in cases such as this, involving only vertical forces, there are only two equations of statics available for determining the external reactions. Therefore, let's go directly to the use of $\Sigma M = 0$. Taking the point on the beam through which R_1 passes, then

$$\Sigma M_{R_1}: \quad (10^K)(6') + (18^K)(18') = (R_2)(30')$$

and

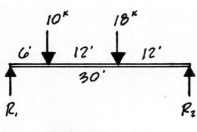

FIGURE 3–13

$$R_2 = \frac{(10^K)(6') + (18^K)(18')}{30'} = 12.8^K$$

Now that we have a value for R_2, we can use the other available equation of statics, $\Sigma F_y = 0$, and solve for R_1.

$$R_1 + R_2 = 28^K$$

$$\therefore R_1 = 28^K - 12.8^K = 15.2^K$$

UNIFORM LOADING

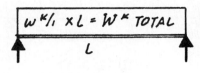

FIGURE 3–14

Most often, structural beams are loaded uniformly throughout their length or over a portion of their length, rather than with concentrated loads. For example, when a beam supports a slab, the load is delivered uniformly throughout the length of the beam on which the slab rests. Such a loading pattern is shown graphically as a block of load, as shown in Fig. 3–14. The magnitude of the uniform load is expressed, generally, in terms of pounds per linear foot or kips per linear foot. The *total* uniformly distributed load is found by multiplying the load per foot by the number of feet over which the load is distributed. For the purpose of finding the external reactions, the resultant total load (W^K) is used, and its position is taken as the center of gravity of the block of loading, which is in the middle of the rectangular block.

Example 3–3 (Fig. 3–15)

We'll now consider the beam shown in Fig. 3–15 and determine the external reactions R_1 and R_2. In this case a uniform load is being delivered to only 20' of the 24'-long beam, at the rate of $1^K/'$. The *total* uniform load is $1^K/' \times 20' = 20^K$. The position of this total load, for the purpose of finding reactions, is in the middle of the block, which is 10' from the left end. In order to find the reactions we'll take moments at R_1. Therefore,

$$\Sigma M_{R_1} = 0 = (20^K)(10') - (R_2)(24')$$

$$R_2 = \frac{200'^K}{24'} = 8.33^K$$

and $\Sigma F_y = 0$:

$$\therefore R_1 = 20^K - 8.33^K = 11.67^K$$

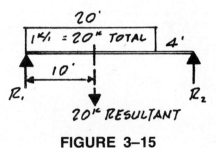

FIGURE 3–15

Example 3–4 (Fig. 3–16)

In the case shown in Fig. 3–16 we have a combination of uniform loading over a portion of the beam, and concentrated loads. In order to determine the external reactions R_1 and R_2, let's take moments at R_2.

$$\Sigma M_{R_2} = (12^K)(6') + (8^K)(12') + (16^K)(16') - R_1 (20')$$

$$R_1 = \frac{424'^K}{20'} = 21.2^K$$

and $\Sigma F_y = 0$:

$$\therefore R_2 = 36^K - 21.2^K = 14.8^K$$

END CONDITIONS

An important consideration at this time is that of the various support, or *end conditions,* that might be used or encountered in the design or analysis of structural members. These will be listed now with a brief discussion of the physical meaning of each.

1. *Simple support.* The examples that have been presented thus far represent "simply supported" beams. A beam that is simply supported is one where the support condition provides no restraint against rotation at the ends when the beam bends under the influence of the load. Therefore, an angular change takes place between the horizontal and a tangent

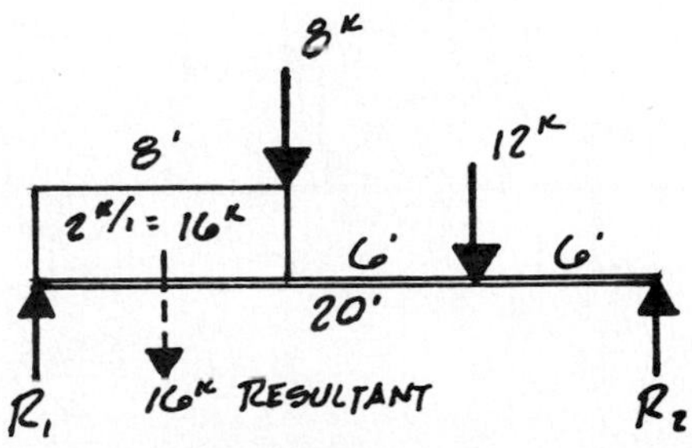

FIGURE 3–16

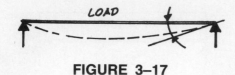

FIGURE 3–17

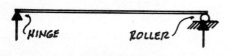

FIGURE 3–18

FIGURE 3–19

to the beam at the end. This is shown in Fig. 3–17, where the dotted line represents the general shape of the bent beam (greatly exaggerated) and the solid horizontal line shows the shape of the member before loading. This idea holds true regardless of the loading pattern. The simple support may be thought of (and is often referred to) as a "pinned" or "hinged" end. The idea of the "hinge" may best exemplify what is happening. The "hinge," much like the hinge on a door, allows total freedom for rotation but allows no translation, either vertically or horizontally. Graphically, the symbol that will be used throughout this text to depict a simply supported member will be arrows, as shown in Fig. 3–17 and the preceding example problems. The supports represented by the arrows may actually be columns, walls, or girders.

It should be mentioned, for the student who may supplement these studies by referring to another text, that variations in the graphic indication may be encountered. For example, some authors prefer to use the symbol shown in Fig. 3–18, which represents "knife-edge" supports. In any case, the principles, of course, are universal. It must be understood that in reality, the amount of rotation that takes place at the support in a simply supported beam is extremely small. Consequently, there are many connection conditions that may not be recognized as simple supports by the beginning student. For example, a wood beam nailed to a wood column or girder is, in fact, a simply supported condition. The nailing will not restrain the very small rotation that takes place. Likewise, a simple bolted connection between steel members is representative of a simple support. Structural members resting on walls are, likewise, simply supported.

2. *Roller support.* The "roller" type of support is a variation of a simple support. The graphic indication for this type of support condition is shown in Fig. 3–19. While the "roller" will allow the end of the supported member to rotate freely, it differs from the "hinge" in that it will not resist any forces in the direction in which it can roll. For example, in Fig. 3–20 a force is shown that has a horizontal component, acting on a beam. Since the "roller" can roll in the horizontal direction, the entire value of the horizontal component must be resisted by the "hinge." Since the "roller" can provide a reaction in the direction perpendicular to that in which it can move, then the reactions to the vertical component will distribute to the supports in accordance with its position.

"Rollers" are often used as supports in long-span roof structures in response to thermal effects on the structure. When a roof structure expands or contracts due to extreme temperature variations, serious damage can be done to the supporting walls if the member is restrained from changing dimension. In such cases the "roller" is actually a graphite or Teflon-coated plate on which the beam bears. This allows the beam to slide freely on the support, thereby avoiding the transfer of stresses to the supporting walls. Oftentimes, in very large scale structures such as highway bridges, absolute assurance must be provided for the beam to slide laterally on its support. In such cases, frictionless supports are necessary and the "roller" is used in the most literal sense, taking the form of a ball bearing in a socket. In most architectural structures, however, the use of a ball in a socket may be more elaborate than is necessary to meet the requirements.

3. *Single and double cantilevers.* Depending on factors involved in the architectural design process, it may be necessary or desirable to extend

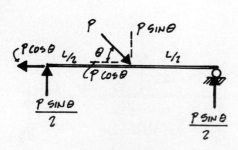

FIGURE 3–20

structural beams beyond the supports. The extension is referred to as a *cantilever*. In general, the cantilever is defined as a member that is supported at one end and free at the other. Where one end of a beam is extended to form a cantilever, this is referred to as a single-cantilever beam (see Fig. 3–21). Where both ends are extended, as shown in Fig. 3–22, this is referred to as a double-cantilever beam. In either case, one of the supports may be a "roller."

One of the important features of a cantilevered beam, which should be understood at this time, is that it affects the rotation that takes place at the support. For this reason, the cantilevered beam must be considered as something different from the simply supported beam, even though the supports are "hinges." For example, the simply supported beam of Fig. 3–23 is *totally* free to rotate at the supports when it bends under the influence of the load. The fact that the beam bends downward at the supports and the member rotates at the supports in the directions shown should be quite obvious. In a cantilevered beam this may not be so obvious. While the double cantilever of Fig. 3–24 is supported on "hinges" (or perhaps a roller in place of one "hinge"), this member is not *totally* free to rotate at the support because the rotations that take place are, to a great degree, dependent on the cantilevers. In fact, if the loads on the cantilevers are very large, or the cantilevers are very long, or both, the beam can conceivably bend, as shown in Fig. 3–25, with the corresponding rotation at the supports. This, of course, is opposite that of the simply supported beam of Fig. 3–23.

It might be well for the student to do some simple experiments, perhaps using a stick of balsa wood resting on supports, in order to help develop a better understanding of the physical behavior of cantilevered beams under various kinds of loading conditions. In any case, the procedure for determining the reactions for cantilevered beams is the same as that shown in previous problems.

4. *Fixed ends*. A *fixed end* condition is exactly the opposite of the simple support. Where the simple support allows complete freedom to rotate at the support, the fixed end allows no rotation. The member is said to be totally restrained at the support. The graphic indication for a fixed end condition is shown on the right-hand side of Fig. 3–26. In this figure the left-hand side is simply supported, and the bent shape of the beam is shown by the dotted line. In comparing this with the bent shape of a beam that is simply supported at both ends, as shown in Fig. 3–27, the physical consequence of the fixed end can easily be seen.

In reality, it is virtually impossible to completely fix the end of a beam. This condition is brought to the attention of the student because even as an abstract idea, the concept of the fixed end forms the basis for certain analytical procedures that will be encountered later in the text (Chapters 16 and 17). The determination of the reactions in a member with a fixed end (or both ends fixed) is somewhat more complex than for a simply supported beam. It is, in fact, beyond the scope of this chapter and will be covered in Chapter 16.

NON-PARALLEL FORCES

While parallel forces represent one form of non-concurrent forces, since they never intersect, many structural configurations will involve something more than just parallel forces. The technique, however, to determine the external reactions required for equilibrium still involves nothing more than

FIGURE 3–21

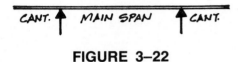

FIGURE 3–22

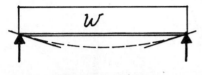

FIGURE 3–23

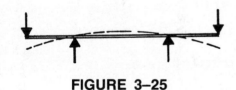

FIGURE 3–24

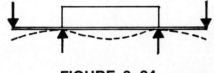

FIGURE 3–25

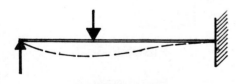

FIGURE 3–26

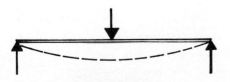

FIGURE 3–27

satisfying the basic equations of statics. We'll now look at several examples involving structures of this kind.

Example 3–5 (Fig. 3–28)

Let's consider the structure shown in Fig. 3–28 and determine the external reactions required for static equilibrium. Due to the 2000# weight, the entire structure will tend to rotate in a clockwise direction. In order for this structure to be in equilibrium there will be reactions at the anchorage points. The cable will be pulling at the anchor and a reaction *(R)* will develop which must be equal and opposite to the pull in the cable. The reaction *R* has horizontal and vertical components, *Z* and *Y*. In addition to the horizontal and vertical effects, there will also be a rotational tendency due to the position of the applied load. Consequently, all the equations of static equilibrium will have to be satisfied.

1. $\Sigma F_x = 0$. This means that $W = Z$. But since both of these are unknowns, this equation is not of much use at this point.
2. $\Sigma F_y = 0$. This means that $Y + X = 2000\#$. Both Y and X are unknown at this point; consequently, this equation does not provide any answers.
3. $\Sigma M = 0$. This tells us that no rotation is taking place with respect to any point. If we select the point where Z and Y intersect as the center of moments, this equation will be helpful (the intersection of W and X could have been used). Previously established principles tell us that Z and Y produce no moment about that point since they pass through the point. Also, the line of action of W passes through the point o and produces no moment about that point. Therefore, taking moments about point o, the moment equation is

$$\Sigma M_o = 0 = (2000\#)(20') - (X)(10')$$

assuming that X acts upward.* Therefore, $X = 4000\#$.

To satisfy $\Sigma F_y = 0$, $Y = 2000\#$ acting downward. By the geometry of the structure, $\tan 45° = Y/Z$ and $Z = 2000\#$ acting to the left.

In order to satisfy $\Sigma F_x = 0$, $W = 2000\#$ acting to the right. The solution for the external reactions is now complete.

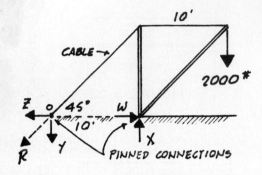

FIGURE 3–28

Example 3–6 (Fig. 3–29)

We'll now determine the external reactions required for equilibrium on the bracket type of structure shown in Fig. 3–29. To do this, we must satisfy the three equations of statics.

1. $\Sigma F_x = 0$. This equation tells us that $A = B$ in magnitude and that they must be opposite in direction, since they are the only horizontal forces involved. But from this equation alone, we cannot determine a numerical value for the magnitude.
2. $\Sigma F_y = 0$. Since there is a roller type of connection on the bottom leg of the bracket, then no vertical force can be resisted at this point. Therefore, all of the vertical forces on the structure must be resisted by the reaction V. Consequently,

$$V = 10^K \uparrow$$

3. $\Sigma M = 0$. If we take moments about the point where A and V intersect, we'll get an expression with the reaction B involved, which can be solved. Assuming the direction of B is to the right, then

$$\Sigma M_{AV} = 0 = (6^K)(6') + (4^K)(18') - B(6')$$

$$\therefore B = \frac{108'^K}{6'} = 18^K \rightarrow$$

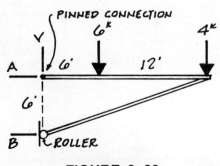

FIGURE 3–29

*When initially writing the moment equation, it is only necessary to make an assumption regarding the direction of the unknown reaction. If the assumption is incorrect, the answer will come out with a negative sign, but the magnitude will be correct. Therefore, if the direction of unknown reactions is not easily determined by inspection, simply make an assumption and the sign of the answer will verify or correct the assumption.

and to satisfy $\Sigma F_x = 0$:

$$A = 18^K \leftarrow$$

The solution for the external reactons is now complete.

PARALLEL FORCE RESULTANT

Occasionally, a problem contains a large number of parallel forces. Therefore, it may be convenient to determine the resultant of this group of forces and deal with it in the solution of a problem.

The resultant of a group of parallel forces is defined in the same way as the resultant of a group of concurrent forces. That is, *the resultant is the single force that has the same effect as the group of forces.* However, it must be considered that with a group of parallel forces, the effect is not only one of translation but also rotation, and the resultant must have the same rotational effect as the group of parallel forces. The magnitude of the resultant of parallel forces is easily determined, but determining the position of the resultant requires a small amount of work.

Example 3–7 (Fig. 3–30)

Consider the problem shown in Fig. 3–30. It is desired to determine the resultant of the given group of parallel forces. The magnitude of the resultant is determined by the algebraic addition of the forces. The forces shown produce a translational tendency in the downward direction with a magnitude of 11^K. Therefore, the resultant must have a downward *direction* and a *magnitude* of 11^K.

The *position* of the resultant (R) can be determined by remembering that the resultant must have the same rotational effect, with respect to any point, as the group of forces. Taking point A as the reference, the forces produce a rotational tendency of*

$$\Sigma M_A = (3^K)(3') + (5^K)(9') + (1^K)(12') = 66'^K$$

All of the forces produce a rotational tendency, with respect to point A, in the clockwise direction. Given that the resultant is directed downward, then it must be to the right of point A in order to produce clockwise rotation. Its distance from point A must be such that $(R)(d) = 66'^K$.

Therefore, $(11^K)(d) = 66'^K$ and $d = 6'$.

Based on the above example, a general expression for determining the location of the resultant from a given reference point may be stated as

$$d = \frac{\Sigma(\text{Forces} \times \text{Distances})}{\Sigma \text{ Forces}}$$

Where d is the distance from the reference point. Whether this distance is measured to the right or left of the reference point can be determined by inspection, remembering that the resultant must produce the same rotational tendency, with respect to the reference point, as the group of forces.

The location of the resultant of parallel forces is generally referred to as the *center of gravity* of the forces. In the following chapter we will elaborate on the center of gravity and related concepts.

*The choice of reference point is totally arbitrary.

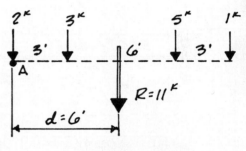

FIGURE 3–30

PROBLEM 3–1

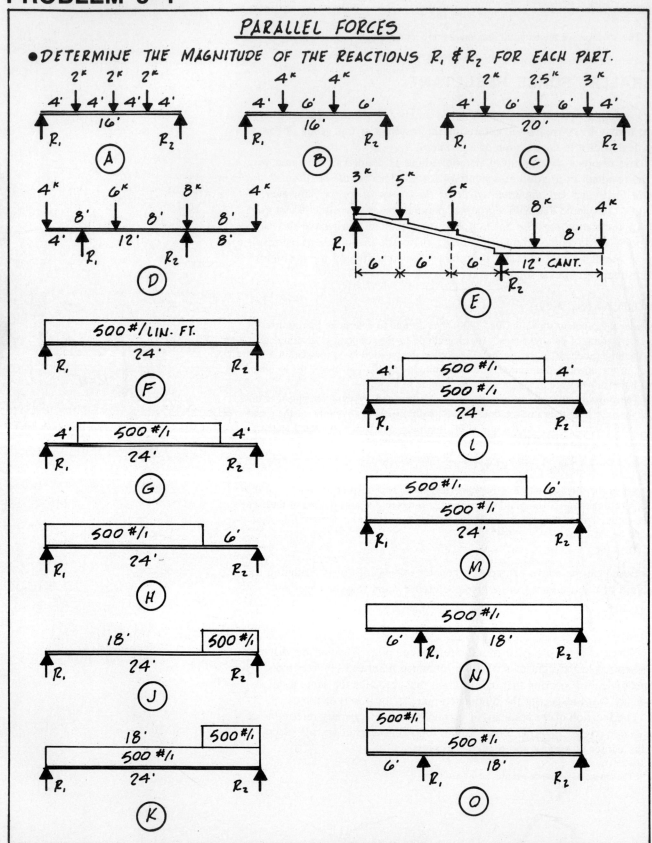

PARALLEL FORCES

• DETERMINE THE MAGNITUDE OF THE REACTIONS R_1 & R_2 FOR EACH PART.

PROBLEM 3–2

PARALLEL FORCES

● DETERMINE THE REACTIONS INDICATED FOR EACH OF THE FOLLOWING:

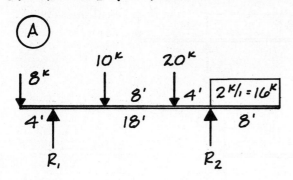

(A)

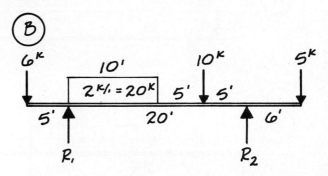

(B)

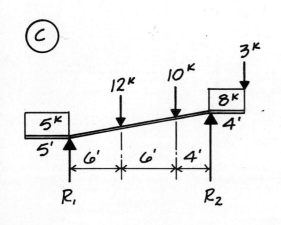

(C)

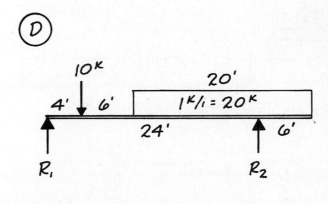

(D)

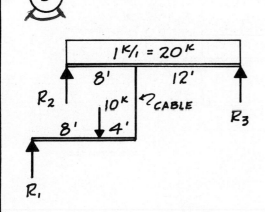

(E)

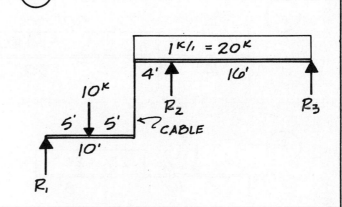

(F)

PROBLEM 3–3

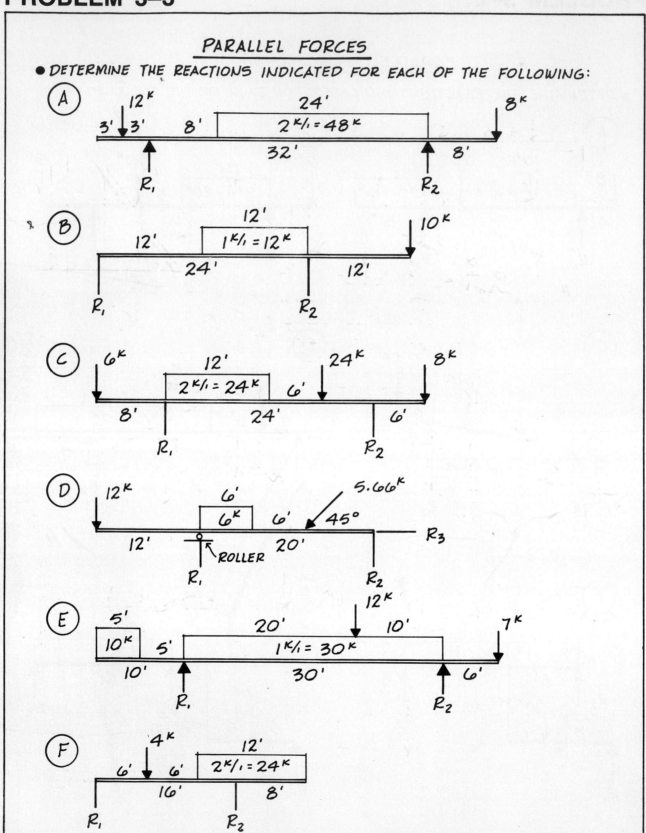

PARALLEL FORCES

● DETERMINE THE REACTIONS INDICATED FOR EACH OF THE FOLLOWING:

A

B

C

D

E

F

PROBLEM 3-4

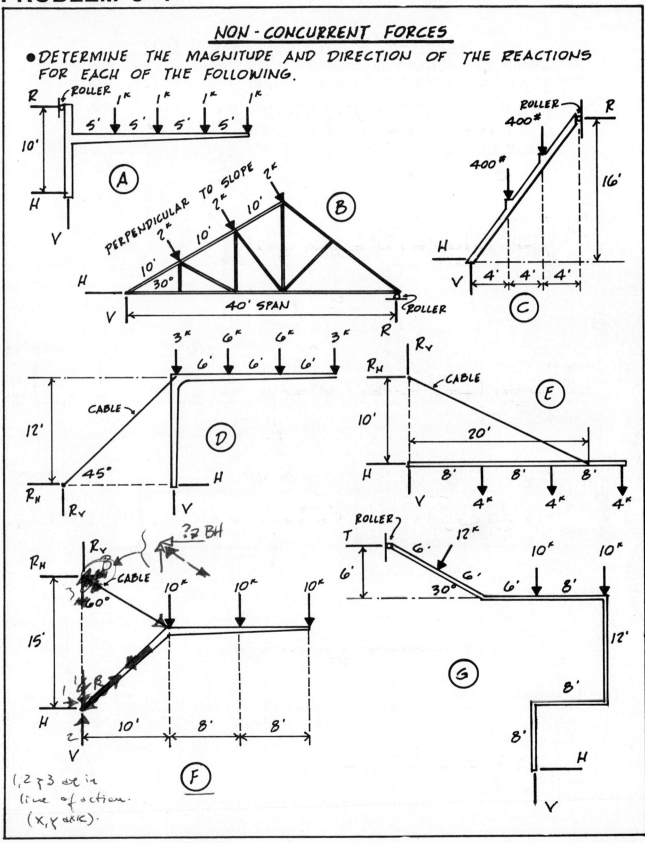

NON-CONCURRENT FORCES

- DETERMINE THE MAGNITUDE AND DIRECTION OF THE REACTIONS FOR EACH OF THE FOLLOWING.

1, 2 & 3 are in
line of action.
(X, Y & K).

PROBLEM 3–5

RESULTANT OF NON-CONCURRENT FORCES

DETERMINE THE MAGNITUDE, DIRECTION, AND POSITION OF THE
RESULTANT FOR EACH OF THE FOLLOWING, WITH RESPECT TO POINT "A".

4

Center of Gravity

INTRODUCTION

In the preceding chapter a process was shown to determine the location of
the resultant force of a group of parallel forces. Frequently, in architectural
problems, the need arises to determine the position of the resultant force of
gravity due to the weight of a body. For example, it would be desirable to
know the location of the *center of gravity* of the stone ledge shown in Fig.
4–1. If the stone ledge is placed so that the position of its resultant weight
is supported by the structure below, it will be stable (provided it is not acted
upon by external forces). This would be true if the center of gravity of the
stone was as far out as the edge of the supporting structure below. If, however, the center of gravity of the stone were positioned outside of the supporting structure, as shown in Fig. 4–2, there would be a tendency to rotate
about point x, since the position of the resultant weight has a lever arm to
this point.

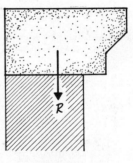

FIGURE 4–1

If the stone were supported by a "point" support, as shown in Fig. 4–
3, and the support coincided with the line of action of the center of gravity
of the stone, no rotation would take place. It may be said that:

> The point about which a body may be balanced and which is the point of
> application of the resultant weight is called the *center of gravity* of the body.

The center of gravity of a body may be thought of as an infinitesimal
point where the tendency to rotate caused by everything to one side of this
point is countered by an equal and opposite tendency due to everything to
the other side of the point.

COMPUTATIONAL PROCEDURES

Determining the location of the center of gravity of a symmetrical and homogeneous body poses no problem since this can be determined by inspec-

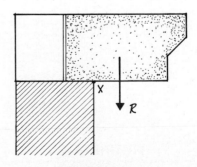

FIGURE 4–2

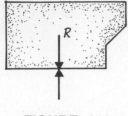

FIGURE 4–3

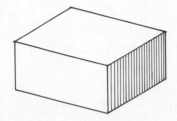

FIGURE 4–4

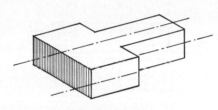

(2 AXES OF SYMMETRY)

FIGURE 4–5

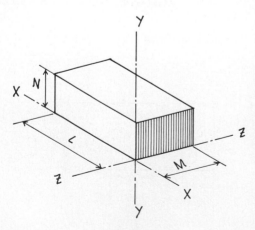

FIGURE 4–6

tion. In the case of symmetrical and homogeneous bodies, the center of gravity lies at the geometric center. This is specifically so in a body such as that shown in Fig. 4–4, which is regular in all directions and has three axes of symmetry. In the case where a body has one or two axes of symmetry, as in Fig. 4–5, then only two coordinates necessary for defining the location of the infinitesimal point called the center of gravity can be determined by inspection, but the third coordinate will require some computation. The following principle will prove useful in determining the location of the center of gravity of three-dimensional bodies:

> For a *homogeneous* body, the center of gravity will lie on the axis, or axes, or symmetry.

Since we are dealing with three-dimensional bodies, the location of the center of gravity must be defined by three coordinates. Conventionally, three axes (*x–x, y–y, z–z*) are used as a reference, and the center of gravity of the object is located from the intersection of these axes. In Fig. 4–6 the axes are shown and are located along edges of the object. The choice of location of the coordinate axes is purely arbitrary. The intersection of the axes could have been at another corner of the object, off of the object, or, for that matter, somewhere within the object.

Another conventional bit of terminology we will use is the names given to the various coordinates measured from the intersection of axes, or the "zero" point. These distances are called *bar x, bar y,* and *bar z,* written as $\bar{x}, \bar{y}, \bar{z}$. Each of these signifies the distance measured from the zero point along the appropriate axis. That is, the distance measured along the *x–x* axis is given as $\bar{x}$, and so on. The object shown in Fig. 4–6 has dimensions of *L* (length), *M* (width), and *N* (height). The location of the center of gravity from the reference axis would be given as follows:

$$\bar{x} = L/2$$

$$\bar{y} = N/2$$

$$\bar{z} = M/2$$

It was mentioned earlier that when dealing with irregular bodies some computation will be required to locate all coordinates of the center of gravity. There are, essentially, two methods available. They are the *Method of Addition* and the *Method of Subtraction*. In the Method of Addition the irregular object under consideration is visualized as broken into regular parts and the results are added. In the Method of Subtraction the irregular object is visualized as a regular object and then those parts which, in fact, do not exist are subtracted from the results. To demonstrate, we'll go to an example problem.

Example 4–1 (Fig. 4–7 through 4–14)

Let's consider the object shown in Fig. 4–7 and determine the location of the center of gravity using the Method of Addition. The object is irregular in shape, but is homogeneous and is made of a material with a density (or unit weight) of $100\#/\text{ft}^3$.

In Fig. 4–8, two views are shown. In View #1 (plan view) it can be readily seen that $\bar{x}$ is determined by inspection since an axis of symmetry exists. In View #2 it is also seen that $\bar{y}$ can be determined by inspection because there exists another axis of symmetry. In View #1 it can also be seen that $\bar{z}$ cannot be determined by inspection because the object is not symmetrical in this direction. The procedure to determine $\bar{z}$ is as follows:

1. Visualize the irregular object broken into regular constituents for which the individual centers of gravity are known by inspection. In the object we are dealing with, parts *A* and *B* are regular in shape, and the center of gravity for each piece may be determined by inspection.

2. Determine the weight of each of the regular parts, which is simply the product of the volume of the part and its unit weight. For the figure we are dealing with:

   Part *A:* $W = (2')(4')(8')(100\#/\text{ft}^3) = 6400\#$

   Part *B:* $W = (2')(3')(6')(100\#/\text{ft}^3) = 3600\#$

3. Sketch the object (or visualize in your mind's eye) so it is oriented with the axis in question placed horizontally on the sketch (such as shown in Fig. 4–9). Show the *resultant* weight of each of the regular parts acting through the center of gravity of the part.

4. Visualize the object as tending to rotate with respect to the axis that is perpendicular to the page (referring to Fig. 4–9, this would be the *x–x* axis). This problem then becomes one of finding the location of the resultant of parallel forces, as presented in Chapter 3. In other words, the resultant weight of this object must produce the same tendency to rotate with respect to the reference point as the summation of the rotational tendencies of the parts. Referring to Fig. 4–9 and 4–10:

$$R\bar{z} = (6400\#)(2') + (3600\#)(7')$$

$$\bar{z} = \frac{12800'\# + 25200'\#}{10000\#} = 3.8'$$

The third coordinate has now been determined and the problem is complete. To summarize:

$$\bar{x} = 4'$$

$$\bar{y} = 1'$$

$$\bar{z} = 3.8'$$

The process just shown is a method where the irregular object is broken down into regular parts and the results of each part are added. This is a Method of Addition. The same problem could have been solved using a Method of Subtraction. The process is very much like that used in the Method of Addition except that the irregular figure is made into a regular shape for which the center of gravity is determined by inspection. Then the results of those parts that do not exist are subtracted from the *assumed* regular shape (refer to Fig. 4–11). The resultant weight of the assumed regular shape $= (2')(10')(8')(100\#/\text{ft}^3) = 16000\#$ (see Fig. 4–12) and its center of gravity is 5' from the reference point, which is the intersection of the axes. From this we must subtract the results of those parts that do not exist. The weight to be subtracted $= (2\ \text{pieces})(2.5')(2')(6')(100\#/\text{ft}^3) = 6000\#$ and the center of gravity

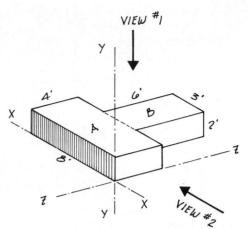

FIGURE 4–7

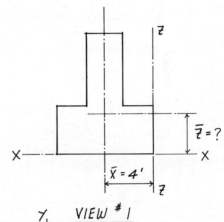

FIGURE 4–8

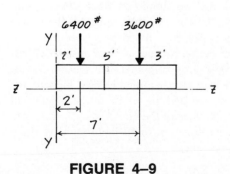

FIGURE 4–9

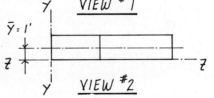

FIGURE 4–10

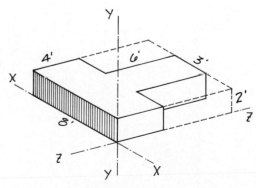

FIGURE 4–11

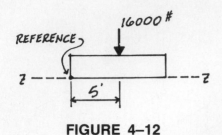

FIGURE 4–12

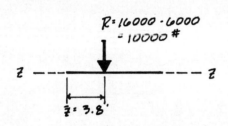

FIGURE 4–13

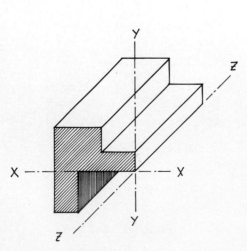

FIGURE 4–14

of the voids is 7' from the reference point (see Fig. 4–13). Consequently (referring to Fig. 4–14):

$$(R)(\bar{z}) = (16000\#)(5') - (6000\#)(7')$$

$$10000\#(\bar{z}) = 80000'\# - 42000'\# = 38000'\#$$

$$\bar{z} = \frac{38000'\#}{10000\#} = 3.8'$$

Each of the methods shown may be used to determine the location of the center of gravity of an irregular object with equal success. This is so even if the object is composed of parts that are not of the same unit weight (non-homogeneous object). Generally, however, the Method of Addition is somewhat simpler when dealing with non-homogeneous objects.

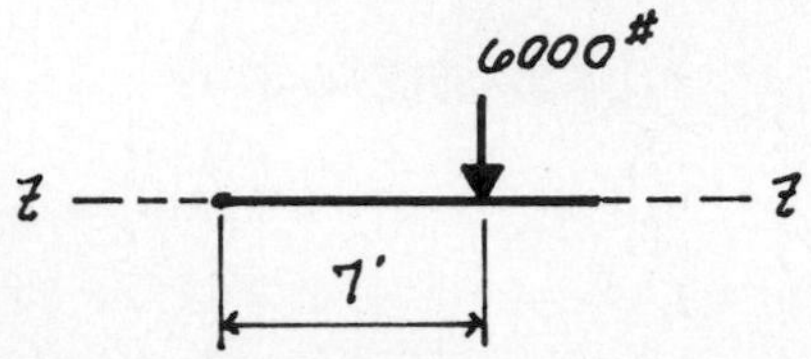

GENERAL EXPRESSIONS

The preceding discussions may be summarized with a single general expression for locating the center of gravity of an irregular object:

$$R\bar{x} = \Sigma Wx$$

where R = total resultant weight
$\bar{x}$ = distance from reference axis to center of gravity of the object (may be $\bar{y}$ or $\bar{z}$ depending on the axis being investigated)
W = weight of regular part which is part of the object
x = distance from reference axis to center of gravity of part being considered

This general expression can be somewhat simplified when dealing with *homogeneous* objects, such as the object used in the previous example. In looking back at the solution to Example 4–1, it can be recognized that the unit weight of the material is common to all terms on both sides of the equality sign. This common term can be factored out and the equation divided by this term, leaving only the expressions that deal with the volume. Consequently, when dealing with homogeneous objects, the general expression for locating the center of gravity becomes:

$$V_t\bar{x} = \Sigma Vx$$

where V_t = total volume
$\bar{x}$ = distance from reference axis to center of gravity of the object
V = volume of regular part which is part of the object
x = distance from reference axis to center of gravity of part being considered

The preceding expression can further be simplified when dealing with homogeneous objects of constant length. For example, consider the object shown in Fig. 4–15. The $\bar{z}$ value is easily determined by inspection since there is an axis of symmetry in this direction because of the constant length. The problem here is to determine $\bar{x}$ and $\bar{y}$ by computational means. Essentially, all that needs to be dealt with here is a two-dimensional view of the shaded area. Further, since the unit weight and the length of the object will be constant to every expression on both sides of the equality sign, these can be factored out and both sides of the equation divided through by these values. The general expression then becomes:

$$A_t\bar{x} = \Sigma Ax$$

where A_t = total area (shaded portion of Fig. 4–15)
$\bar{x}$ = distance from reference axis to center of gravity of the area

FIGURE 4–15

A = area of regular part which is part of the total area
x = distance from reference axis to center of gravity of partial area
    being considered

STRUCTURAL STABILITY

At the beginning of this chapter the idea of stability and the relationship to the center of gravity was shown in the discussion of the stone ledge and the location of its center of gravity relative to the supporting element. These ideas have a wide variety of application.

Example 4–2 (Figs. 4–16 and 4–17)

Consider the reinforced concrete wall of Fig. 4–16 acted upon by a lateral wind force of $20\#/ft^2$. It would be desirable to know if the wall, due to its own weight, will resist the tendency to overturn created by the lateral wind force. Since the wind force is considered to be acting uniformly over the entire surface of the wall (such as the "block" of load on a uniformly loaded beam), the *resultant* lateral wind force will be used. The value of this resultant is simply the product of the unit force and the area over which this force is acting, or

$$R = (20\#/ft^2)(10')(12') = 2400\#$$

Referring to Fig. 4–17, the position of this force is at the middle of the block of loading. It should be recognized that if the wall is to overturn, then it must rotate about point x. The resultant wind force produces an "overturning moment" of $(2400\#)(5') = 12000'\#$ in the counterclockwise direction. This is resisted by the weight of the wall, the resultant of which will act through the center of gravity of the wall. This will produce a rotational tendency (with respect to point x) in the clockwise direction. Given that reinforced concrete has a unit weight of $150\#/ft^3$,

$$P = (150\#/ft^3)(1')(10')(12') = 18000\#$$

This force acting through the center of gravity of the wall produces a clockwise moment $= (18000\#)(.5') = 9000'\#$. The wind produced an overturning moment of $12000'\#$, which is greater than the resistance provided by the weight of the wall; consequently, the wall will be rotating about point x in the counterclockwise direction. The wall will turn over. This can be taken care of by some sort of positive anchorage of the wall to the ground, or by making the wall thicker and therefore heavier. There is also the possibility of building the wall with some curvature or with wing walls, as shown in Fig. 4–18. This will alter the location of the center of gravity of the wall, thereby creating a larger lever arm to resist overturning. It should be recognized from the arrangements of Fig. 4–18 that the center of gravity may be a point in space, related to, but not part of, the body itself. The architectural designer should realize that the concept of center of gravity and its relationship to the idea of structural stability can play an important role in those decisions that deal with the development of form.

CENTROID OF AREAS

It was pointed out previously that when dealing with an object that is homogeneous and of constant length, the problem of finding the center of gravity is reduced to finding the center of gravity of the cross-sectional area of the object. For a variety of purposes we will, in succeeding discussions, be required to find the center of gravity of cross-sectional areas, which are two-dimensional problems. That is, we will not be concerned with the location of a third axis. When dealing with two-dimensional problems (where the intersection of only two axes is required), the center of gravity of the cross-sectional area is known as the *centroid*.

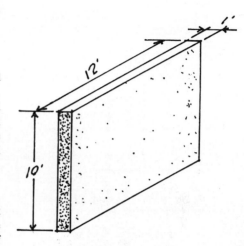

FIGURE 4–16

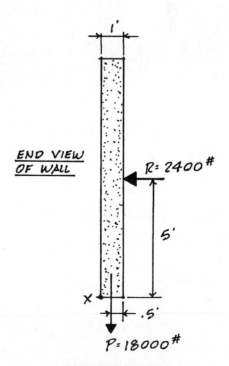

FIGURE 4–17

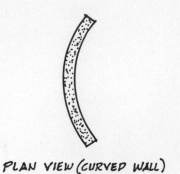

PLAN VIEW (CURVED WALL)

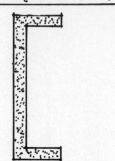

PLAN VIEW (WING WALL)

FIGURE 4–18

Considering that, generally, cross sections of structural members have at least one axis of symmetry, then the problem of locating the centroid is a relatively simple matter. The general expression used for this purpose is:

$$A_t \bar{x} = \Sigma A x$$

where A_t = total cross-sectional area

$\bar{x}$ = distance from reference axis to the centroid of the area

A = area of the regular part which is part of the total area

x = distance from the reference axis to the centroid of the partial area being considered

Example 4–3 (Fig. 4–19)

It is desired to determine the centroidal location of the T-shaped cross section of Fig. 4–19. Since there is an axis of symmetry, the centroid will be on that axis. This is determined by inspection. The problem becomes one of determining the location of the intersecting axis, shown as $\bar{x}$. In Fig. 4–19, the top of the cross section has been chosen as the reference, but this is an arbitrary choice. Using this and the general expression

$$A_t \bar{x} = \Sigma A x$$

$\underline{A_t}$

$$\text{(3'')(12'')} = 36 \text{ in.}^2$$
$$\text{(2'')(10'')} = \underline{20 \text{ in.}^2}$$
$$A_t = 56 \text{ in.}^2$$

$\underline{\Sigma A x}$

$$(36 \text{ in.}^2)(1.5'') = 54 \text{ in.}^3$$
$$(20 \text{ in.}^2)(8'') = \underline{160 \text{ in.}^3}$$
$$\Sigma A x = 214 \text{ in.}^3$$

Therefore,

$$56 \text{ in.}^2 \, (\bar{x}) = 214 \text{ in.}^3$$

$$\bar{x} = \frac{214 \text{ in.}^3}{56 \text{ in.}^2} = 3.82 \text{ in.}$$

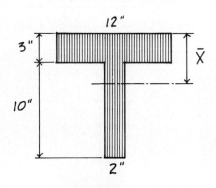

FIGURE 4–19

Hopefully, it is recognized that this procedure is identical to that used for finding the center of gravity of a three-dimensional object, except that we have only been concerned with an *area,* rather than a mass.

In the introduction to this chapter it was stated that "the center of gravity of a body may be thought of as an infinitesimal point where the tendency to rotate caused by everything to one side of this point is countered by an equal and opposite tendency due to everything to the other side of the point." Although somewhat of an abstraction (since an area has no weight), the centroidal axis may be thought of in the same way. That is, the *centroidal axis* is that axis where the tendency of the *area alone* of everything to one side to rotate is countered by an opposite tendency due to everything on the other side of the axis. Referring to Fig. 4–20, this means that the *product* of $A_1 \bar{x}_1$ must be equal to the product of $A_2 \bar{x}_2$ where $\bar{x}_1$ and $\bar{x}_2$ are the distances from the centroid of the entire cross-section to the centroids of the areas that lie on either side of this axis. It must be emphasized that, in Fig.

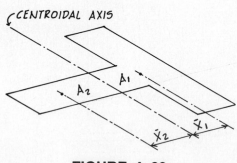

FIGURE 4–20

4–20, the areas A_1 and A_2 *are not* equal to each other. These areas can only be equal to each other, coincidentally, when dealing with an area that is symmetrical about both axes, such as Fig. 4–21. This same statement is true concerning the values of $\bar{x}_1$ and $\bar{x}_2$.

The tendency for an *area alone* to rotate about an axis in the plane of that area is known as its:

$$\text{Static Moment} = A\bar{x} = Q$$

This bit of terminology is something that will be used considerably in the solution of future problems, and while the ideas presented in the preceding discussion may seem somewhat abstract, the physical significance of these ideas will, it is hoped, become apparent.

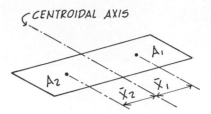

FIGURE 4–21

Example 4–4 (Fig. 4–22)

We will now determine the location of the centroidal axis of the cross-sectional area of Fig. 4–22. In this case, both the Methods of Addition and Subtraction will be used. In order to have appropriate data for this problem, reference must be made to Data Sheet D–2 in the Appendix.

Since the cross section shown has a vertical axis of symmetry, then the problem becomes one of finding $\bar{x}$. Using the top of the figure as the reference:

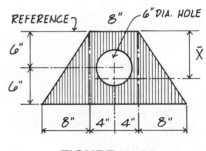

FIGURE 4–22

ΣAx

$$2[(48 \text{ in.}^2)(8'')] = \quad 768.0 \text{ in.}^3$$

$$(96 \text{ in.}^2)(6'') = \quad 576.0 \text{ in.}^3$$

$$-\bigcirc \quad (28.3 \text{ in.}^2)(6'') = \quad \underline{-169.8 \text{ in.}^3}$$

$$1174.2 \text{ in.}^3$$

and $A_t = 163.7 \text{ in.}^2$

$$\therefore \bar{x} = \frac{1174.2 \text{ in.}^3}{163.7 \text{ in.}^2} = 7.17''$$

The problem could have been solved strictly by using the Method of Subtraction. The cross section could have been visualized as a solid rectangle $12'' \times 24''$ and then the triangles and circle subtracted. The student can verify this for the sake of practice.

UNIFORM UNIT STRESS

If a group of 100 fibers were subjected to a load of 1000#, as shown in Fig. 4–23, then each individual fiber would carry 10#. If there were more fibers, then the individual stress taken by each would be less. All structural materials may be thought of as being made of "fibers," with each individual fiber carrying its share of the total load. The average stress per "fiber" may be found by dividing the total load by the number of fibers. Conventionally, for standardization and ease of computation, we do not work with these fibers in an infinitesimal sense, but rather in a wholesale way. That is, we deal with *one square inch* of cross section at a time, and the individual stress is not expressed as pounds per fiber, but as pounds per square inch of cross section. Such an expression is referred to as the *unit stress*. Therefore, if a bar $2'' \times 2''$, as shown in Fig. 4–24, is subjected to a load of 10000#, the unit stress is 10000# divided by 4 in.^{2} = 2500#/in.^{2} The unit stress is most often expressed in lbs/in.^{2}, but such units as lbs/ft^2 and tons/ft^2 are sometimes used depending on the scale of the item with which we are dealing.

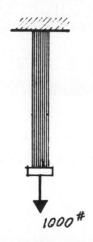

1000#

FIGURE 4–23

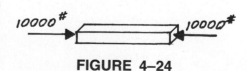

10000#   10000#

FIGURE 4–24

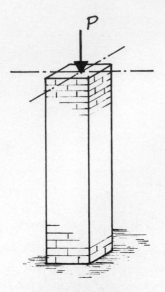

FIGURE 4–25

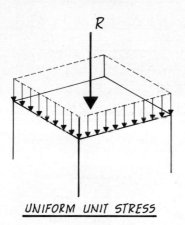

UNIFORM UNIT STRESS

FIGURE 4–26

AXIAL LOADS

If a concentrated load is placed on the centroid of a cross section, as shown in Fig. 4–25, the resulting unit stresses will be *uniformly* distributed throughout the entire cross section. Conversely, if a cross section is subjected to a uniformly distributed unit stress, as shown in Fig. 4–26, then the position of the resultant will be at the centroid of the cross section. When a member is subjected to a concentrated load placed at the centroid, or a *uniform unit stress,* the member is said to be *axially loaded.*

We may summarize the preceding discussions with two very important principles:

1. Any load that is applied on the centroid of the cross section of a structural member will produce uniformly distributed unit stresses throughout the cross section (the member must be of a homogeneous material).

2. The uniform unit stress produced by a load placed on the centroid of a section is found by dividing the load by the cross-sectional area. Expressed mathematically:

$$f = P/A$$

where f = unit stress (sometimes called the "fiber" stress)

P = total load

A = cross-sectional area

When dealing with structural materials that are being stressed due to applied loads, we are interested in the level of unit stresses being produced and the capacity of the material to withstand these stresses. All materials of construction have an "ultimate" stress that can be applied, which will bring about failure of the material. It should be obvious that when designing a structure, we would not want to stress the materials to their ultimate capacity, under the anticipated loadings. Instead, we use an "allowable" unit stress in the design process. Ultimate unit stresses for the various structural materials are determined by testing. Allowable unit stresses for the various materials are normally based on the ultimate stresses with factors of safety applied. The issues of ultimate stress and allowable stress will be dealt with in more detail in subsequent chapters which deal with the structural design process.

PROBLEM 4–1

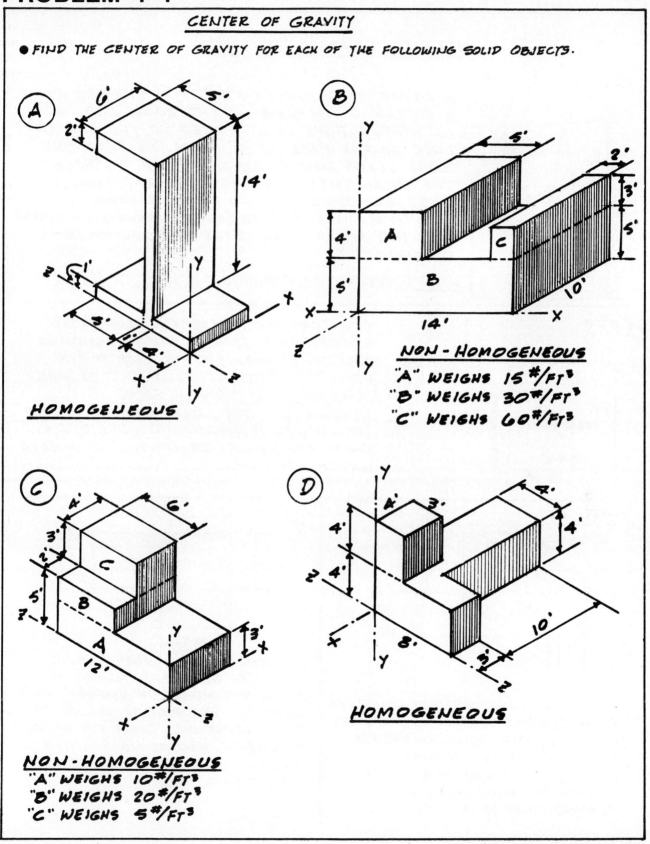

CENTER OF GRAVITY

● FIND THE CENTER OF GRAVITY FOR EACH OF THE FOLLOWING SOLID OBJECTS.

A 14' 6' 5' 2' 1' 3' 4'

HOMOGENEOUS

B 5' 2' 3' 5' 4' 5' 14' 10'

NON - HOMOGENEOUS
"A" WEIGHS 15#/FT³
"B" WEIGHS 30#/FT³
"C" WEIGHS 60#/FT³

C 4' 6' 3' 2' 5' 7' C B A 3' 12'

NON - HOMOGENEOUS
"A" WEIGHS 10#/FT³
"B" WEIGHS 20#/FT³
"C" WEIGHS 5#/FT³

D 4' 3' 4' 4' 4' 4' 8' 10' 5'

HOMOGENEOUS

PROBLEM 4–2

STRUCTURAL STABILITY

(I)

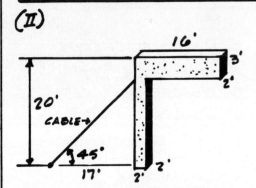

A BLOCK OF GRANITE (175 #/FT³) SHAPED AS A PYRAMID IS 5'x5' AT THE BASE AND IS 16' HIGH. A ROPE IS ATTACHED TO THE CENTER-LINE OF ONE FACE AT 10' FROM THE BASE. THE ROPE LEADS AWAY AT AN ANGLE OF 30° BELOW THE HORIZONTAL, AS SHOWN. WHAT PULL (P) MUST BE APPLIED TO THE ROPE IN ORDER TO JUST OVERTURN THE PYRAMID? ASSUME FRICTION WILL NOT ALLOW IT TO SLIDE BEFORE OVER-TURNING

<u>NOTE</u>: VOLUME OF PYRAMID = ⅓ A_BASE X HEIGHT.

(II)

IT IS DESIRED TO ERECT A REINFORCED CONCRETE "HALF ARCH" AS SHOWN. ASSUMING THAT THE ANCHORAGE OF THE ARCH TO THE GROUND PROVIDES NO RESISTANCE TO OVER-TURNING, WHAT TENSILE FORCE IS NECESSARY IN THE CABLE IN ORDER TO KEEP THE ARCH FROM OVERTURNING DUE TO ITS OWN DEAD WEIGHT? REINFORCED CONCRETE WEIGHS 150 #/FT³.

(III)

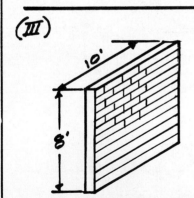

A WALL MADE OF 8" THICK CONCRETE BLOCK (60# PER SQ.FT. OF SURFACE AREA) AND IS 8' HIGH. WHAT UNIFORMLY DISTRIBUTED WIND PRESSURE (IN P·S·F·) WILL CAUSE THE WALL TO OVERTURN? ASSUME THE WIND FORCE IS APPLIED PERPENDICULAR TO THE WALL.

(IV)

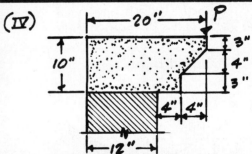

A STONE LEDGE, AS SHOWN, IS SUPPORTED BY A SOLID WALL. THE STONE HAS A UNIT WEIGHT OF 200 #/FT³. WHAT LOAD "P" PER FOOT OF LENGTH OF STONE LEDGE WILL CAUSE THE STONE TO ROTATE? ASSUME NO POSITIVE ANCHORAGE OF STONE TO THE SUPPORTING WALL.

PROBLEM 4-3

CENTROIDS

● FOR EACH OF THE FOLLOWING, DETERMINE THE LOCATION OF THE CENTROID ($\bar{x}$), MEASURED FROM REFERENCE AXIS A-A.

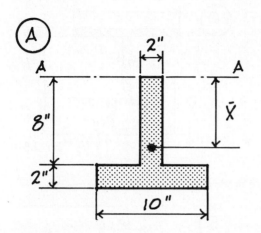

(A)

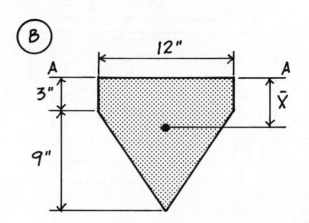

(B)

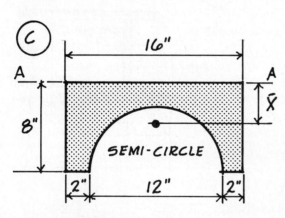

(C) SEMI-CIRCLE

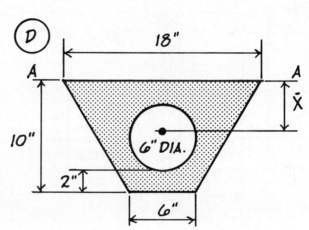

(D) 6" DIA.

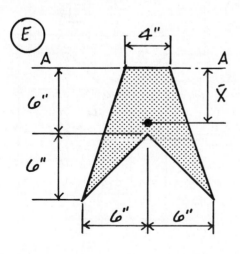

(E)

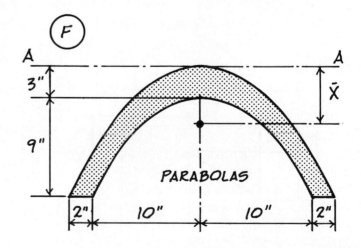

(F) PARABOLAS

PROBLEM 4-4

CENTROIDS

● DETERMINE THE LOCATION OF THE CENTROID ($\bar{X}$) FROM THE GIVEN REFERENCE FOR EACH OF THE FOLLOWING CROSS-SECTIONS.

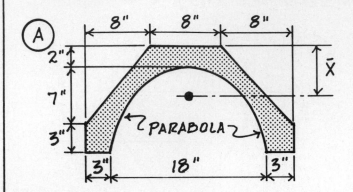

(A)

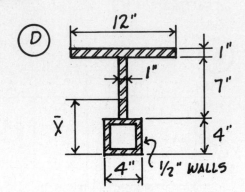

(D)

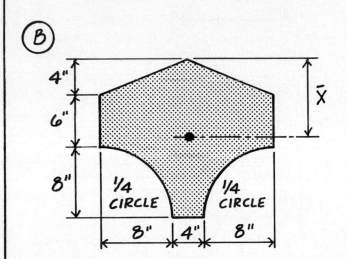

(B)

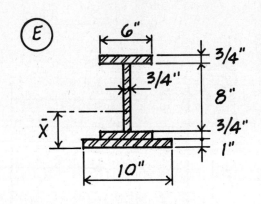

(E)

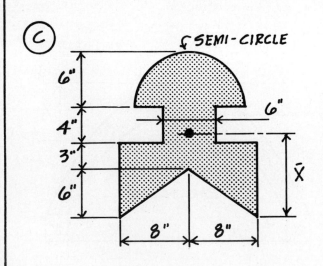

(C)

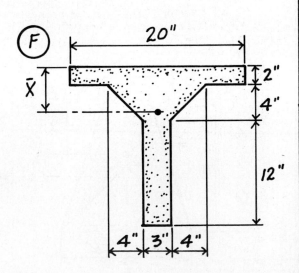

(F)

PROBLEM 4–5

UNIFORM UNIT STRESS

(I)

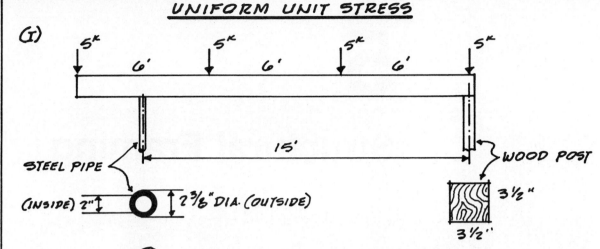

STEEL PIPE

(INSIDE) 2" 2 3/8" DIA. (OUTSIDE)

WOOD POST

3 1/2"

3 1/2"

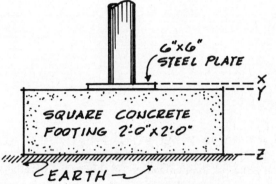

6"x6" STEEL PLATE

X
Y

SQUARE CONCRETE FOOTING 2'-0" x 2'-0"

Z

EARTH

FOR THE ARRANGEMENT SHOWN: DETERMINE THE BEAM REACTIONS, THE UNIT STRESSES IN THE STEEL PIPE AND WOOD POST AND THE UNIT STRESSES ON THE CONTACT AREAS AT PLANES X, Y, Z

<u>NOTE:</u> NEGLECT WEIGHT OF CONSTRUCTION MEMBERS.

(II)

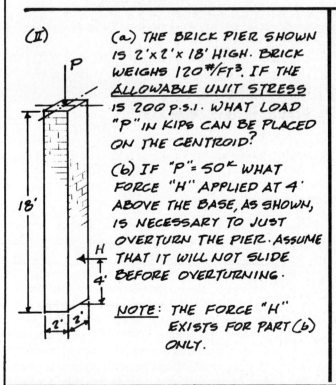

P

18'

H

4'

2' 2'

(a) THE BRICK PIER SHOWN IS 2' x 2' x 18' HIGH. BRICK WEIGHS 120 #/FT³. IF THE <u>ALLOWABLE UNIT STRESS</u> IS 200 P.S.I. WHAT LOAD "P" IN KIPS CAN BE PLACED ON THE CENTROID?

(b) IF "P" = 50ᴷ WHAT FORCE "H" APPLIED AT 4' ABOVE THE BASE, AS SHOWN, IS NECESSARY TO JUST OVERTURN THE PIER. ASSUME THAT IT WILL NOT SLIDE BEFORE OVERTURNING.

<u>NOTE:</u> THE FORCE "H" EXISTS FOR PART (b) ONLY.

(III) THEORETICALLY, HOW HIGH CAN A SOLID BRICK PIER (NO MORTAR) BE BUILT BEFORE IT WILL CRUSH UNDER ITS OWN WEIGHT? THE BRICK HAS A UNIT WEIGHT OF 120 #/FT³. ASSUME AN <u>ULTIMATE (CRUSHING) COMPRESSIVE UNIT STRENGTH</u> OF 4000 P.S.I.

REALISTICALLY, A BRICK PIER MADE OF MATERIAL HAVING THE ABOVE ULTIMATE STRENGTH WOULD BE DESIGNED FOR AN <u>ALLOWABLE</u> UNIT STRESS OF 200 P.S.I. WHAT IS THE MAXIMUM HEIGHT FOR THIS STRESS CONSIDERING ONLY THE WEIGHT OF THE PIER?

WHY DO YOU THINK SUCH A LARGE FACTOR OF SAFETY IS INVOLVED FROM THE ULTIMATE STRENGTH TO THE ALLOWABLE STRESS?

5

Structural Framing

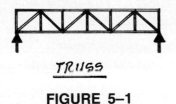

TRUSS

FIGURE 5–1

INTRODUCTION

Some of the basic terminology of skeletal and bearing wall structures was presented in Chapter 1; in this chapter, the language will be extended. We shall deal with such issues as load collection on supporting members, lateral stability of low-rise and high-rise skeletal structures, bearing wall behavior, and related issues.

The first step in these studies must necessarily be a brief survey of various floor and roof systems.

FLOORS AND ROOFS

In this section, we will consider several ways of supporting floors and roofs when using the most common structural framing materials; that is, structural steel, reinforced concrete, and wood.

Steel Construction

In skeletal steel construction, the most common way of supporting a floor (or roof) deck is with beams, trusses, or bar joists. A beam, as defined in Chapter 1, is simply a horizontal member that supports a floor (or roof) deck. Most commonly, in steel design, wide flange shapes are used for this purpose (see Fig. 1–9). However, trusses or bar joists are sometimes used. A truss is simply a beam made of small pieces, as indicated in Fig. 5–1. Trusses will be studied in some detail in the following chapter. Let it suffice, here, to say that a truss is capable of relatively long spans while requiring minimum amounts of material, as compared to a solid beam. Trusses may require greater depth than a wide flange beam to perform a given function, and while the amount of material used may be less than a wide flange

TYPICAL METAL DECK PROFILE

FIGURE 5–2

beam, fabrication costs are generally higher (on a pound-for-pound basis) since they are made of many small pieces.

A bar joist is really nothing more than a small truss generally used on a very close spacing. It is referred to as a "bar joist" because many of the members are made of "bar stock" (i.e., solid round sections of steel) as opposed to structural shapes (such as a steel angle) which would normally be used to make a truss.

The choice of using beams, trusses, or bar joists would depend on the magnitude of loading, span, and the spanning ability of the floor deck, which must deliver the load that it carries to these supporting members. Wide flange steel beams and joists are considered standard items since they are premanufactured and are available in a variety of depths and weights to satisfy a variety of loading conditions.

In steel construction, a very common method of spanning between supporting members is through the use of metal deck. Metal deck is simply sheet metal that has been formed to give it strength. A typical cross section of structural metal decking is shown in Fig. 5–2. The spanning capability of metal deck is dependent upon the gauge of metal used and the depth of the corrugation. Metal decking is available in a variety of depths and gauges to satisfy a variety of load and span combinations. When used on a floor of a building, metal deck must be filled with concrete, as shown in Fig. 5–3, in order to give it the appropriate stiffness required to carry the floor loads. When used on a roof (where the required live load is normally less than that of a floor), the metal deck is sometimes simply covered with a rigid insulating material, as shown in Fig. 5–4.

In addition to the possibility of metal decking to span between supporting members, pre-cast concrete and cast-in-place concrete is sometimes used. Several typical shapes in pre-cast concrete are shown in Fig. 1–11. When cast-in-place concrete is used on a steel frame, the structure is often designed as a "composite construction." In this type of construction, both materials (structural steel and concrete) are designed to act in unison due to the bond between them. The concrete is then considered to have structural capability rather than simply being a dead weight on the steel beam. A cross section through a composite beam is shown in Fig. 5–5. The combination of the two materials into a single structural unit leads to a savings in structural steel. One of the advantages of the kind of composite construction shown in Fig. 5–5 is that the concrete, which is normally required as fireproofing for the steel, is also called upon to provide structural strength, thereby serving a double function. One of the disadvantages is that concrete is a very heavy material, and much of the work it does is due to its own dead weight. Consequently, the method indicated in Fig. 5–5 has largely been replaced by the process shown in Fig. 5–6. In this process, fireproofing is provided for the steel beam by the use of a "spray-on" technique. The spray-on coating is generally about 3/4-in. thick, which satisfies most fireproofing requirements; therefore, a great deal of concrete dead load is eliminated. Composite action is then achieved between the concrete floor slab and the steel beam, through the use of "shear studs." The shear studs (which, physically, are like steel mushrooms welded to the top of the beam) are needed to resist the tendency of the slab to slide with respect to the beam when the unit is deformed under the influence of a load. This is an example of horizontal shear (defined in Chapter 1) that must be resisted so that the slab and beam will act as an integral structural unit, as shown in Fig. 5–7.

The few systems described are the most basic; they are commonly used in steel construction for floors and roofs. There are many variations of these

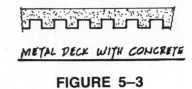

METAL DECK WITH CONCRETE

FIGURE 5–3

METAL DECK WITH INSULATION

FIGURE 5–4

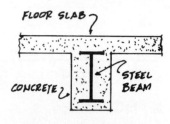

COMPOSITE CONSTRUCTION

FIGURE 5–5

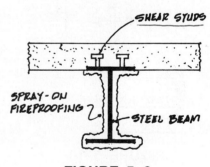

FIGURE 5–6

WITHOUT STUDS, SLAB SLIPS ON BEAM WHEN DEFORMED

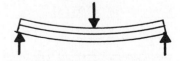

SLAB AND BEAM ACT TOGETHER WHEN SHEAR STUDS ARE USED.

FIGURE 5–7

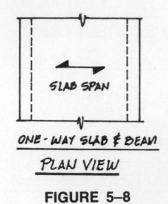

ONE-WAY SLAB & BEAM
PLAN VIEW

FIGURE 5–8

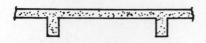

ONE-WAY SLAB & BEAM
SECTION

FIGURE 5–9

TWO-WAY SLAB
PLAN VIEW

FIGURE 5–10

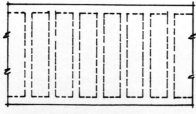

PAN-JOIST SYSTEM
PLAN VIEW

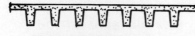

PAN-JOIST SYSTEM
SECTION

FIGURE 5–11

systems and there are much more elaborate systems, which are sometimes employed when conditions call for large, column-free spaces or other unusual conditions. The coverage of the large number of possibilities is beyond the purpose or intent of this chapter.

Reinforced Concrete Construction

Cast-in-place concrete construction differs from that of structural steel (or any structure made of "sticks") because it is, initially, a viscous liquid that will take any form, depending on the molds into which it is poured. Consequently, the nature of the structural systems in concrete are not at all like those in steel or, for that matter, wood, although there are certain behavioral similarities.

Because concrete is placed as a liquid, it is not possible to distinguish between the floor "deck" and the supporting members, as it was in a steel structure. All of the elements in a cast-in-place concrete structure are integral. In reinforced concrete construction there are several systems that can be considered standard. The choice of one method over another depends on a variety of factors such as load, span, availabilty of forms (the molds into which concrete is poured), etc.

Reinforced concrete floor systems may be divided into two kinds: one-way and two-way systems. Very simply, a one-way system is one where the loads are delivered to the supporting members in one direction; a two-way system is one where loads are delivered to supporting members in two directions. To elaborate on this, one of the most common one-way floor systems in reinforced concrete is the beam and slab. This is shown in plan view in Fig. 5–8, and a section is shown in Fig. 5–9. In this system, as in all cast-in-place concrete systems, the slab and beams are poured and act as an integral unit. The slab delivers load to the beams in one direction, as indicated by the slab span shown in Fig. 5–8. The direction of the slab span is dictated primarily by the direction of the reinforcing steel. In the case shown, the slab will span between the beams. It should be noted that in Fig. 5–8, supporting beams are shown only on two edges. If beams were supporting the slab on four edges, the slab may actually be spanning in two directions, as shown in Fig. 5–10. For two-way action to take place, the reinforcing steel must be placed in both directions indicated, and certain proportions of the bay must be maintained. That is, where dimensions between supporting members are in excess of a 2:1 ratio, the slab is considered to be out of the range of any significant two-way action and will, for all practical purposes, deliver loads by spanning in the short direction. Merely as an indication of the scale being discussed here, a one-way slab is generally not a good choice for spans in the neighborhood of 20 ft or more. Two-way slabs tend to be somewhat more efficient than one-way slabs; consequently, the spans may be larger than 20 ft.

Another very common one-way system is the ribbed slab, which is more often referred to as the *pan-joist system*. It is called the *pan-joist* because it is formed with metal forms that are available in a variety of dimensions. A plan view and section of a typical pan-joist system is shown in Fig. 5–11. This system is good for spans up to 40–45 ft, but spans in the low 30's are generally most economical. As a further indication of scale, the ribs shown in Fig. 5–11 may vary in depth from 6–20 in. and they are spaced at 20 or 30 in. apart. The depth and spacing are based on the span and the loads that must be carried.

A fairly common two-way system in reinforced concrete is the *waffle slab*. The waffle slab has closely spaced ribs running in two directions, as shown in Fig. 5–12. Like the pan-joist system, the waffle slab is generally

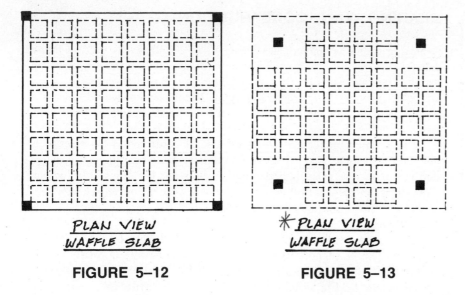

PLAN VIEW
WAFFLE SLAB

FIGURE 5-12

*PLAN VIEW
WAFFLE SLAB

FIGURE 5-13

SECTION

FIGURE 5-14

formed with metal forms that are available in a variety of sizes to satisfy load and span requirements. Waffle slabs may be supported at all four edges, as shown in Fig. 5–12, but more commonly, beams at the edges are not used, as shown in Fig. 5–13. In this kind of framing scheme, solid concrete is poured around the columns equal to the depth of the ribs, as shown in Fig. 5–14. This is necessary to avoid a problem known as "punching shear." This is caused by the large intensity of load collecting at the column, which produces a tendency for the column to punch through the slab. It should be noted that the waffle slab, like all two-way systems, is most efficient when used in a square bay. Under such conditions a waffle slab may be used for spans (in both directions) of about 15–50 ft.

Another two-way system commonly used in reinforced concrete is the *flat plate*. The flat plate is simply a slab of uniform thickness without beams at the edges or shear panels around the columns, as shown in Fig. 5–15. Reinforcing steel is placed in both directions to produce two-way action. Because of the absence of shear panels around the columns, the flat plate is normally considered a light-duty system; it is used for spans up to about 26–28 ft. The flat plate requires very simple formwork and also gives the advantage of minimum structural depth due to the uniform thickness.

A variation of the flat plate, where heavier duty is required, is the *flat slab*. In the flat slab, shear panels are used at the columns, as shown in Fig. 5–16. The flat slab may be used efficiently for spans up to about 30–34 ft.

Wood Construction

The most common support for a floor or roof in wood construction is wood joists spaced very closely. The spacing of these members is normally 16 in. or 24 in. apart, depending on the loads they must carry. A typical plan view is shown in Fig. 5–17. The deck used to span between the joists is normally plywood. Sometimes, on roofs, a tongue-and-groove wood decking is used. A typical section of this kind of deck is shown in Fig. 5–18. Tongue-and-groove decking is available in thicknesses that are greater than those of plywood decking. This allows longer spans between the supporting joists. Since tongue-and-groove decking is usually made of high-quality material, it is commonly used where the underside of a roof deck is to be exposed and finished.

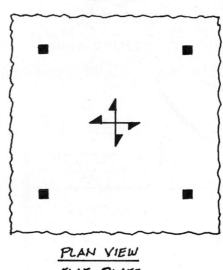

PLAN VIEW
FLAT PLATE

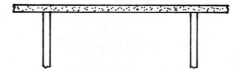

SECTION
FLAT PLATE

FIGURE 5-15

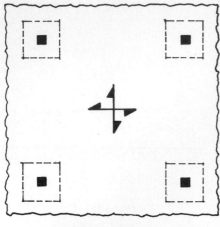

PLAN VIEW
FLAT SLAB

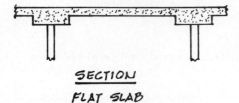

SECTION
FLAT SLAB

FIGURE 5–16

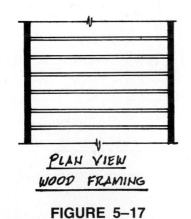

PLAN VIEW
WOOD FRAMING

FIGURE 5–17

TONGUE & GROOVE DECKING

FIGURE 5–18

BEARING WALL CONSTRUCTION

A bearing wall was defined, in Chapter 1, as a wall that is used in lieu of columns to collect loads from floors or roofs and deliver them to the foundations. Bearing walls are linear supports, as opposed to columns, which are point supports for a structural system. In this section we will elaborate on this definition in the light of certain behavioral aspects of the bearing wall.

For a variety of reasons—such as today's construction practices, seismic forces, load-carrying capacity of materials, etc.—bearing wall construction in masonry or wood is best suited for low-rise buildings up to perhaps three stories in height. In this section we will concern ourselves with some of the qualities of masonry bearing walls, which are very common for low-rise buildings.

In masonry bearing wall structures, such as that shown schematically in Fig. 5–19, linear supports (beams, joists, or trusses) are normally used to collect the loads from the floor or roof deck and deliver them to the supporting walls. Such a system delivers concentrated loads to various points on the bearing wall. This does not make the best use of the strength of a bearing wall if the beams are spaced a great distance apart (perhaps more than 2–3 ft). This is so because these concentrated loads stress the wall unevenly, thereby not making maximum use of the potential strength of the bearing wall material. The intensity of the stress at the point where the load is delivered is extremely high; it diminishes as it is distributed through the wall on the way to the foundation. This phenomenon is shown in Fig. 5–20. According to the load distribution pattern shown in this figure, it should be clear that a bearing wall is most efficient, from the standpoint of load distribution, when the load it carries is as close as possible to being uniformly distributed. The load distribution pattern due to concentrated loads, as shown in Fig 5–20, may be generally taken as a 45° distribution from the point of delivery. Point loads, therefore, leave certain areas unstressed, thereby not making the most use of the materials involved in the wall. On the other hand, uniform load distribution, as shown in Fig. 5–21, makes best use of the strength of the wall. It should also be pointed out that bearing walls are not only best when the loads on them are uniformly distributed, but also when the magnitudes of these loads are as great as the wall can

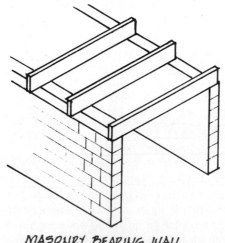

MASONRY BEARING WALL

FIGURE 5–19

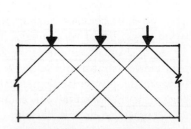

BEARING WALL ELEVATION

FIGURE 5–20

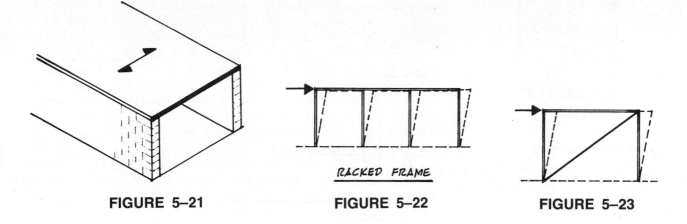

FIGURE 5–21	FIGURE 5–22	FIGURE 5–23

safely carry. Such an arrangement on a bearing wall provides the greatest stability against the effects of much of the lateral force to which low-rise bearing wall structures may be subjected. In areas where the magnitude of lateral loads may be quite high due to high-velocity winds or seismic activity, masonry bearing walls may need to be reinforced with steel.

FRAME STABILITY

When a skeletal structure is made with simple connections, there is a tendency for the frame to rack, as shown in Fig. 5–22, when subjected to a lateral force. This is so because simple connections will not resist rotation between members, thereby allowing the original angle (90° in this case) to change. Because of this, the entire frame will be displaced in the direction of the lateral force. This is a problem in simply connected "stick" structures, such as wood or steel. In looking at the racked frame of Fig. 5–22, it should be obvious that serious damage would happen to any finishes within the panels. Therefore, certain measures must be taken to resist the displacement of the frame. There are several ways of doing this, each of which has potentially strong architectural implications. One procedure involves the use of "X-bracing." In Fig. 5–23 a single-panel frame is shown with a lateral force being applied. Considering all joints to be simply connected (bolted or nailed), there will be no resistance to angular change between members, and the frame will rack, as shown by the dotted lines. If the joint at the upper right was not allowed to be displaced, the frame would be stable against the lateral force. To do this, it is only necessary to tie this point back to something immovable; this would be the junction at the lower left corner. This tie, or brace, would be in a state of tension, and the horizontal component of the tensile force would be equal and opposite to the applied lateral force. Of course, the lateral force can be coming from the other direction, as shown in Fig. 5–24. The same technique is used here; the final configuration would be an X, as shown in Fig. 5–25. It was mentioned that these ties would be in a state of tension. However, only one half of the X would be needed, as shown in Fig. 5–26, if the bracing member were designed to resist compression as well as tension. In Fig. 5–26, if a lateral force were coming from the left, then the brace would be in a state of compression. Conversely, if the force were coming from the right, the brace would be acting in a state of tension. Where possible, however, the X-brace made of members that can only resist tension is a better choice because less material would be required. Tension members can be steel cables or rods of small

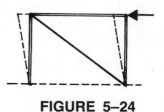

FIGURE 5–24

FIGURE 5–25

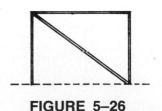

FIGURE 5–26

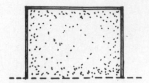

FIGURE 5-27

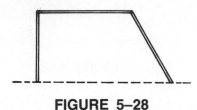

FIGURE 5-28

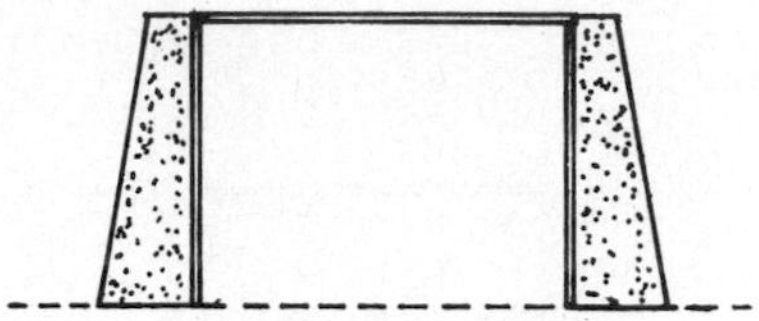

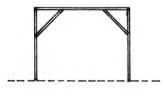

FIGURE 5-29

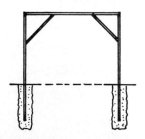

FIGURE 5-30

FIGURE 5-31

diameter. Such members cannot resist compressive stresses. Therefore, if a brace must be designed to resist compression, larger cross sections are required to resist the tendency of the member to buckle.

Another possible way to brace a frame is with the use of a "diaphragm." This is simply an infill made of some substantial material (obviously not glass), as shown in Fig. 5-27. In order for this frame to deform due to the influence of a lateral force, the diaphragm must change shape. If properly designed, most materials will resist this change, thereby offering stability to the frame. Materials commonly used for this purpose are brick, concrete blocks, or plywood.

There are many other ways to stabilize a frame to resist deformation due to the application of lateral forces. For example, it is possible to provide a sloping support member, as shown in Fig. 5-28, or buttresses for the frame, as shown in Fig. 5-29. Another possibility is to provide "knee braces," as shown in Fig. 5-30. Knee braces, in the arrangement shown, will resist changes in the original angle between the beam and the columns, thereby making the frame stiffer. However, there will still be some movement due to pinned connections at the base of the columns. This can be taken care of by providing rigid connections at the base of the column. One type of structure where this is done is generally referred to as a *pole structure*. In this kind of structure the columns are fixed at the base by sinking them to some depth and anchoring them in concrete, as shown in Fig. 5-31.

We have only discussed single-panel frames to this point. However, when there are multiple panels, such as shown in Fig. 5-32, the principles are precisely the same. In such cases, only one panel needs to be braced to stabilize the frame. This may be done by X-bracing, as shown in Fig. 5-32, or by the use of a diaphragm infill, as shown in Fig. 5-33. It is also possible to brace more than one panel, as shown in Fig. 5-34. In this sort of scheme, each of the braces will take a share of the total lateral force.

The concepts of frame stability discussed to this point have dealt with ways to stabilize a frame in the vertical plane. There still remains the problem of stabilizing a "stick structure" in the horizontal plane. In Fig. 5-35, a plan view of a very simple three-bay "stick structure" is shown which is subjected to lateral forces. It would be a reasonably simple matter to stabilize the end frames in the vertical plane. Any of the schemes previously mentioned could satisfy the requirements. However, how can the *interior* frames be stabilized against lateral displacement? If this were an open space (without interior walls that could be used to stabilize the frames), lateral displacement of the frames would take place, as shown in Fig. 5-36. The architectural consequences of such behavior should be apparent. Assuming that this must be a totally open interior space, how can these frames be kept from racking? Actually, the solution is rather simple, using the same concepts already presented for providing stability in the vertical plane. This involves nothing more than bracing the roof (or floor) in the horizontal plane, as shown in Fig. 5-37. With a pattern such as this, all horizontal forces applied at the roof level will be transferred to the end frames, which can be braced. Since the lateral forces can come from either direction, the

FIGURE 5-32

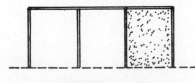

FIGURE 5-33

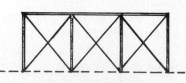

FIGURE 5-34

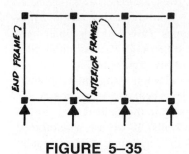

FIGURE 5–35

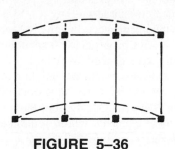

FIGURE 5–36

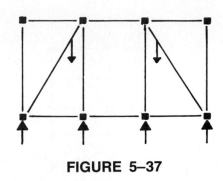

FIGURE 5–37

final bracing pattern will look like that shown in Fig. 5–38.

A multistory building frame differs from one- or two-story frames in that the design of the main structural members may be heavily influenced by lateral forces, especially as the height of the building increases relative to the width. In general, this occurs when the building height is something in the order of two or more times the least plan dimension. However, many of the same techniques already presented may be used to provide lateral stability for tall buildings.

When subjected to lateral wind forces, a "pin-connected" frame will deform in the manner shown in Fig. 5–39. In order to stabilize such a frame, X-bracing may be used as shown in Fig. 5–40. Essentially, this forms a vertical truss that carries the lateral forces to the ground. This, however, may not be feasible except at the ends of the building, since interior partitions may need openings where the X-braces are required. If this is the case, vertical bracing at the interior frames may be eliminated by delivering the lateral forces through a bracing system in the horizontal plane, as shown in the plan view of Fig. 5–41. This sort of scheme, however, generally presents problems because of the number of pieces to be fabricated and installed. The more usual technique is to design the floor system as a horizontal diaphragm. This requires a rigid floor system with positive attachment to the supporting girders.

Although a symmetrical and uniform pattern for bracing is recommended (especially for earthquake forces), it is possible, when necessary, to place bracing at random, as shown in Fig. 5–42. The important thing to remember is that each level must somehow be braced.

Bracing for a multistory frame may be provided by a solid wall that has good structural properties such as reinforced concrete, masonry, or reinforced masonry. These are called "shear walls," which are essentially diaphragms in the vertical plane (see Figs. 5–43 and 5–44). The choice of solid walls or exposed X-bracing clearly has design implications.

In addition to X-bracing or solid shear walls, there are other ways to stabilize a building against lateral forces. For example, knee braces, shown in Fig. 5–45, offer the advantage of bracing without interfering with openings in interior partitions. A negative aspect of this scheme is that the horizontal component of the force in the knee brace resolves itself in the span of the column (like a point load on a beam), introducing significant additional stresses to the column and thus requiring more material for the column. Also, while knee braces will stiffen the frame, they are not as effective as X-braces. Use of knee braces is infrequent in high-rise buildings.

The K-braced pattern, as shown in Fig. 5–46, is another way to brace a building. However, the members will obviously interfere with a space in which there are no partitions where they can be hidden. This system of bracing also has the disadvantage of requiring a greater amount of material

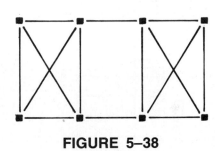

FIGURE 5–38

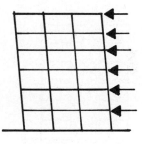

FIGURE 5–39

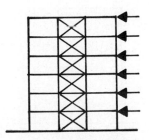

END ELEVATION

FIGURE 5–40

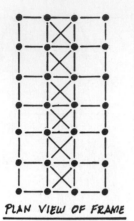

PLAN VIEW OF FRAME

FIGURE 5–41

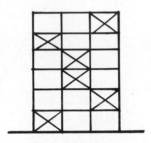

FRAME ELEVATION

FIGURE 5–42

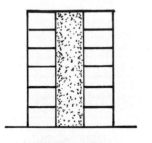

END ELEVATION
SINGLE BAY SHEAR WALL

FIGURE 5–43

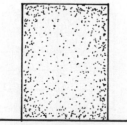

END ELEVATION
SHEAR WALL

FIGURE 5–44

than other systems and can present detailing problems (such as the connections to the girders).

Supplementary bracing systems of the types shown to this point may not always be required. It should be recognized that high-rise buildings generally have at least one "utility core." The utility core contains those elements necessary for the building to function (elevators, stairs, ducts, plumbing chases, etc.). Building codes demand a great deal of fire protection for these elements. Most often, these fireproofing requirements lead to enclosure of these elements within reinforced concrete walls. If we use a reinforced concrete enclosure for the services involved in a building, this essentially creates a large vertical shaft (vertical cantilever) of great structural strength. Therefore, why not use this vertical cantilever to resist, or at least help resist, the lateral loads? To do so makes a great deal of sense, and the placement of the utility core within the plan of a building should include careful evaluation of structural and functional criteria. To elaborate on these ideas, consider the plan view of the frame shown in Fig. 5–41. Only the bare frame is shown in this diagram. Of course, it will be necessary to provide elevators, stairs, etc., to service this multistory building. There may be many ways to do this, depending on the function of the building. One possibility, in a schematic sense, is to place these strong vertical cantilevers at the ends of the building, as shown in Fig. 5–47 (if the building proportions are such that two utility cores are feasible). This serves the same function as any other bracing pattern, and if the floor system is designed to act as a horizontal diaphragm, the whole building will be braced. Continuing with the same idea, but with a single utility core (which is probably more common), it may be possible to arrange floor girders so they deliver the lateral forces directly to the core, as shown in Fig. 5–48. The core is then designed to be stiff enough to take these loads safely to the ground. In this kind of scheme the floor system must be designed for diaphragm action.

Finally, it is not uncommon to design a rigid frame structure for the purpose of resisting lateral forces. This means that the connections between beams and columns will be rigid (as opposed to simple) connections. Rigid connections provide resistance to lateral forces because they maintain the original angle between girder and column (generally 90°). In steel construction, rigid connections are often all welded. In cast-in-place concrete construction, rigid connections are inherent due to the integral nature of the joint between girders and columns.

TUBE STRUCTURES

The previous discussion has dealt with multistory building frames that are called "braced frames" or "rigid frames." A comparatively new develop-

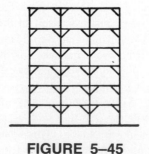

FIGURE 5–45

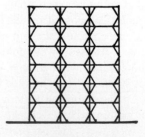

FIGURE 5–46

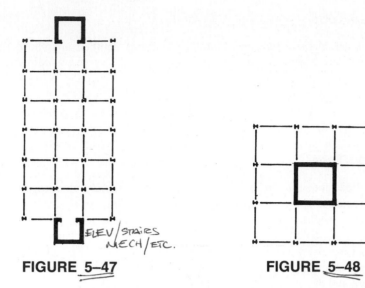

ELEV/STAIRS MECH/ETC.

FIGURE 5–47

FIGURE 5–48

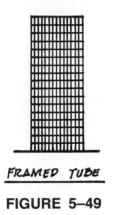

FRAMED TUBE

FIGURE 5–49

ment in structural systems for high-rise buildings is the *tube structure*. The tube structure is designed and constructed in such a manner that the outer walls form a very rigid "tube" that resists lateral forces efficiently. Such systems have been used extensively in past years for very tall buildings. Several types of tubular structures have been developed. These can be categorized into five major groupings, but variations of the systems and combinations of two or more categories are possible.

Framed Tube

The framed tube consists of very closely spaced exterior columns rigidly connected to spandrel beams. The framed tube is, in essence, a bearing wall structure that is extemely rigid and capable of resisting all the lateral forces to which a tall building may be subjected, without dependence on inner parts of the frame or the interior core. An example of a framed tube structure is the 110-story World Trade Center in New York City. A schematic elevation of a typical framed tube is shown in Fig. 5–49.

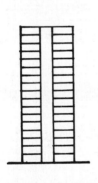

TUBE - IN - TUBE

FIGURE 5–50

Tube-in-Tube

The tube-in-tube structure is simply an extension of the framed tube. In this kind of tube structure the inner core is designed to interact with the outer framed tube, thereby taking a share of the horizontal forces as well as gravity loads. A section through a typical tube-in-tube structure is shown in Fig. 5–50.

Shell Tube

Another way of making a tube structure is by the use of continuous steel plates as structural members to form the exterior walls. This forms a diaphragm that carries both gravity and lateral loads. An efficient tube can be created in this way, provided openings in the plates are kept to a minimum. An example of a shell tube structure is the Standard Oil Building in Chicago. A schematic elevation of this kind of tube structure is shown in Fig. 5–51.

SHELL TUBE

FIGURE 5–51

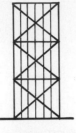

TRUSSED TUBE

FIGURE 5–52

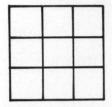

SCHEMATIC PLAN VIEW
BUNDLED TUBE

FIGURE 5–53

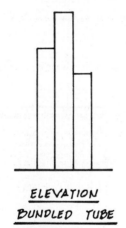

ELEVATION
BUNDLED TUBE

FIGURE 5–54

Trussed Tube

In the trussed tube structure, the main load-carrying members are placed diagonally, forming a vertical truss on the outside walls of the building. These diagonals are designed to carry gravity loads as well as lateral forces. The trussed tube has been found to be an extremely efficient form for very tall buildings, such as the John Hancock Center in Chicago. A schematic elevation of this kind of tube structure is shown in Fig. 5–52.

Bundled Tube

The bundled tube structure is an extension of the framed tube concept. The system consists, basically, of a framed tube with shear walls spanning between exterior walls. This greatly increases the efficiency of the basic framed tube. The bundled tube idea allows each module in a building to be designed as an independent tube. This system was used for the Sears Building in Chicago, where nine tubes were "bundled" together at the lower part of the building and then reached various heights. A schematic plan view and elevation of a bundled tube are shown in Fig. 5–53 and 5–54.

The tube structure concept has provided for great efficiency in very tall buildings, but this is not necessarily so for buildings of moderate height. A chart showing optimum systems for various heights of multistory buildings is given on Data Sheet D–19, in the Appendix.

LOAD COLLECTION

In Chapter 3 the process for determining reactions for structural members subjected to loads was presented. In this section we will deal with ways that loads are delivered to the members, and the loading patterns that are developed. Before doing this, it is necessary to discuss briefly some of the graphic techniques used for structural framing plans.

In Fig. 5–55 a framing plan of a typical bay in structural steel is shown. The lines indicating the structural members are conventionally single lines, as shown in the figure. Note that the lines do not touch the supporting members. That is, the lines indicating beams do not touch the girders, and the girders do not touch the columns. This is the standard graphic indication for "simply supported" members. If the intent was for the connections between members to be rigid (welded, as opposed to a simple bolted connection), the lines would touch the supporting members, as shown for the rigid frames of Fig. 5–56. If cantilevers are used, they must be shown as continuous over the support, indicated in Fig. 5–57.

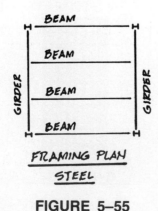

FRAMING PLAN
STEEL

FIGURE 5–55

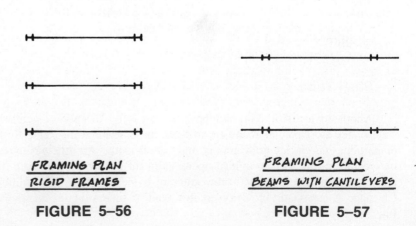

FRAMING PLAN
RIGID FRAMES

FIGURE 5–56

FRAMING PLAN
BEAMS WITH CANTILEVERS

FIGURE 5–57

In cast-in-place reinforced concrete construction, the conventional method for indicating supporting members is with double lines, as indicated in Fig. 5–58. The lines are shown dotted because they are beams below the slab. Since cast-in-place concrete has rigid joints between the members, the lines are drawn to the supports. In Fig. 5–58 a one-way beam and slab system is indicated. Graphic indications of other concrete systems have already been shown in Figs. 5–8 through 5–16.

Graphic indications in wood construction are much like those used for steel construction. That is, single lines are used to indicate the members. In Fig. 5–59 a typical framing plan indication is shown for a wood skeletal structure. The lines stop short of the supporting members, indicating simple connections. In wood framing, all members, for all practical purposes, are simply supported. Nailing or bolting of members at their supports, will not restrain the members from rotating with respect to each other. In Fig. 5–60 a segment of a framing plan for a wood bearing wall structure is shown (such as in residential construction). Conventionally, the bearing walls are shown shaded and the joists, which they are supporting, are shown as simply supported.

With some of the basic means for indicating various situations graphically, we will now proceed to the means for determining the loads on structural members.

One-Way Systems

In Fig. 5–61 a typical interior bay of a building is shown. The floor system used here is a concrete slab supported by steel beams. Considering that the slab is supported at the middle of the bay by the beam called B–1, the slab will deliver its load to the supporting members through one-way action. That is, the slab will tend to span in the direction shown by the symbol ⟶ . It will tend to choose this route simply because of the proportions involved. In the case under consideration, the slab dimensions between supports are 10′ × 20′, or a ratio of 1:2. Given such proportions, a slab will tend to deliver its load to the supporting member in the short direction or "the path of least effort." Therefore, when the proportions of a slab are in the order of 1:2 or less, the slab will behave as a one-way slab. As the proportion increases beyond a 1:2 ratio, most of the load will be delivered in the short direction, but some of the load will travel in the longer direction to the supporting members, provided that reinforcing steel is placed appropriately in both directions.

✳ In the problem of Fig. 5–61 we want to determine the load on the beam B–1 and the girder G–1. The live load is given as 40 p.s.f., which is a function of the use of the space and is given by building codes. Generally, the 40 p.s.f. load given here would be applicable to residential use (hotel, dormitory, etc.). In addition to the live load, the floor must carry its own weight. In this bay, a 4″-thick reinforced concrete slab is being used. The dead weights of various building materials are given on Data Sheet D–3. Considering that reinforced concrete weighs 150#/ft³, each square foot of slab, 4″ thick, is one third of a cubic foot and will weigh 50# (see Fig. 5–62). Therefore, the dead load of the reinforced concrete slab is 50 p.s.f. Depending upon the building, there would probably be more dead load involved. For example, there may be a floor finish involved. Also, there may be a ceiling and lighting fixtures hung from the underside of the floor, and so on. However, for the sake of simplicity, let us consider that the only dead load involved is the concrete slab.

Considering that the one-way slab will span between B–1 and B–2, visualize a typical 1′-wide strip of slab (in this case there would be 20 such

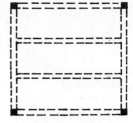

FRAMING PLAN
REINF. CONCRETE

FIGURE 5–58

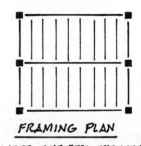

FRAMING PLAN
WOOD SKELETAL STRUCTURE

FIGURE 5–59

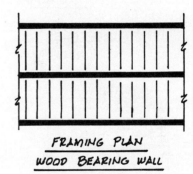

FRAMING PLAN
WOOD BEARING WALL

FIGURE 5–60

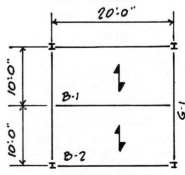

INTERIOR BAY
LIVE LOAD : 40 P.S.F.
DEAD LOAD : 50 P.S.F.
TOTAL LOAD : 90 P.S.F.

FIGURE 5–61

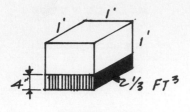

FIGURE 5–62

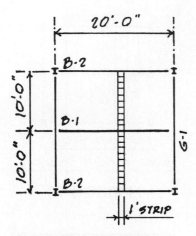

FIGURE 5–63

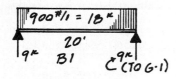

FIGURE 5–64

strips) as shown in Fig. 5–63. The 1'-wide strip is further divided into 20 ft², as shown, and each of these square feet is carrying a 90# load to the support. Each of these square feet will deliver the load through the "path of least effort"; that is, to the closest support. With this in mind, 5 ft² on each side of B–1 will deliver the load to B–1. Therefore, a typical 1' length of B–1 will be supporting 10 ft² of slab; the load on this 1' length of B–1 will be (10 ft²) (90#/ft²) = 900#.

Since every linear foot of B–1 is loaded in precisely the same manner, the load on B–1 will be 900#/lin ft, which is a uniformly distributed load. The graphic representation of this is shown in Fig. 5–64. The end reactions for B–1 are 9000# each, which is determined by inspection. These end reactions are provided by the girders G–1. Therefore, there is a 9000# action on G–1 produced by the beams B–1 that are framing in. Since this is a typical *interior* bay, there will be a B–1 framing in from each side, thereby producing a total <u>concentrated load</u> on G–1 of 18000# (see Fig. 5–65). The beams marked B–2 will be framing directly into the columns and will not affect G–1 (see Fig. 5–66).

It should be pointed out at this time that although the total load is the same magnitude on B–1 and G–1, the member required for G–1 would be larger. This is so because the concentrated load will produce more flexure (bending) than the uniform load of equal magnitude. This will be seen mathematically in a later chapter, when we deal with the design of structural members. Let it suffice for now to simply understand that point loads in the center of the span of a beam will produce greater requirements than uniformly distributed loads of equal magnitude (see Fig. 5–67).

Two-Way Systems

When the proportions of a slab are greater than 1:2, the "path of least effort" is not such a clear issue. In fact, the slab will tend to deliver loads in both directions. This is known as *two-way action*. Reconsider the framing plan of Fig. 5–61, but this time without the member B–1 (see Fig. 5–68).

Before going any further it should be emphasized, that two-way action in a concrete slab will only occur when the reinforcing steel within the concrete slab is placed in both directions. If the reinforcing steel is placed in one direction only one-way action will occur regardless of the proportions of the slab. The same idea is true if the concrete slab is poured on a metal deck. The ribs on the metal deck will dictate the direction of the span regardless of the proportions (see Fig. 5–69).

Returning to the two-way slab, the issue becomes one of determining the manner in which such a slab distributes loads to the supporting members. The contributory area (that is, the area of slab delivering load to a

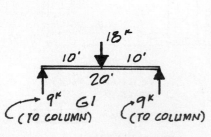

FIGURE 5–65

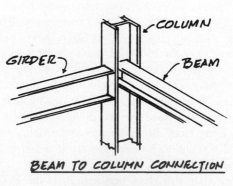

BEAM TO COLUMN CONNECTION

FIGURE 5–66

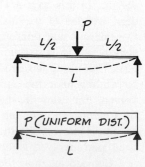

FIGURE 5–67

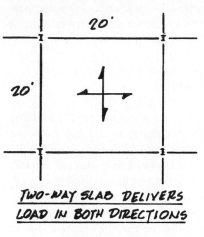

TWO-WAY SLAB DELIVERS
LOAD IN BOTH DIRECTIONS

FIGURE 5–68

TYPICAL METAL DECK PROFILE

FIGURE 5–69

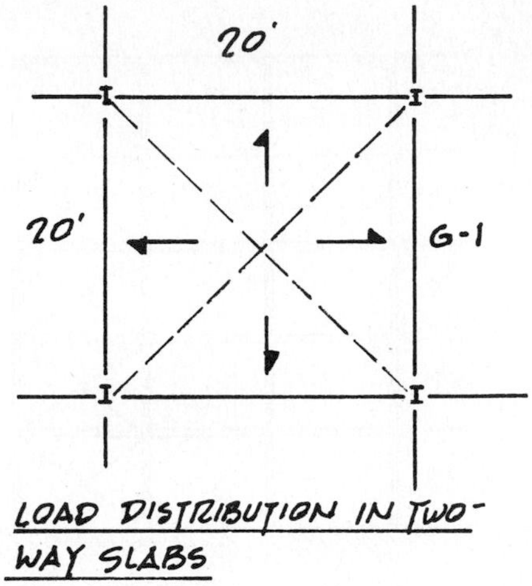

LOAD DISTRIBUTION IN TWO-
WAY SLABS

FIGURE 5–70

beam) is determined by drawing diagonals on the framing plan, from the corners of the bay, at an angle of 45°, as shown in Fig. 5–70. The loads in the triangular areas encompassed by these 45° lines will travel in the directions indicated. The loading on the member G–1 (as well as all other beams in this example) will be triangular in distribution, varying from a maximum rate of loading at the center to zero at the ends, as shown in Fig. 5–71. Therefore, with a total load of 90#/ft² and a bay on each side, the loading on G–1 varies from 1800#/lin ft to zero, as shown in Fig. 5–71. The *total* load on this member is determined by simply computing the area of the triangle

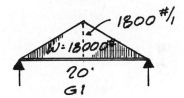

FIGURE 5–71

$$W = 1800\#/' \ \frac{\times \ 20'}{2} = 18000\#$$

and the load is triangularly distributed. Compared to the original scheme of Fig. 5–61, where the load on G–1 was also, coincidentally, 18000# as a point load, the two-way system would have some benefit because G–1 would be receiving a somewhat distributed load (triangular variation) rather than the entire load concentrated in the middle of the span. The determination of loads on supporting members where a square bay is involved (such as the example of Fig. 5–70) is a rather straightforward problem. Each supporting member will simply carry one quarter of the total bay, and the weight will be distributed in a triangular pattern, as shown. However, where the bay is not square, but of such proportions that two-way action can still take place, the process for determining the contributory area is still quite simple. Referring to Fig. 5–72, draw diagonals at 45° angles, from the corners of the bay until they intersect. Then connect the intersection of the diagonals, as shown. The areas bordered by these lines will deliver loads to the supporting members, as shown. Still considering that the total floor load is 90 p.s.f. and that this is a typical interior bay (equal contributory areas on each side), then the loadings on G–1 and G–2 are shown in Fig. 5–73. Considering the previous discussions, it may be said that when dealing with two-way systems, the square bay represents the optimum arrangement of supporting members. For example, consider a 30 ft × 30 ft bay. All members would be equally loaded with a triangular load. The total load would be 20250# on each of the supporting members, which is heavier than the loads on G–1 of the previous example, but lighter than the loads on G–2. The optimization here would come primarily from standardization and from

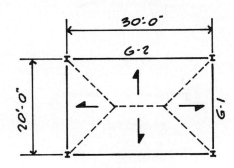

FIGURE 5–72

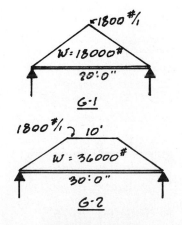

FIGURE 5–73

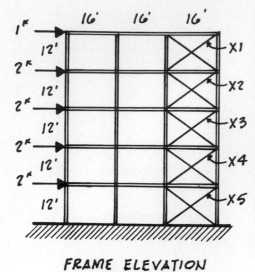

FRAME ELEVATION

FIGURE 5-74

the fact that less columns would be required for a given total building dimension. This statement should be considered as somewhat of a generalization, and each particular problem should be evaluated separately in order to determine the optimum arrangement of elements.

FORCES IN DIAGONAL BRACING

Earlier in this chapter we discussed the various ways to brace a building subjected to lateral forces. Since diagonal bracing is a fairly common method for stabilizing a structure made of "sticks," and which is essentially "pin-connected," it seems that it would be useful to briefly discuss the way in which forces in braces are determined. The computational aspects of finding the forces in diagonal bracing are, in fact, extraordinarily simple; they make use of the fundamental laws of static equilibrium.

For example, let's determine the forces in the diagonal braces, due to wind acting on the structural frame, as shown in Fig. 5-74. It should be mentioned at this point that it is a matter of convention to show wind forces as concentrated loads acting at the floor levels. This is based on the assumption that wind forces will apply pressure uniformly to the face of a building. The concentrated loads, therefore, are simply the resultant of the uniform loads acting between the floor levels. The intensity of uniformly distributed wind loads acting on the surface of a building is largely a matter that is determined based on building code recommendations and, as one would expect, varies with the height of the building. For very tall buildings, wind tunnel tests are normally required in order to determine the behavior of the building due to the dynamic quality of wind forces.

In our problem, we are dealing with a low-rise building. By whatever means, the wind loads have been determined and are shown as point loads at the floor levels. Let's now consider that the X-braces shown are made of members that can only resist tension (such as steel cables or thin rods), and that the structural frame cannot resist any of the lateral force. The braces marked X1 through X5 will go into a state of tension when the wind forces come from the direction shown, and the unmarked braces will go slack. When the wind comes from the other direction, the unmarked braces go into tension and X1 through X5 will go slack. The force in each brace is determined by the very simple fact that the horizontal component of the force in any brace, and at any level, *must be equal to, and opposite to* the summation of the applied lateral forces above that level. Therefore, the *horizontal component* of the force in X1 must be 1^K and, based on the geometry, the force in the brace will be 1.25^K (based on 3:4:5 right-triangle geometry). The forces in all the braces have been found by the same process and are summarized, as follows:

$$X1 = 1.25^K$$
$$X2 = 3.75^K$$
$$X3 = 6.25^K$$
$$X4 = 8.75^K$$
$$X5 = 11.25^K$$

The important point to remember when determining the force in a diagonal brace (in a building designed so that the braces take all lateral forces) is that the *horizontal component* of the force in a brace at any level must be *equal and opposite to the summation* of all horizontal forces above that

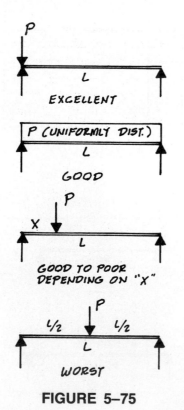

FIGURE 5-75

level. This is so whether the bracing pattern is uniform, which is recommended, or not uniform, such as that shown in Fig. 5–42.

EQUIPMENT LOADS

Often, structural members must support very heavy equipment, such as that used in a manufacturing facility or in a mechanical equipment room. The equipment is usually mounted on supporting members in such a way that a concentrated loading pattern is produced. Considering the earlier discussion regarding the relationship between concentrated loads and uniformly distributed loads, heavy equipment mountings (which produce concentrated loads) should be located as close to the columns as possible. Such an arrangement will place the greater portion of the load away from the center of the supporting beam. It can be said that concentrated loads placed far away from the middle of a supporting beam will produce less bending than the same value of load placed at the center, although shearing forces will be higher. Figure 5–75 shows the general relationship involved in the various loading arrangements discussed. It should be realized that the more a concentrated load can be distributed, the better it will be for the beam. This idea is shown in Fig. 5–76, which, in essence, is approaching the uniform loading pattern. The proper distribution of loads is not only important from the standpoint of the equipment weight, but also from the standpoint of vibration, which is generally present. The weights of various kinds of equipment, and the method in which the equipment must be mounted, should be furnished by the manufacturers.

WALL LOADS

In addition to supporting dead and live loads delivered from a floor surface, beams must sometimes carry loads from walls they are supporting. This generally occurs on a spandrel beam or spandrel girder (see Fig. 5–77). Sometimes the wall system used is such that the wall may take some of its own load directly to the columns. In spite of this, it is generally assumed that the total wall weight is supported uniformly by the spandrel beam. In order to determine the load on a member due to a wall, one only needs to know the weight of the material being used for the wall. Consider, for example, that it is desired to know the wall loading on S–1 in Fig. 5–77. The wall is made of brick and is 12″ thick. Considering that brick (clay masonry) has a unit weight of $120\#/ft^3$, each linear foot of S–1 is supporting 10 ft^3 of brick (see Fig. 5–78), or 1200#. Since each linear foot is considered to be providing the same support, then the loading is shown as a uniform load with a magnitude of 1200# per linear foot. In order to complete the loading diagram for this member, the load coming in from the floor would have to be added.

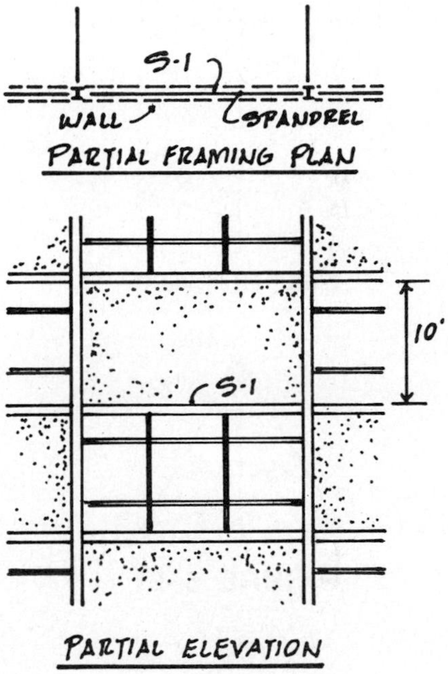

Lintels

Often it is necessary or desirable to create openings in masonry walls for windows or doors. In order for the masonry to carry across the top of the opening, a structural member known as a *lintel* is employed. A lintel is really nothing more than a beam used to carry masonry across an opening. In order to determine the loading on a lintel, the "arching" action of masonry is taken into account. Referring to Fig. 5–79, the loading on the lintel

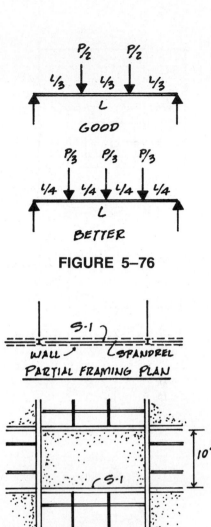

FIGURE 5–76

FIGURE 5–77

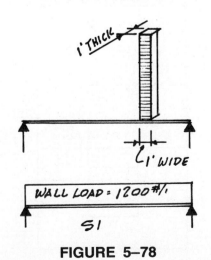

FIGURE 5–78

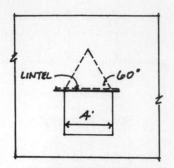

FIGURE 5–79

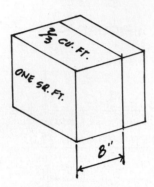

FIGURE 5–80

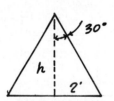

FIGURE 5–81

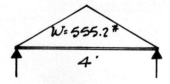

FIGURE 5–82

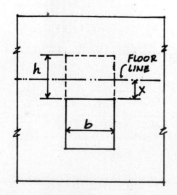

FIGURE 5–83

is taken as a triangular area above the opening. This triangle is generally constructed with 45° to 60° angles from the corner of the opening. The 60° angle will, of course, take in a larger area than a 45° angle; consequently, the design of the lintel will be slightly on the safe side. The choice of angle to use is primarily a matter of judgment. For the sake of this discussion, the 60° triangle will be used, as shown in Fig. 5–79. The brick outside of this triangle is considered to carry itself across the opening through arch action. In order to determine the loading on the lintel of Fig. 5–79, simply compute the area of the 60° triangle and multiply this value by the unit weight of the material being used. In this example let's consider that the wall is brick (clay masonry) and is 8 in. thick. Given that the unit weight of clay masonry is 120#/ft³ (as mentioned previously), then each square foot of wall surface with an 8-in. thickness represents two thirds of a cubic foot and weighs 80# (see Fig. 5–80). Since clay brick generally comes in 4-in. thicknesses, it is usually convenient to remember that each 4 in. increment of thickness weights 40# per square foot of surface.

In order to determine the load on the lintel for the problem being considered, it is first necessary to determine the height of the triangle (see Fig. 5–81).

$$h = \frac{2}{\tan 30°} = 3.46'$$

The *area* of the triangle is

$$A = \frac{1}{2}(4)(3.46) = 6.92 \text{ ft}^2$$

The total load is then

$$W = (6.92 \text{ ft}^2)(80\#/\text{ft}^2) = 553.6\#$$

and the load is distributed in a triangular pattern, as shown in Fig. 5–82. Where a floor load is coming into a masonry bearing wall and the point of application of the floor load is closer to the top of the opening than the dimension of the width of the opening (b), as shown in Fig. 5–83, then the general assumption is that the lintel must carry the floor load above it and a weight of brick determined by a square area which is the product of (h) (b) where h is equal to b.

CONCLUSION

The concepts presented in this chapter are general and should not be considered as all-encompassing. They are presented as generalizations for the student of architecture in order to provide an overview of the issues involved. It is hoped that this presentation will provide a basis for further research where specific information is required. The author wishes to emphasize that there are virtually an infinite number of "systems" that can be developed by the creative architect. The "systems" presented in this chapter should, therefore, merely be considered as convenient vehicles for the presentation of certain issues. They should by no means be considered the only structural configurations possible.

PROBLEM 5–1

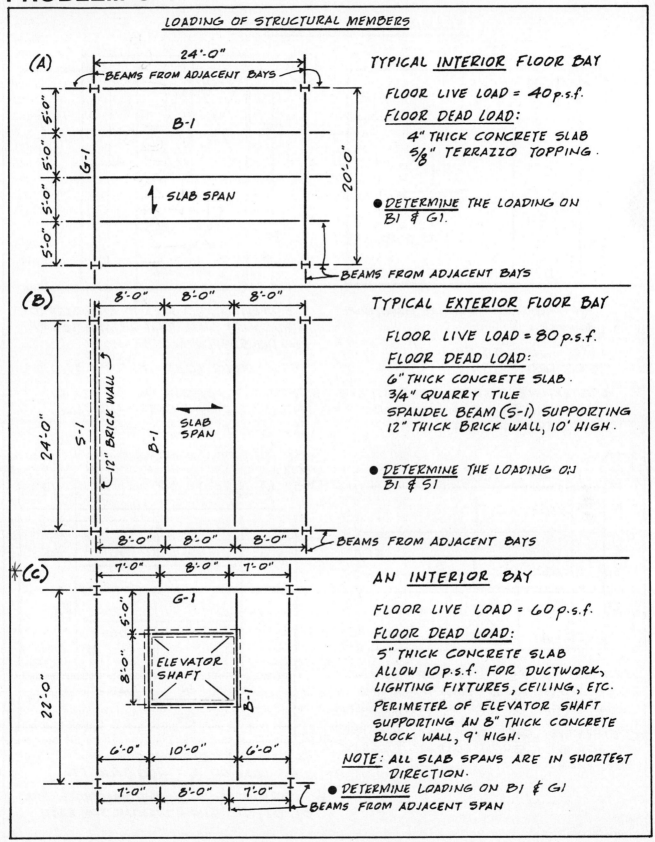

LOADING OF STRUCTURAL MEMBERS

(A)

24'-0"
BEAMS FROM ADJACENT BAYS
5'-0" 5'-0" 5'-0" 5'-0" 5'-0"
G-1
B-1
SLAB SPAN
20'-0"
BEAMS FROM ADJACENT BAYS

TYPICAL **INTERIOR** FLOOR BAY

FLOOR LIVE LOAD = 40 p.s.f.

FLOOR DEAD LOAD:

  4" THICK CONCRETE SLAB
  5/8" TERRAZZO TOPPING.

● **DETERMINE** THE LOADING ON B1 & G1.

(B)

8'-0" 8'-0" 8'-0"
24'-0"
S-1
12" BRICK WALL
B-1
SLAB SPAN
8'-0" 8'-0" 8'-0"
BEAMS FROM ADJACENT BAYS

TYPICAL **EXTERIOR** FLOOR BAY

FLOOR LIVE LOAD = 80 p.s.f.

FLOOR DEAD LOAD:

6" THICK CONCRETE SLAB.
3/4" QUARRY TILE
SPANDEL BEAM (S-1) SUPPORTING 12" THICK BRICK WALL, 10' HIGH.

● **DETERMINE** THE LOADING ON B1 & S1

✳(C)

7'-0" 8'-0" 7'-0"
22'-0"
5'-0"
8'-0"
G-1
ELEVATOR SHAFT
B-1
6'-0" 10'-0" 6'-0"
7'-0" 8'-0" 7'-0"
BEAMS FROM ADJACENT SPAN

AN **INTERIOR** BAY

FLOOR LIVE LOAD = 60 p.s.f.

FLOOR DEAD LOAD:

5" THICK CONCRETE SLAB
ALLOW 10 p.s.f. FOR DUCTWORK, LIGHTING FIXTURES, CEILING, ETC.
PERIMETER OF ELEVATOR SHAFT SUPPORTING AN 8" THICK CONCRETE BLOCK WALL, 9' HIGH.

NOTE: ALL SLAB SPANS ARE IN SHORTEST DIRECTION.

● **DETERMINE** LOADING ON B1 & G1

PROBLEM 5-2

FRAMING PLAN ANALYSIS

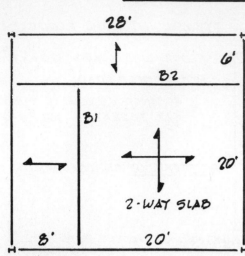

Ⓐ 28'

B2 6'

B1

20'

2-WAY SLAB

8'   20'

CONCRETE SLABS SPAN IN DIRECTIONS
SHOWN.

LIVE LOAD = 75 P.S.F.
DEAD LOAD = 75 P.S.F.

● DETERMINE LOADING AND REACTIONS
FOR B1 & B2 - SHOW LOADING
DIAGRAM FOR EACH.

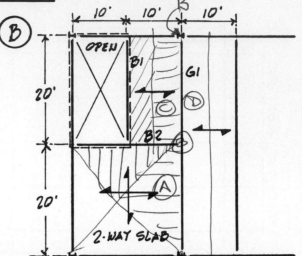

Ⓑ 10' 10' 10' Ⓑ

OPEN B1 Ⓐ

20' G1 Ⓓ

B2 Ⓒ

20'

2-WAY SLAB Ⓐ

THE OPEN 10'x 20' AREA IS ENCLOSED BY
A MASONRY WALL THAT WEIGHS 800 #/,
ON THE SUPPORTING BEAMS.

TOTAL FLOOR LOAD = 160 P.S.F.

● DETERMINE LOADING ON B1, B2, G1
SHOW LOADING DIAGRAM FOR EACH.

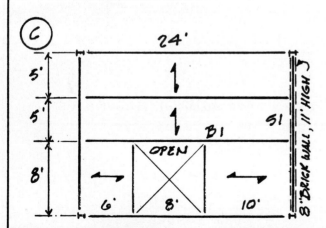

Ⓒ 24'

5'

5'

B1   S1

8'

OPEN

6'   8'   10'

8" BRICK WALL, 11' HIGH

DEAD LOAD = 50 P.S.F., LIVE LOAD = 80 P.S.F.
S1 SUPPORTS AN 8" THICK BRICK WALL
WHICH WEIGHS 80 P.S.F OF SURFACE AREA.

● DETERMINE LOADS AND REACTIONS FOR
B1 AND S1 - SHOW LOADING DIAGRAM
FOR EACH.

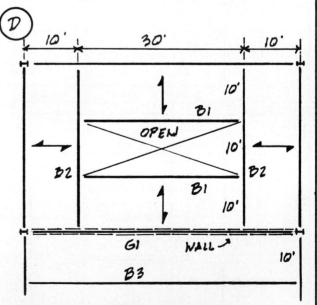

Ⓓ 10' 30' 10'

B1   10'

B2 OPEN 10' B2

B1   10'

G1   WALL 10'

B3

TOTAL FLOOR LOAD = 160 P.S.F
WALL LOAD ON G1 = 400 #/LIN.FT.

● DETERMINE LOADING AND REACTIONS FOR
B1, B2, G1 - SHOW DIAGRAM FOR EACH.

PROBLEM 5–3

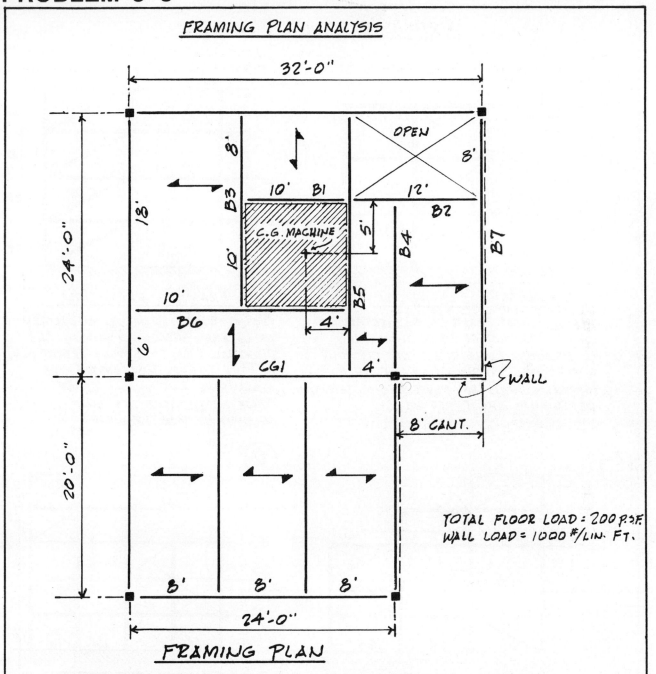

FRAMING PLAN ANALYSIS

FRAMING PLAN

ONE-WAY SLABS AS INICATED (←→), DELIVER LOADS TO SUPPORTING BEAMS.

MACHINE WEIGHT = 20ᵏ ACTING THROUGH THE CENTER OF GRAVITY (C.G.) AS INDICATED. THE MACHINE IS SUPPORTED ONLY AT THE FOUR CORNERS. THERE IS NO SLAB UNDER THE MACHINE

• DETERMINE THE LOADING ON B1, B2, B3, B4, B5, B6, B7, CG1 SHOW LOADING DIAGRAM FOR EACH.

TOTAL FLOOR LOAD = 200 P.S.F.
WALL LOAD = 1000 #/LIN. FT.

PROBLEM 5–4

VERTICAL BRACING

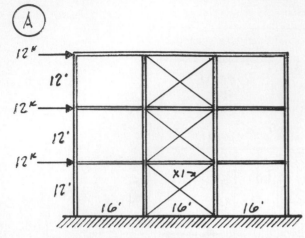

FRAMING ELEVATION

THE 3-STORY FRAME IS SUBJECTED TO LATERAL FORCES AS SHOWN. THE X-BRACING RESISTS ALL OF THE WIND FORCE. THE BRACING IS MADE OF A MATERIAL THAT CAN ONLY RESIST TENSION. DETERMINE THE FORCE IN X1.

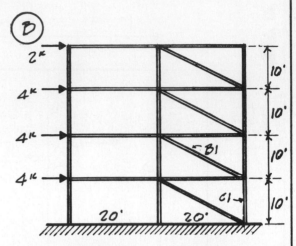

FRAMING ELEVATION

THE 4-STORY FRAME IS SUBJECTED TO LATERAL FORCES AS SHOWN. THE DIAGONAL BRACES SHOWN RESIST ALL OF THE FORCE. DETERMINE THE MAGNITUDE AND CHARACTER OF THE FORCE IN B1 AND THE LOAD ON COLUMN C1 DUE TO WIND.

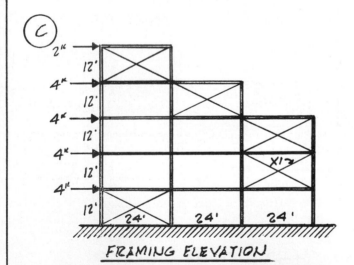

FRAMING ELEVATION

THE X-BRACES SHOWN RESIST ALL OF THE LATERAL WIND FORCES. THE BRACES ARE MADE OF A MATERIAL THAT CAN ONLY RESIST TENSION. DETERMINE THE FORCE IN X1.

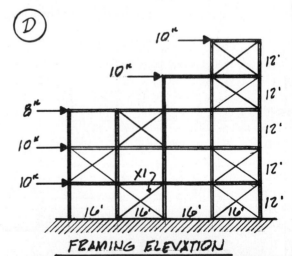

FRAMING ELEVATION

IF THE X-BRACING SHOWN RESISTS ALL OF THE LATERAL FORCE, AND THE BRACES CAN ONLY RESIST TENSION, WHAT IS THE FORCE IN X1.

6
Trussed Structures

INTRODUCTION

Essentially, a truss is a large beam made of small "sticks" arranged in triangular patterns, as shown in Fig. 6–1. The primary advantage of a truss is that it is capable of extremely long spans while carrying heavy loads. Also, a truss uses minimal amounts of material as compared to the amount that would be required for a solid beam under the same load and span conditions. In addition, the truss has the advantage of being made of relatively small pieces that can be put together to satisfy virtually any span requirement.

Trusses can, usually with considerable economic feasibility, cover spans of 30 ft to 400 ft, though they have been used for spans greater than this where overriding factors made it necessary.

It is the primary purpose of this chapter to present methods of analysis in order to determine the stresses in the members of which a truss is made. This will form the necessary background for the design of trussed structures. Before proceeding to the analytical methods involved, it will be helpful to develop the terminology used for trussed structures, as well as to familiarize the reader with some of the standard truss configurations.

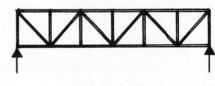

FIGURE 6–1

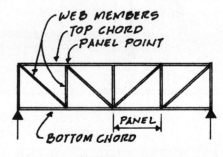

FIGURE 6–2

NOMENCLATURE

In Figs. 6–2 and 6–3 the names of the various parts of trussed structures are given. The member on top of the truss, regardless of whether it is flat or sloped, is referred to as the *top chord*. Consequently, the linear member on the bottom of the truss is called the *bottom chord*. All of the members that knit the top and bottom chords together are referred to as *web members*, regardless of whether they are oriented diagonally or vertically. The horizontal dimension between points where web members tie into the chords is known as a *panel*; the points where the web members tie into the chords are

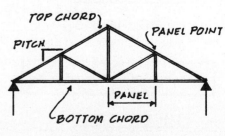

FIGURE 6–3

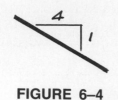

FIGURE 6–4

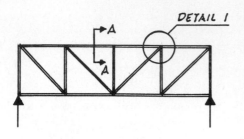

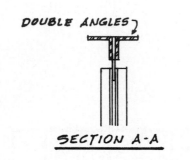

FIGURE 6–5

known as the *panel points*. When dealing with a sloped top chord, such as that shown in Fig. 6–3, the slope is referred to as the *pitch,* and is expressed as a ratio of vertical rise for each increment of horizontal dimension. For example, if the top chord rises 3 in. for every 12 in. of horizontal dimension, this is expressed as a ratio of 3:12 or 1:4, as shown in Fig. 6–4. This ratio is commonly referred to as the "rise to run."

TRUSS DETAILS

To this point the figures shown have been single-line diagrams. Such schematic representations will suffice for the purposes of structural analysis. However, it should be realized that the "sticks" of which trusses are made do, in fact, have dimensions. Consider that the trussed structure of Fig. 6–5 is made of steel members. It is of great importance in a trussed structure that the *centroidal lines of all members coming together at a panel point have a common intersection.* This is because all members in a truss must be subjected to only pure axial load (tension or compression) for the truss to work as expected. From previously established principles, this means that the resultant force in all members must be at the centroidal axis of the member. If the centroidal axes and, therefore, the resultant forces did not have a common intersection at the panel point, there would be a tendency for rotation at the panel point due to the eccentricity of the forces. This would upset the goal to have only pure axial load in truss members. It may, therefore, be said that in the detailing of a truss, one of the primary considerations is that all forces coming into a panel point are *concurrent.*

With these ideas in mind, the typical details for the truss of Fig. 6–5, made of steel, would be as shown in Section A–A and Detail 1. In order to eliminate eccentricity it is not only important that the members intersect at a common point as shown in Detail 1, but also that the cross section is symmetrical. In structural steel this is generally achieved through the use of double-angle members for the chords as well as the web members, as shown in Section A–A of Fig. 6–5. In order to tie all these members together at the panel point, a steel plate is generally employed, known as a *gusset plate.* All members are then either bolted or welded to the gusset plate. The welded connection is most commonly used for this purpose. Regardless of the materials used to make a truss, it is important that eccentricity be avoided at the joints. With this in mind, the designer should develop details in order to achieve this goal.

TRUSS TYPES

In order to familiarize the reader with the possibilities for patterns of triangulation and general truss configurations, we will take a brief look at some trusses that have been developed and, over the years, have become standards.

Warren Truss (Figs. 6–6 and 6–7)

The Warren Truss is a flat-top-chord truss with the web members arranged as shown. The vertical members may be eliminated as shown in Fig. 6–7. The decision to have vertical members basically depends on the question of scale—that is, the span of the truss, depth, and required distance between panel points. The distance between panel points in a trussed structure is a function of the floor or, more generally, roof system that is being

FIGURE 6–6

FIGURE 6–7

supported by the truss. This will be discussed in some detail shortly. Considering structural steel as the material being used, the Warren Truss is most economical for spans of about 40–120 ft.

Flat Pratt Truss (Fig. 6–8)

The Flat Pratt has a flat top chord with the web members arranged as shown in the figure. Considering structural steel as the material, the range of most economical span is essentially the same as that of the Warren Truss: that is, about 40–120 ft.

It should be pointed out that for both the Warren Truss and the Flat Pratt Truss, when the spans exceed about 40–50 ft it may be necessary to slope the top chord slightly, as shown in Fig. 6–9, in order to permit water to run off, where a roof is being supported. This slope only needs to be very slight, perhaps in the order of $\frac{1}{4}$ in. per foot. Although for the longer spans this may result in a substantial rise to the center, these would still be considered primarily flat-chord trusses.

Pitched Pratt Truss (Fig. 6–10)

The Pitched Pratt Truss has its web members arranged as shown. This configuration in structural steel is economical for spans not in excess of about 100 ft. The slope employed for the top chord should be about 5:12 to 6:12, as shown in the figure.

Howe Truss (Fig. 6–11)

The Howe Truss is similar to the Pitched Pratt Truss except that the sloped web members are opposite in direction from those in the Pratt Truss. The largest economical span is also about 100 ft. The choice of a Howe Truss or a Pitched Pratt is dependent on the spans, magnitude of loads, and the various patterns of loading that may be imposed on the truss. Only careful mathematical analysis can reveal which of the two would be most economical for a given situation.

Fink Truss (Fig. 6–12)

The Fink Truss is economical for spans in the range of 100–125 ft, when made of structural steel. The arrangement of the web members leads to the economical edge over the Howe or Pratt, because in the Fink Truss most of the members are in tension, and those acting in compression are relatively short. Tension members require less cross-sectional area to resist a force than do compression members of equal length. This is because the danger of buckling in a compression member necessitates different methods of design.

The few truss types discussed represent the most commonly used configurations possible. The basic idea is that the truss achieves stability through triangulation. With this in mind, the designer can develop virtually any configuration to satisfy conditions of span, loading, patterns of loading, type of roof, climate, etc.

LOADING OF ROOF TRUSSES

It was mentioned earlier in this chapter that all members in a truss must be subjected to pure axial load only. Generally, members subjected to pure

FLAT PRATT TRUSS

FIGURE 6–8

FLAT PRATT TRUSS
WITH SLOPED TOP CHORD

FIGURE 6–9

PITCHED PRATT TRUSS

FIGURE 6–10

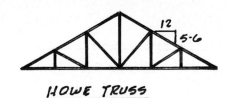

HOWE TRUSS

FIGURE 6–11

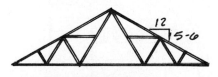

FINK TRUSS

FIGURE 6–12

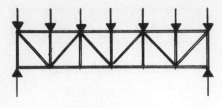

FIGURE 6-13

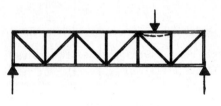

FIGURE 6-14

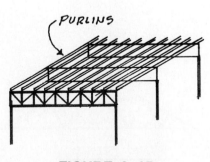

FIGURE 6-15

axial tension or compression will require less material than members subjected to bending stresses. In order to achieve this end, the loads coming into the truss must be delivered to the panel points, as shown in Fig. 6–13. Loads coming into the panel points will create only axial stresses in the members. This will be seen in the next section of this chapter, which deals with methods of analysis. If the loads were introduced to the truss between the panel points, the condition shown in Fig. 6–14 would be created. Bending stresses would be introduced to the top chord, which would be detrimental to the goal of economical design.

In order to control the delivery of loads to the truss so that the loads are delivered as concentrated loads to the panel points, supplementary members known as *purlins* are used (see Fig. 6–15). The roofing system is designed to span between purlins; the purlins, in turn, bring the loads to the truss. This system of controlling loads is of primary importance in determining the panel dimensions of a truss. The purlins can be spaced only as far apart as the roof system can span. Since the roof system spans between purlins and not directly to the trusses, this allows a fairly wide spacing of trusses. The purlin must then be designed to span between the trusses. The purlins may be of any cross section necessary, but generally, in steel design, channel sections or wide flange sections are most commonly used. In essence, the purlin is nothing more than a uniformly loaded beam with the truss playing the role of a girder.

In order to achieve the most economical design for the conditions involved, the designer must make decisions regarding the type of roof to be used, the spacing between trusses, the pitch of the trusses, and the variables involved in the loading patterns, such as gravity, wind, moving loads, and the possible effects due to combinations of these loads. Since there are virtually an infinite number of possibilities when considering these factors, the most economical arrangement is oftentimes difficult to determine. Each designer must exercise good judgment in determining those factors that have priority and, from this point, proceed to develop the most suitable design.

METHODS OF ANALYSIS

Once a truss configuration has been established, based on the variety of factors previously discussed, the next step is to design the members of which the truss is to be made. It has already been mentioned that the detailing involved in a truss is such that all members will be in a state of pure axial stress, either tension or compression. In order to design the individual members properly, we must determine the magnitude and character of the stresses. We'll now look at the methods of analysis, which include two closely related analytical methods (one an extension of the other) and a graphic method.

The analytical methods are called the *Method of Joints* and the *Method of Sections*. Both of these methods are based on an idea called the "free body" concept. While this is the first mention of this concept in this text, it must be pointed out that the ideas involved will be used extensively from this point on, regardless of the configuration of the structure. The truss represents an ideal opportunity for the initial presentation of the "free body" concept, but it will be seen in subsequent chapters that this concept has universal application in the study of equilibrium and structural analysis. It is vital, therefore, that the reader develop a thorough understanding of the free body concept.

THE "FREE BODY" CONCEPT

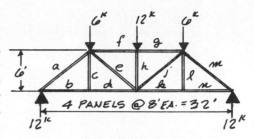

FIGURE 6–16

In the analysis of structures subjected to loads, the chief concern, to this point, has been the determination of the values of *external reactions* necessary to equilibrate the loads. This, of course, is a very necessary step in the design of a structure. The next step will be to determine the *internal stresses* in a structural member, due to the application of external forces. In order to determine the internal stresses in a structural member, a very useful tool known as the "free body diagram" will be employed. The entire free body concept is based on one very sensible notion:

> If a structure is in equilibrium, then *every part* of that structure is in equilibrium.

This means that if a part of a structure is visualized as being isolated from the structure of which it is a part, then all three equations of *static equilibrium* ($\Sigma F_x = 0$, $\Sigma F_y = 0$, $\Sigma M = 0$) as they relate to the isolated part must be satisfied.

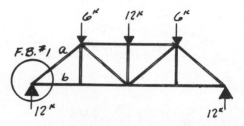

FIGURE 6–17

As previously mentioned, the idea of the free body is presented in this chapter because the concept is best demonstrated when dealing with the trussed structure, although the idea is applicable to any structure, regardless of configuration. For the purpose of making the free body concept as clear as possible, we will immediately move to the analysis of trusses by the first of three methods.

METHOD OF JOINTS

Before designing the members that make up the truss of Fig. 6–16, it is necessary to determine the stresses to which the individual members will be subjected. In order to determine these *internal* stresses, a series of free body diagrams will be employed. The first step in the design of any structure is to determine the *reactions*. In the problem of Fig. 6–16 this is easily done by inspection, since the structure and loading are symmetrical. Therefore

$$R_1 = R_2 = 12^K$$

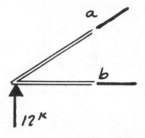

FIGURE 6–18

After the equilibrating reactions are determined, isolate each joint in succession as a *free body diagram*, starting at a joint where there is a support reaction. For example, consider the joint shown as Free Body #1 in Fig. 6–17. The joint is isolated from the remainder of the structure, as shown in Fig. 6–18. *All external forces* acting on the isolated joint must be shown in the free body diagram. In Free Body #1 the 12^K reaction is shown. Assuming that the truss is properly made, all members will be subjected to either *axial* tension or compression. When the free body is "cut loose" from the parent structure, the members that are cut release their *internal* axial stresses; these are shown as *external* forces on the free body diagram. Referring to Free Body #1 in Fig. 6–18, we have the 12^K reaction, the unknown axial stress in member *a,* and the unknown axial stress in member *b.* All of these forces have a common intersection, which is simply a set of concurrent forces (refer to Chapter 2). Considering the notion that this joint must be in a state of equilibrium, we will be concerned with satisfying the two equations of static equilibrium that deal with translation ($\Sigma F_x = 0$, $\Sigma F_y = 0$). Referring to Fig. 6–19, we shall now proceed to equilibrate this joint of Free Body #1:

Since there is a 12^K upward force, there must be 12^K furnished in the

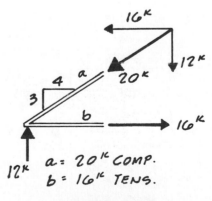

FIGURE 6–19

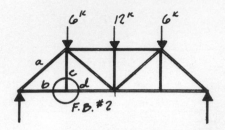

FIGURE 6-20

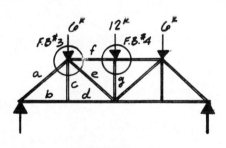

FIGURE 6-21

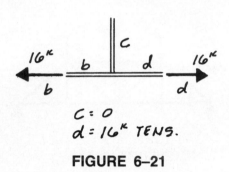

FIGURE 6-22

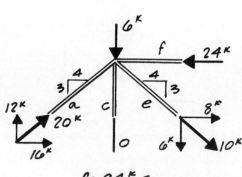

FIGURE 6-23

downward direction for $\Sigma F_y = 0$. In Free Body #1 the only force that can furnish vertical resistance is the axial force of member a. Therefore, the vertical component of a must be 12^K. Based on the slope of member a, the parallelogram of forces will be made of right triangles in the proportion of $3:4:5$. Therefore, given that the vertical component *must* be 12^K, then the horizontal component must be 16^K (this is the 4 side of the $3:4:5$ triangle). The resultant of these components is 20^K; this is the axial stress in member a. The determination of the direction of the resultant is a relatively simple matter. Since this is an axial force, there are only two possibilities for the direction. That is, either *down and to the left* or *up and to the right*. In order to satisfy $\Sigma F_y = 0$, it was determined that the 12^K component *must* be downward. Therefore, the direction of the resultant *must* be *down and to the left* and the horizontal component must be heading to the left. The determination of the character of the stress (tension or compression) is also a relatively simple matter. If the resultant axial force is *pushing into the joint*, then the member is in a state of compression. Conversely, if the resultant force is *pulling away from the joint*, then it is in a state of tension. In the problem being considered, member a is in a state of compression. Still referring to Fig. 6–19, it was found that there is a 16^K horizontal component going to the left. In order to satisfy $\Sigma F_x = 0$, there must be 16^K going to the right. There is only one unknown remaining, and this is the stress in member b. Therefore, the stress in member b is 16^K, and it is in a state of tension.

Referring to Fig. 6-20, let's evaluate Free Body #2, which is shown isolated from the structure in Fig. 6–21. The stress in member b was already found to be 16^K tension. Nothing has happened between Free Body #1 and Free Body #2 to change this. Consequently, member b is shown as 16^K tension. *Since it is in a state of tension, it must be shown pulling away from the joint under investigation.* The solution for unknown stresses in members coming into this joint now follows:

$\Sigma F_y = 0$: There are no vertical forces applied to this joint. Therefore, the stress in member c must be zero. The member is included in this truss in order to reduce the free span of the bottom chord and thereby provide an extra measure of stability. Without this "zero stress" member the bottom chord, which will generally be very slender, would be extremely flexible in both the vertical and horizontal directions.

$\Sigma F_x = 0$: Member b is in a state of tension (pulling away from the joint) by an amount of 16^K. The only unknown remaining is the stress in member d. It *must* be providing a 16^K force going to the right, which is pulling away from the joint and, therefore, in a state of tension. The solution for stresses in members coming into the joint of Free Body #2 is now complete.

The solution for stresses in members coming into the joints of Free Body #3 and Free Body #4 (as indicated in Fig. 6–22) is shown in Figs. 6–23 and 6–24. Because of the symmetry in the truss under investigation, the stresses in the members on the right-hand side will be the same as those to the left of the centerline.

The procedure for determining the magnitude and character of stresses in truss members by the Method of Joints may be summarized as follows:

1. Determine the reactions.
2. Isolate each joint as a free body and sketch the free body diagram showing all forces acting on the isolated joint. Consider the stress in all members that have been cut to be external forces acting on the free body.

3. Start and proceed with this process dealing only with joints where there are no more than two unknowns. Since we are dealing with a coplanar concurrent force problem at each joint, there are only two equations of static equilibrium available.

    For example, in the preceding problem, if the joint of Free Body #3 were investigated *before* the joint of Free Body #2, it would be found that four members were coming into that joint and three of them would be unknown. There would be one more unknown than there are equations of statics, and the stresses in these members could not be determined.

4. Determine the magnitude of stresses in the members by equilibrating each joint against translation.

5. Determine the character of stress in each member, considering the following:
   (a) If the axial force in the member is *pulling away* from the joint being investigated, the member is in a state of *tension*.
   (b) If the axial force in the member is *pushing into* the joint, the member is in a state of *compression*.

    If a joint exists with more unknowns than there are equations of static equilibrium, the truss cannot be completely analyzed by the methods presented. Such a truss is said to be "statically indeterminate." We will not consider statically indeterminate trusses.

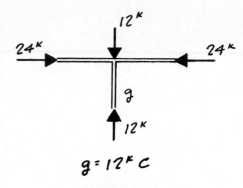

FIGURE 6–24

METHOD OF SECTIONS

In analyzing the stresses in a truss by the Method of Joints, it is necessary to analyze each joint in succession, as previously demonstrated. Sometimes it is desirable or necessary to determine the stress in a member without analyzing each joint. For example, suppose it is desired to know the magnitude and character of the stress in member f, as shown in Fig. 6–25. This is the same truss that has been analyzed by the Method of Joints, and by that method two joints had to be analyzed before dealing with the joint that includes member f. Actually, the stress in member f can be found directly (without determining the stress in any other members) by a process known as the Method of Sections. The ideas involved in solutions by this method are very much the same as in the Method of Joints, except that the free body diagram involved will include a much larger piece of the structure than simply a joint.

    To determine the stress in member f, a section of the structure, as shown in Fig. 6–25, is visualized as being "cut loose" as a free body diagram (see Fig. 6–26). Two important points about the free body concept must be remembered:

1. The free body shown in Fig. 6–26 must be in a state of static equilibrium since it is part of a structure that is in a state of static equilibrium.
2. The members that have been "cut" (members d, e, f in Fig. 6–26) *release* their internal forces, which are treated as external forces on the free body.

    Therefore, the forces acting on the free body must satisfy the equations of static equilibrium. It should be clear that the diagram of Fig. 6–26 is a *non-concurrent* force system and, therefore, both translation and rotation will have to be considered ($\Sigma F_x = 0$, $\Sigma F_y = 0$, $\Sigma M = 0$).

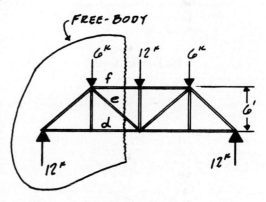

FIGURE 6–25

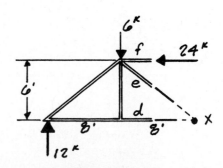

FIGURE 6–26

In order to determine the stress in member f using the free body of Fig. 6–26, consider that the section is in a state of equilibrium against rotation. This means that the section is not rotating with respect to *any* point, If moments are taken about point x, the lines of action of two unknown forces will pass through this point and will not appear in the moment equation. The force released in member f will have a lever arm to point x; this will appear in the moment equation as the only unknown. The solution is as follows:

$$\Sigma M_x = 0 = (12^K)\,(16') - (6^K)(8') - f\,(6')$$

$$f = \frac{144'^K}{6'} = 24^K C$$

In writing this moment equation, the direction of the force in member f was initially assumed to be pushing into the section. The positive answer verifies this assumption; therefore, the member is in a state of compression. The method for determining the character of a stress when analyzing by the Method of Sections is the same as that when using the Method of Joints. That is, if the direction of the force is pushing into the cut member, the member is in compression; if the direction is pulling away from the cut member, the member is in a state of tension.

The section of Fig. 6–26 may be used to find the stresses in members d and e. Referring to Fig. 6–27, the section *must* be in equilibrium against translation in both the vertical and horizontal direction ($\Sigma F_y = 0$, $\Sigma F_x = 0$). The solution follows:

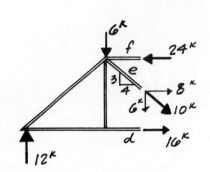

FIGURE 6–27

1. $\Sigma F_y = 0$: On this section there is a 12^K upward force and a 6^K downward force. Consequently, there must be an additional 6^K force in the downward direction. There is only one force that can furnish this, and that is the vertical component of member e, which must be 6^K. The geometry of member e is that of a 3:4:5 triangle. Therefore, the horizontal component must be 8^K to the right, and the resultant (which is the axial force in member e) is 10^K tension.

2. $\Sigma F_x = 0$: Member f has a 24^K force to the left, and member e furnishes an 8^K force to the right. For equilibrium against translation in the horizontal direction, a 16^K force to the right is required. The force in member d must furnish this. Therefore, member $d = 16^K$ tension.

GRAPHIC METHOD

When analyzing the stresses in a trussed structure by the method of joints, the forces at each joint may be determined graphically by the use of the force (or vector) diagram (see Chapter 2). When analyzing each joint *individually* with a force diagram, the order in which the forces are drawn at each joint is not important. This was pointed out in Chapter 2.

However, there is a process by which a single force diagram can be created for the entire truss and the order in which things are done is quite important. Referring to Fig. 6–28, the process is as follows:

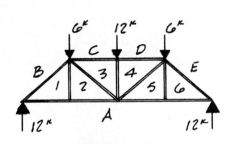

FIGURE 6–28

1. Draw the truss (as a single line drawing) to scale, accurately.

2. Mark the spaces between external forces with a letter, and proceed with this in a *clockwise* direction. Conventionally, this is started in the space between external forces along the bottom chord, although the lettering can begin in any space.

3. Number the triangular spaces in the web *from left to right*.
4. Referring to Fig. 6–29 for the entire drawing, begin by drawing the force diagram for all external forces in the following manner:

   (a) Going from the space marked *A* to the space marked *B*, we pass through a 12^K upward force. Draw this force to scale. The tail of this vector is the point *A*, the head represents point *B*.

   (b) Going from *B* to *C*, a 6^K force is encountered in the downward direction. Draw this force to scale starting at point *B* on the force diagram (this will be superimposed on the initially drawn 12^K vector). The head of this force will be point *C* on the drawing.

   (c) Continue to draw all external forces in the manner described. When this is complete, the head of the last vector must come back to the starting point—that is, point *A*. If this *does not* happen, an error has been committed in the determination of the reactions.

   (d) Still referring to Fig. 6–29, between the space marked *A* and triangle 1 (on the drawing of the truss) there is a horizontal member. Therefore, on the force diagram draw a horizontal line through point *A*. Between the space marked *B* and triangle 1 there is a sloping member. Therefore, on the force diagram draw a line of the same slope through point *B*. This line will intersect the horizontal line drawn through point *A*. This intersection becomes point 1 on the force diagram.

   Note that in order to define the location of point 1 on the force diagram it was necessary to use two expressions, each containing the number 1 (which represents the triangular space marked 1) and a point *that has already been located* on the force diagram. In the case under consideration the expressions were *A*–1 and *B*–1.

   (e) Continuing with this process, find point 2 on the force diagram which will represent the space marked 2 on the drawing of the truss. Again, in order to locate the point, two expressions must be used with the number 2 and a known point. There are only two expressions that will fit this requirement: That is, *A*–2 and 1–2. Going from *A*–2 there is a horizontal member. Therefore, draw a horizontal line through point *A* on the force diagram. Going from 1–2 there is a vertical member, so draw a vertical line through point 1. The horizontal line drawn through *A* will hit point 1; consequently, points 1 and 2 will have a common location on the force diagram. This indicates that there is no distance on the force diagram between points 1 and 2. The significance of this is that the member between spaces 1 and 2 on the truss is a "zero stress" member, which was discovered when analyzing this truss by the method of joints.

   (f) Continuing with the process, it is now necessary to locate point 3 on the force diagram. The only two expressions available, each containing a known point and both containing the unknown point 3, are 2–3 and *C*–3. From *C*–3 we pass through a horizontal member. Therefore, draw a horizontal line through point *C* on the force diagram. Going from 2–3 there is a sloping web member. Draw a line of the same slope through the known point 2. This line and the horizontal line drawn through point *C* can only intersect at one point; this intersection is point 3.

   (g) In order to locate point 4, the only two expressions that can be used are 3–4 and *D*–4. Going from 3–4 there is a vertical member, so draw a vertical line through point 3. Going from *D*–4 we encounter a horizontal member, so draw a horizontal line through point *D*.

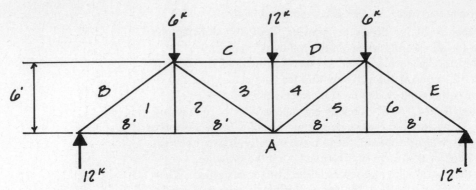

TRUSS - SCALE: 1/8" = 1'-0"

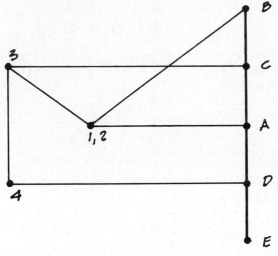

A-1 = 16ᵏ T
B-1 = 20ᵏ C
1-2 = 0
3-2 = 10ᵏ T
A-2 = 16ᵏ T
3-4 = 12ᵏ C

GRAPHIC SOLUTION - SCALE: 1" = 10ᵏ

FIGURE 6-29

These two lines can only intersect at one point; this becomes point 4.

Since this truss is symmetrical, there is no need to go further with the drawing. All members on the right side of the centerline will be stressed exactly as those on the left side.

Now that the force diagram is completed, it is a relatively simple matter to determine the stresses in the individual members. These are determined by scaling the lines between points at the scale by which the force diagram for external forces was drawn. For example, the stress in the member between the space marked A and the space marked 1 may be determined by scaling the line between points A and 1 on the force diagram. The stresses in all members may be found by this procedure.

5. The determination of the character of stress in members of a truss which has been analyzed graphically is reasonably simple. The procedure for determining the character of stress is as follows:

(a) Considering each joint individually, go around each joint in a clockwise direction, considering each member encountered individually. For example, consider the joint at the left end of the truss (see Fig.

6–30). If we wish to know the character of stress in the member between space B and space 1, then, going in a clockwise direction, this member would be encountered in going from B to 1. Referring to the force diagram, the direction from B to 1 is down and to the left. This indicates the direction of the force in this member; that is, referring to Fig. 6–30, the force is heading down and to the left and is *pushing into* the joint being considered. Therefore, this member is in a state of compression.

(b) Still referring to Fig. 6–30, the next member encountered while proceeding on a clockwise course around this joint is the member between space 1 and the space marked A. Referring again to the force diagram, the direction from 1 to A is from left to right. Therefore, the direction of the force is to the right, which is pulling away from the joint. This indicates that the member between spaces 1 and A is in a state of tension.

(c) Proceed in the manner outlined to determine the character of stress in all members.

CONCLUSION

The trussed structure is a particularly important structural configuration because of its ability to span large distances while consuming minimum amounts of material. However, it should be realized that because of the many pieces involved in the construction of a truss, there will be a great deal of labor involved in the fabrication. Thus, the cost of material saved may be outweighed by fabrication costs. A decision to create large column-free spaces through the use of trusses must be carefully weighed in relation to the function of the spaces and the requirements necessitated by the function.

Aside from the architectural considerations, and more important for the purpose of this book, the truss is very useful for exemplifying some of the fundamental principles involved in the analysis and design of structures. The truss is an extremely suitable type of structure for use in the explanation of the free body concept. It also provides for rigorous exercise in the application of the laws of static equilibrium. Since the free body concept and the laws of static equilibrium are applicable to *any structure*, it is recommended that all problems in this chapter be done so that a strong background in the fundamental principles of structural equilibrium will be developed.

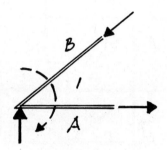

FIGURE 6–30

PROBLEM 6–1

<u>TRUSSES</u>

(1) DETERMINE:
- REACTIONS A & B (WITHOUT RELATION TO MEMBER STRESS)
- STRESS IN EACH MEMBER BY THE <u>METHOD OF JOINTS</u> USING THE KNOWN REACTIONS AS NECESSARY.

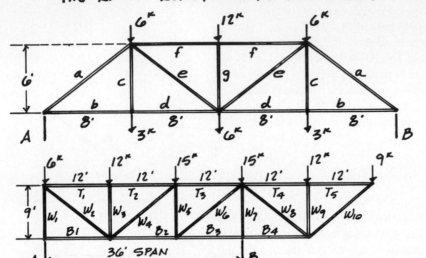

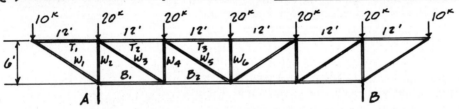

(2) SAME AS ABOVE, EXCEPT USE <u>METHOD OF SECTIONS</u>

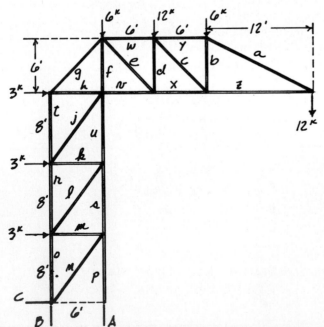

PROBLEM 6–2

TRUSSES

(I)

12ᵏ

20'

b

a

12ᵏ

c

d

20'

e

f

12ᵏ

g

h

20'

j

k

A   15'   D

B   C

(II)

12ᵏ

a   5ᵏ

b   c

d   10'

e   8ᵏ

f   g

h   10'

j   5ᵏ

k   l   m

10'

A   D

B   5'   5'   C

FOR EACH OF THE ABOVE, DETERMINE THE REACTIONS AND DETERMINE THE
MAGNITUDE AND CHARACTER OF STRESS IN ALL MEMBERS BY THE METHOD
OF JOINTS.

(III)

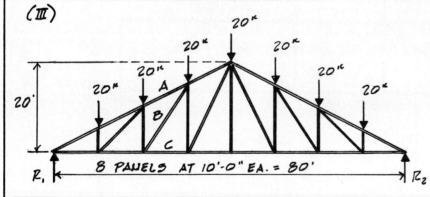

20'

20ᵏ   20ᵏ   20ᵏ   20ᵏ   20ᵏ   20ᵏ   20ᵏ

A

B

C

R₁   8 PANELS AT 10'-0" EA. = 80'   R₂

DETERMINE THE MAGNITUDE
AND CHARACTER OF STRESS
IN MEMBERS A, B, C BY
THE METHOD OF SECTIONS.

(IV)

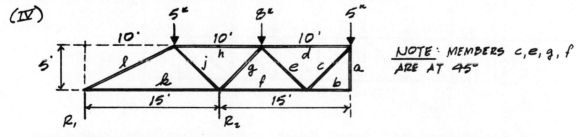

10'   10'   10'

5ᵏ   8ᵏ   5ᵏ

5'

l   j   h   d

k   g   f   e   c   a

b

15'   15'

R₁   R₂

NOTE: MEMBERS c, e, g, f
ARE AT 45°

(1) DETERMINE REACTIONS R₁ & R₂ FOR THE CANTILEVERED TRUSS SHOWN.
(2) DETERMINE MAGNITUDE AND CHARACTER OF STRESSES IN ALL MEMBERS BY
THE METHOD OF JOINTS.

PROBLEM 6–3

TRUSSES

(I)

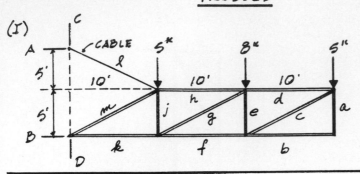

(A) DETERMINE THE REACTIONS A,B,C,D.

(B) DETERMINE THE MAGNITUDE AND CHARACTER OF STRESS IN ALL MEMBERS BY THE METHOD OF JOINTS.

(II)

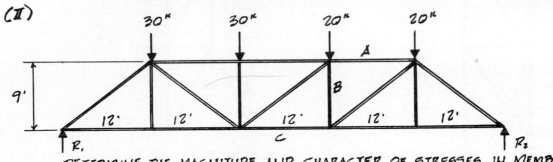

DETERMINE THE MAGNITUDE AND CHARACTER OF STRESSES IN MEMBERS A,B,C BY THE METHOD OF SECTIONS.

(III)

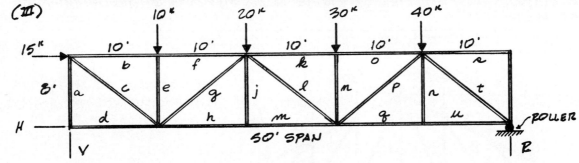

(1) DETERMINE MAGNITUDE AND DIRECTION OF H,V,R.
(2) DETERMINE STRESS IN a,b,c,d,e BY THE METHOD OF JOINTS
(3) DETERMINE THE STRESS IN k,l,m BY THE METHOD OF SECTIONS

(IV)

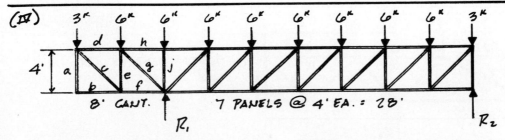

USING THE METHOD OF JOINTS, DETERMINE THE MAGNITUDE AND CHARACTER OF THE STRESS IN THE DESIGNATED MEMBERS.

PROBLEM 6–4

<u>TRUSSES</u>

(1) DETERMINE REACTIONS AND STRESS IN MEMBERS. —
GRAPHIC SOLUTION.

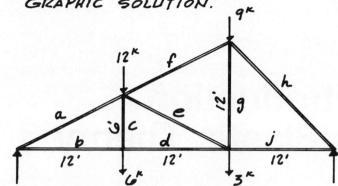

<u>ANSWERS</u>

a = 35.8^k C
b = 32^k T
c = 6^k T
d = 32^k T
e = 20.1^k C
f = 15.6^k C
g = 12^k T
h = 19.8^k C
j = 14^k T

(2) DETERMINE REACTIONS AND STRESSES IN MEMBERS —
GRAPHIC SOLUTION

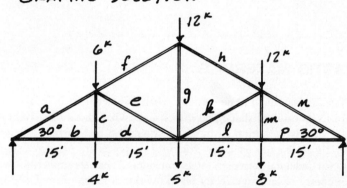

<u>ANSWERS</u>

a = 42^k C   k = 20^k C
b = 36.4^k T   l = 45^k T
c = 4^k T   m = 8^k T
d = 36.4^k T   n = 52^k C
e = 10^k C   p = 45^k T
f = 32^k C
g = 20^k T
h = 32^k C

(3) FIND REACTIONS & STRESSES IN MEMBERS — METHOD OF JOINTS OR
SECTIONS, AS CONVENIENT.

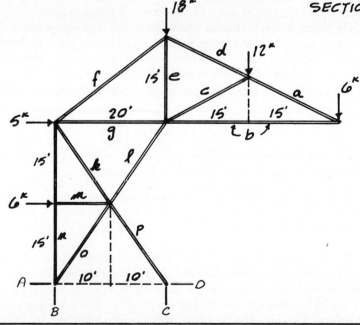

<u>ANSWERS</u>

a = 13.4^k T   h = 79.5^k T
b = 12^k C   k = 73.8^k C
c = 13.4^k C   l = 64.7^k C
d = 26.8^k T   m = 6^k C
e = 48^k C   n = 79.5^k T
f = 30^k T   o = 59.4^k C
g = 12^k T   p = 79.2^k C

7

Introduction to Elastic Theory

ELASTIC MATERIALS

An elastic material is one which, when subjected to external forces, will return to its original shape and dimensions when these forces are removed.

With this definition in mind, it should be recognized that none of the materials commonly used for structural purposes are *perfectly* elastic. However, for practical purposes, it is considered that the materials will exhibit the property of being perfectly elastic within a range of applied forces. This range is referred to as the *elastic range*.

When a force is applied to an elastic material, the dimensions of the material will change. A piece of material will lengthen when subjected to a tensile force and will shorten when subjected to a compressive force, as shown in Fig. 7-1. The change in the original dimension L of the test specimen will be referred to as deformation D as shown on the figure. Generally, however, the change in dimension is expressed on a unit basis and called *strain* (ϵ). The unit strain may be found by dividing the *total deformation* by the *original length* of the specimen. Expressed mathematically,

$$\epsilon = D/L$$

It should be noted that D and L have the same units (generally inches) and, consequently, the unit of strain is *inches per inch*.

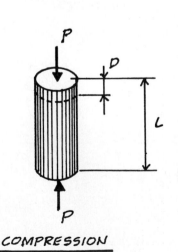

TENSION

COMPRESSION

FIGURE 7–1

HOOKE'S LAW

In the seventeenth century an English scientist named Robert Hooke discovered that a *linear* relationship exists, up to a limit known as the *proportional limit*, between the amount of unit stress ($f = P/A$, see Chapter 4) to which a material is subjected and the amount of strain that takes place. This relationship is known as *Hooke's Law*, which states that

For an elastic material, stress is proportional to strain within a certain limit.

The limit is, as already mentioned, called the proportional limit. This limit is often referred to as the *elastic limit;* we will use the latter expression from this point on. It should be clear, however, that there is a difference in meaning.

The proportional limit is that point beyond which there is a deviation in the linear relationship between stress and strain. The elastic limit is the point beyond which a material will no longer return to its original dimensions when the load is removed. For the materials we will be concerned with, the value of the proportional limit and the elastic limit are virtually the same.

STRESS/STRAIN DIAGRAMS

The relationship between stress and strain for a particular material is generally expressed graphically on a *stress/strain diagram*. A generalized diagram is shown in Fig. 7–2. The elastic limit which has already been defined is at the end of the linear relationship between stress and strain. At some point beyond the elastic limit, large strains will occur without significant increase in stress. This point is known as the *yield point*.

The *ultimate strength* of a material is defined as the maximum unit stress obtained before breaking the specimen.

In the appendix, Data Sheets 4–10 show stress/strain relationships for several materials. Exercises that will require the use of these data sheets are presented at the end of this chapter.

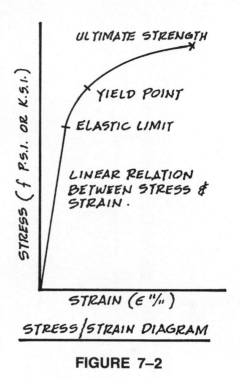

FIGURE 7–2

MODULUS OF ELASTICITY

Each material has its own unique stress/strain diagram. The straight-line portion (the elastic range) will always exist, but it will be steeper for stronger materials. In other words, for a given unit stress, less strain will take place in stronger materials. In order to distinguish between materials, we deal with a property known as the *modulus of elasticity (E)*, which is simply the slope of the straight-line portion of the stress/strain diagram. Expressed mathematically,

$$E = \frac{f}{\epsilon}$$

Note that the units of E are p.s.i. or k.s.i. since the units of f are p.s.i. or k.s.i.; strain, ϵ, is a pure number.

In Fig. 7–3 the initial portion of the stress/strain diagram for a common grade of structural steel is shown. The modulus of elasticity of steel is shown to be 30,000 k.s.i. (or 30,000,000 p.s.i.). Actually, a variety of tests will show E for steel to vary from 29,000 k.s.i. to 30,000 k.s.i.

The modulus of elasticity for structural grades of wood will vary generally from 1,000,000 p.s.i. to 2,000,000 p.s.i. depending on the species being tested. The modulus of elasticity of concrete will vary from about 2,000,000 p.s.i. to 5,000,000 p.s.i. depending on the strength of the concrete mix.

The modulus of elasticity may properly be thought of as an index of strength of the material, and can only be determined by means of an actual test. Generally, ductile materials which are homogeneous and isotropic are

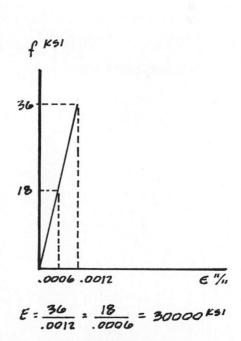

FIGURE 7–3

subjected to tensile tests to determine their properties. An "isotropic" material is one that exhibits the same physical properties in all directions. Brittle materials, such as concrete, are tested in compression.

For most materials, except those that are brittle and considered to have no resistance to tension, the modulus of elasticity is considered to be the same in tension and compression.

DEFORMATION

In order to determine the deformation that will take place in material subjected to stress, the relationships between stress and strain will be used. Referring to Fig. 7–4, which is an elastic material to be stressed in tension (within the elastic range of the material), it is desired to know the total deformation *(D)* that will take place due to the application of the force *P*. Consider that the specimen has a cross-sectional area equal to *A*.

Therefore, with these factors in mind,

$$E = \frac{f}{\epsilon} \text{ and } \epsilon = \frac{f}{E}$$

Furthermore, $\epsilon = D/L$

$$\therefore D/L = \frac{f}{E} \text{ and } f = P/A$$

$$\therefore D/L = \frac{P}{AE}$$

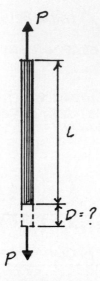

FIGURE 7–4

and

$$\boxed{D = \frac{PL}{AE}}$$

which is the basic equation used to determine the *total* deformation (or change in dimension).

The application of this equation is a simple matter. For example, consider the steel bar of Fig. 7–5 and the following data:

$$L = 10'' \text{ (original length)}$$

$$A = 1 \text{ in.}^2$$

$$P = 20^K$$

$$E = 30,000 \text{ k.s.i.}$$

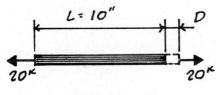

FIGURE 7–5

Then the total deformation:

$$D = \frac{PL}{AE} = \frac{(20^K)\,(10'')}{(1 \text{ in.}^2)\,(30,000 \text{ k.s.i.})} = .0067''$$

Under the applied tensile force, the deformation represents an increase in the original dimension. If the force applied were compressive in character, the deformation would represent a decrease in the original dimension, assuming that the modulus of elasticity were the same in compression as in tension.

The consequences of deformation due to applied forces are most apparent when dealing with extremely large scale structures. For example, in a very tall building the columns, which are very long and are accumulating very large loads to be delivered to the foundations, may deform several inches. This can be very important in buildings of large scale and must be considered in the detailing and fabrication procedures. In low-rise buildings this is generally not a serious issue, since the length of the columns is not great and therefore, deformation due to compressive stresses may be negligible.

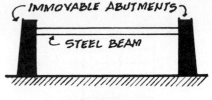

FIGURE 7–6

EFFECTS OF TEMPERATURE ON STRUCTURES

The materials of construction will expand when subjected to a rise in temperature and contract when temperatures are lowered. This tendency to change dimension when subjected to temperature changes can sometimes cause difficulty in a structural member when it is restrained from responding to the dimensional change. If the member cannot change in length, there are stresses introduced due to the restraint. For example, if the member shown in Fig. 7–6 were subjected to a uniform temperature increase and it were fixed between two immovable abutments, a compressive stress would be created in the section. In essence, what is happening here is the same as if the member were allowed to expand and then was shortened with a compressive force, as shown in Fig. 7–7. Conversely, if the member were subjected to a uniform decrease in temperature, a tensile force would be developed, as shown in Fig. 7–8. If the member is free to expand or contract under the influence of temperature change, no stresses will be developed, but a change in dimension will take place.

The amount of dimensional change that will take place is determined by the use of a factor known as the *coefficient of expansion*. Each material has such a coefficient, which is determined by experimentation. The coefficient of expansion (which is the same for contraction) expresses the change in dimension, per unit of original length, per degree of temperature change. The coefficients are extremely small numbers and may be expressed in terms of the Celsius or Fahrenheit temperature scale. The Fahrenheit scale will be used here. Coefficients of expansion are given for several materials on Data Sheet D–11 in the Appendix. The change in dimension of a structural member may be determined by the following:

$$\boxed{D = c\Delta t L}$$

where D = total deformation in inches
  c = coefficient of expansion ($''/''/^0F$)
  Δt = change in temperature (^0F)
  L = original length in inches

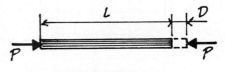

FIGURE 7–7

While it may seem a bit strange, it should be noted that this expression does not include anything that has to do with the cross-sectional area. In fact, regardless of the area, a member will contract or expand uniformly if every square inch of cross section is subjected to the same change in temperature.

If the member is restrained from changing dimension, the stress introduced may be found by a modified version of the above equation:

$$D = c\Delta t L \qquad (7\text{–}1)$$

It was previously established that

$$\epsilon = D/L \text{ or } D = \epsilon L \qquad (7\text{–}2)$$

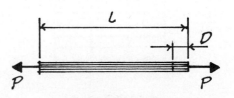

FIGURE 7–8

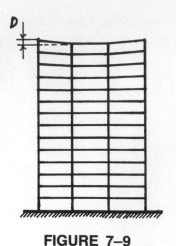

FIGURE 7–9

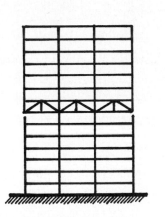

FIGURE 7–10

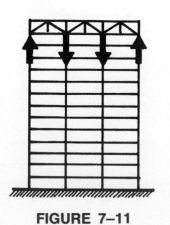

FIGURE 7–11

and

$$\epsilon = \frac{f}{E}, \therefore D = \frac{f}{E} L \qquad (7\text{–}3)$$

Substituting Eq. (7–3) into Eq. (7–1)

$$\frac{f}{E} L = c\Delta t L$$

and

$$\boxed{f = Ec\Delta t}$$

where f = unit stress (p.s.i. or k.s.i.)

  E = modulus of elasticity (p.s.i. or k.s.i.)

  c = coefficient of expansion

  Δt = change in temperature (°F)

Deformation, or induced stresses due to temperature change, can be a problem in high-rise structures or long-span structures; in these cases, design, detailing, and fabrication procedures can be affected.

The problem in high-rise buildings is primarily due to differential deformation between exterior columns, whose exposed faces may be subjected to drastic temperature changes, and interior columns, which are subjected to no significant temperature changes. Figure 7–9 shows, in exaggerated form, what happens when exterior columns are subjected to a drastic rise in temperature. The effect of this expansion on glass, interior finishes, and other architectural elements should be obvious. A drastic decrease in temperature will cause a shortening of the exterior columns, which will also have a serious effect on architectural finishes. There are several ways to prevent the kinds of problems that temperature changes may produce in high-rise structures. The most obvious of these calls for a proper job of detailing, which will allow the structure to move independently of the finish work. This means that where glazing, interior partitions, or other architectural elements tie into the structural frame, appropriate space must be left so that movement of the frame will not destroy the finishes.

Where a building is so tall that movements produced by temperature changes would be in excess of what could feasibly be handled by detailing, other techniques must be employed. One way to handle this is to produce what is known as a "thermal break" at the mid-height of the building, as shown schematically in Fig. 7–10. In this scheme, the loads on the exterior columns from the upper portion of the building are transferred to the interior columns at mid-height. This allows a "break" in the exterior columns; since temperature deformation is dependent on length, this technique can be quite effective in reducing the total deformation in any one section of the building. Essentially, by reducing the deformation affecting any level, the problems are reduced to a point where they can be handled by detailing procedures.

Another way to handle temperature change problems is to provide a very stiff cap for the structure, sometimes called a "hat truss," shown in Fig. 7–11. This very stiff structure is essentially a double cantilevered truss that is designed to restrain movement of the exterior columns, while the interior columns act as supports for the truss. The action of the forces involved is as shown by the arrows, in the case of expansion of the exterior columns. The arrows would, of course, be reversed in the case of contraction of the exterior columns. When using the hat truss scheme, the idea is to restrain the columns from dimensional change. In doing this, large stresses may be introduced which must be accounted for in the analysis and design of the structure.

Temperature change may also cause problems in long-span structures as shown in Fig. 7–12. If the roof structure expands, the columns to which it is attached will move outward, causing problems to the wall finishes that are tied into the columns. If the supports are masonry walls, the problems can be very serious. Since masonry walls are brittle, they cannot undergo the movement dictated by the roof structure without cracking. Cracks in the walls are not only unsightly and distressing to the users of the building, but can also allow water to penetrate the walls, thereby causing further damage. Common preventive measures for this include the use of a "roller" connection at one end of the roof structure and the use of graphite- or Teflon-coated bearing plates, which will allow the structure to slide freely on its support when expansion or contraction takes place. The roller connection usually takes the form of a slotted connection, shown schematically in Fig. 7–13. The "slot" is made in the direction of the expansion or contraction, allowing movement in this direction but restraining movement perpendicular to this, as well as providing a vertical tie to the support.

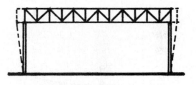

FIGURE 7–12

ALLOWABLE STRESS

Earlier, the expressions "yield point" and "ultimate strength" were defined. it would obviously be nonsensical to design the members of a structure so that the yield strength of the material is reached under the influence of loads. The members would deform excessively, and since the yield point is generally beyond the elastic limit, much of the deformation would be permanent, even if the loads are removed. To guard against this, structural elements are designed with a *factor of safety*. The factor of safety is provided by using a stress below the elastic limit as a maximum allowable stress. This process of design is known as *allowable stress design*, or sometimes, *working stress design*. For example, the most commonly used grade of structural steel has a yield strength of 36,000 p.s.i. By the allowable stress design method, a member subjected to bending would be designed so that the stresses would not exceed 24,000 p.s.i. (this may, in some cases, be reduced to 22,000 p.s.i., depending on the shape of the cross section being used). Therefore, there would be a factor of safety of 1.5 against yielding of the material. The allowable stresses to be used for the various structural materials are primarily based on experience and judgment. The recommended values to be used for allowable stress design are generally provided by manufacturers of the materials through associations that represent them.

ULTIMATE STRENGTH

An alternative to the allowable stress design method is known as *ultimate strength design*. In this approach, the *anticipated* loads on a structural member are modified by a "load factor." These *increased* loads are then used in the computations, and the structure is designed to its ultimate strength. This method of design is now quite common for reinforced concrete structures.

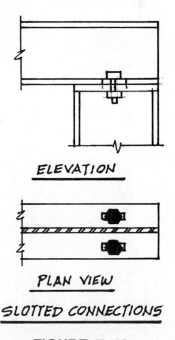

ELEVATION

PLAN VIEW

SLOTTED CONNECTIONS

FIGURE 7–13

FIGURE 7–14

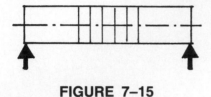

FIGURE 7–15

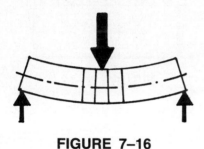

FIGURE 7–16

ELASTIC BENDING

If a structural beam were subjected to a load, bending would occur, as shown in Fig. 7–14. If vertical lines that were considered to be representative of plane sections (or "slices") of the beam were etched on the unloaded member, it would be seen that these sections, which are initially plane, as shown in Fig. 7–15, will remain plane when the load is applied and bending takes place, as shown in Fig. 7–16.

Numerous experiments have revealed that initially plane sections remain plane during bending, *regardless of the shape of the cross section,* with the result that *compressive shortening* and *tensile elongation* of "fibers" are proportional to their distances from a line known as the "neutral axis." This phenomenon is shown in Fig. 7–17, which is a blown-up view of part of the beam of Fig. 7–16. What is being shown here is that the *strains* in a member subjected to bending vary proportionally with the distance from the neutral axis. There will be zero strain at the neutral axis, where no dimensional change takes place between planar sections, and maximum strain at the outermost fibers.

With these ideas in mind, let's now cut a free body diagram from the beam shown in Fig. 7–18. The same ideas concerning the free body which were discussed in the preceding chapter are applicable here as well. To repeat these ideas:

1. The part that has been cut loose as a free body must be in a state of equilibrium, since it is part of a structure that is in a state of equilibrium.
2. Internal forces are visualized a being "cut loose" and are treated as external forces on the free body diagram.

Referring to Fig. 7–18, the force P produces reactions R_1 and R_2. The part of the member cut loose as a free body includes the reaction R_1. In

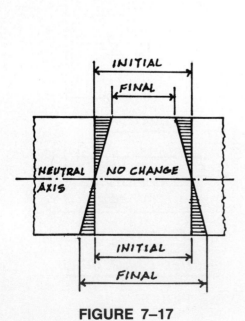

FIGURE 7–17

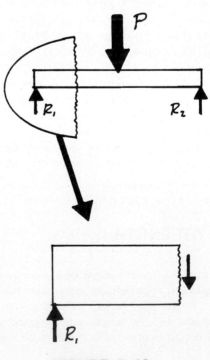

FIGURE 7–18

order to satisfy $\Sigma F_y = 0$, there must be a force on the cut section equal and opposite to R_1. This is a shearing force which exists internally and is treated as an external force on the free body. The shearing stresses are an important consideration in the design of structural members; the evaluation of shearing unit stresses will be treated in some detail in Chapter 10. In this section we will concern ourselves with the evaluation of the compressive and tensile stresses produced by bending. Consequently, the shearing force on the free body will be omitted from the sketches used in the remainder of this chapter.

Returning to the free body, now shown in Fig. 7–19, it can be seen that the reaction R_1, because of its *position* relative to the cut section, will produce a tendency for the section to rotate about point n, which is on the neutral axis of the beam. This tendency to rotate caused by a bending situation is referred to as a *bending moment*. Now, in order to satisfy $\Sigma M = 0$ there must be a moment produced which is equal in magnitude and opposite in direction to the bending moment. This is called the *resisting moment*. Let's consider what is happening to produce this resisting moment.

It was shown earlier that in a beam subjected to bending, initially plane sections will remain plane. This indicated that the strains varied proportionally with the distance from the neutral axis. Considering that Hooke's Law is applicable to the member under discussion, then with a proportional variation in strain there will also be a proportional variation in the *unit stresses* produced in the cross section, as shown in Fig. 7–20. There will be zero unit stress at the neutral axis with a maximum compressive unit stress (f_c) at the top fibers and a maximum tensile unit stress (f_t) at the bottom fibers. These unit stresses will have resultants R_c and R_t, shown in Fig. 7–21. In order to satisfy $\Sigma F_x = 0$, R_c must be equal to R_t. The two resultants are referred to as the *internal couple*, with a couple being defined as two forces, equal in magnitude and opposite in direction.

Still referring to Fig. 7–21, the resultants R_c and R_t have lever arms $\bar{Y}_1$ and $\bar{Y}_2$. The resisting moment furnished by the internal couple with respect to the neutral axis is

$$Resisting\ Moment = R_c\bar{y}_1 + R_t\bar{y}_2$$

It should be noted that although the resultant forces are opposite in direction, they produce a rotational tendency of the same sense with respect to the neutral axis.

The values of the lever arms $\bar{y}_1$ and $\bar{y}_2$ are dependent on the area that the stress is acting upon, since the resultants will act at the center of gravity of the "wedge" of stress. It should be recognized that the wedge of stress will be irregular when acting upon an irregular area. An example of this is shown in the T-shaped cross section of Fig. 7–22. While the wedge of stress (that is, the three-dimensional configuration) is irregular, the intensity of stress will still vary proportionally with the distance from the neutral axis, as shown in Fig. 7–23. This is so because the strains vary proportionally with the distance from the neutral axis, *regardless of the shape of the cross section*.

What the preceding discussion shows is that a bending moment is produced in a member spanning between supports. In order to counter this bending moment, the member must furnish a resisting moment. The resisting moment is provided by the internal couple, which is a product of the internal proportionally varying stresses. We will now consider the evaluation of the maximum stresses produced in a member due to a bending moment.

Referring to Fig. 7–24, which is a section subjected to a bending moment, the internal stresses produce resultants that create a resisting moment

$$Resisting\ Moment = R_c\bar{y}_1 + R_t\bar{y}_2 \qquad (7\text{–}4)$$

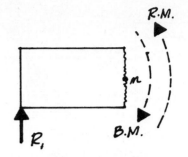

FIGURE 7–19

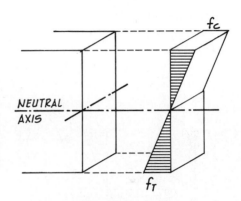

FIGURE 7–20

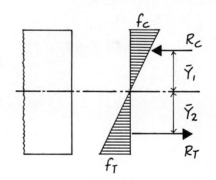

FIGURE 7–21

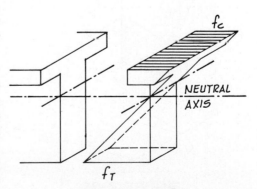

FIGURE 7–22

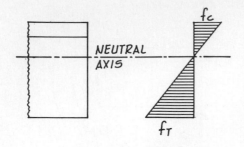

FIGURE 7-23

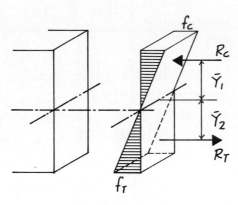

FIGURE 7-24

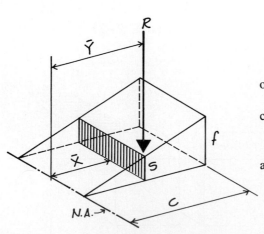

FIGURE 7-25

For the sake of convenience, we will use one of the stress wedges of Fig. 7–24 and orient it, as shown in Fig. 7–25, in order to define the various terms. Referring to Fig. 7–25, the resultant force may be taken as the product of the *average* unit stress (S) and the plane area (A)

$$\therefore R = SA \qquad (7\text{--}5)$$

By similar triangles

$$\frac{S}{f} = \frac{\bar{x}}{c} \text{ and } S = \frac{f\bar{x}}{c} \qquad (7\text{--}6)$$

Substituting equation (7–6) into Eq. (7–5)

$$R = \frac{fA\bar{x}}{c} \qquad (7\text{--}7)$$

Substituting equation (7–7) into equation (7–4) and differentiating between the compression side and tension side

$$Resisting\ Moment = \frac{f_c A_c \bar{x}_1 \bar{y}_1}{c_c} + \frac{f_t A_t \bar{x}_2 \bar{y}_2}{c_t} \qquad (7\text{--}8)$$

and the combination of the group of geometrical factors $A\overline{xy}$ is called the *moment of inertia (I)*. Therefore, rewriting Eq. (7–8)

$$Resisting\ Moment = \frac{f_c I_1}{c_c} + \frac{f_t I_2}{c_t} \qquad (7\text{--}9)$$

For a homogeneous material, with the modulus of elasticity in compression equal to the modulus of elasticity in tension, the ratio of $\frac{f_c}{c_c} = \frac{f_t}{c_t}$ (see Fig. 7–26) and Eq. (7–9) becomes

$$Resisting\ Moment = \frac{f}{c}(I_1 + I_2) = \frac{f}{c}(I_{total}) \qquad (7\text{--}10)$$

To simplify matters, we will refer to the resisting moment as, simply, the *moment (M)*. Then the basic equation which deals with flexural conditions becomes

$$\boxed{M = \frac{fI}{c}} \qquad (7\text{--}11)$$

Until this point we have used the expression "neutral axis" a number of times. The location of the neutral axis will now be determined.

It has previously been shown that the forces that constitute the internal couple must be equal to each other in order to satisfy $\Sigma F_x = 0$. Therefore,

$$R_c = R_t$$

and, according to Eq. (7–7),

$$R = \frac{fA\bar{x}}{c}$$

$$\therefore \frac{f_c A_c \bar{x}_1}{c_c} = \frac{f_t A_t \bar{x}_2}{c_t} \qquad (7\text{--}12)$$

and, for a homogeneous material,

$$\frac{f_c}{c_c} = \frac{f_t}{c_t}$$

Then, in Eq. (7–12),

$$A_c \bar{x}_1 = A_t \bar{x}_2$$

Referring to Chapter 4, it should be recalled that this is true at the centroid of the area. Therefore the neutral axis is at the centroid of the cross section.

SUMMARY (ELASTIC BENDING)

It would be useful at this point to collect the ideas presented in the last few pages and restate them in compact form.

Therefore, consider the following:

1. A structural member subjected to bending (no direct axial stress).
2. the member is made of a homogeneous material with stress-strain characteristics, within the elastic range, of:
   (a) constant ratio $E = f/\epsilon$.
   (b) modulus of elasticity in compression (E_c) = modulus of elasticity in tension (E_t).

Then, on any plane-section area in bending, the following principles apply:

1. *Unit Stress Distribution.* Strains (ϵ) are proportional to distances from neutral axis (axis of zero unit stress).

   Stress/strain ratio $\dfrac{f}{\epsilon} = E$ is constant.

   ∴ Unit stresses (f) must be proportional to distances from the neutral axis.

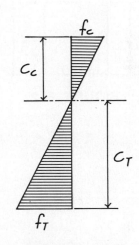

FIGURE 7–26

2. *Location of Neutral Axis.* $R_c = R_t$ (for $\Sigma F_x = 0$).
   For proportional unit stress distribution

   $$R = \frac{fA\bar{x}}{c}$$

   $$\therefore R_c = \frac{f_c A_c \bar{x}_1}{c_c} = R_t = \frac{f_t A_t \bar{x}_2}{c_t}$$

and

$$\boxed{A_c \bar{x}_1 = A_t \bar{x}_2}$$

which can only be true of an axis at the centroid of the section.

∴ The neutral axis is at the centroid of the area.

3. *Resisting Moment.* Bending Moment = Resisting Moment (for $\Sigma M = 0$).

For proportional unit stress distribution:

$$M = \frac{fI}{c}$$

This expression will be referred to as the "bending stress equation" throughout.

MOMENT OF INERTIA

Previously, the expression *moment of inertia* was introduced as the name given to the collection of geometrical factors $A\overline{x}\overline{y}$ where

    A = the planar area on which the "wedge" of stress is acting

    $\overline{x}$ = distance from the neutral axis to the centroid of the planar area

    $\overline{y}$ = distance from the neutral axis to the resultant (R) of the "wedge" of stress

In order to use the bending stress equation, it will be necessary to determine a numerical value for the moment of inertia for any given cross section. In light of the derivation of the bending stress equation, it should be remembered that we are interested in the total moment of inertia of the cross section, with the neutral axis being the reference from which $\overline{x}$ and $\overline{y}$ are measured.

To demonstrate, let's determine the general expression for the moment of inertia of the rectangular section shown in Fig. 7–27. Visualizing a proportionally varying stress acting on the section, with the neutral axis being the axis of zero unit stress, as shown in Fig. 7–28, and evaluating the expression

$$I = A\overline{x}\overline{y}$$

we can determine the moment of inertia. In this case we can deal with one half of the section, since it is symmetrical about the neutral axis, and double the answer to account for the other half.

    Therefore

    $A = (B)(D/2)$

    $\overline{x} = D/4$

    $\overline{y} = (2/3)(D/2)$ since the resultant is acting at the *center of gravity* of the triangular wedge (see Data Sheet D–12 for location of resultant forces).

$$I = A\overline{x}\overline{y} = 2[(B)(D/2)\,(D/4)(2/3)(D/2)] = \frac{BD^3}{}$$

This is the general expression for th⸙
section with respect to the ne⸙
moment of inertia of a vari⸙
the Appendix. The d⸙
with respect to a⸙   ⸙ma-
tion should be⸙   ⸙es. It is
recommend⸙   ⸙a sheet as an
exercise.⸙   ⸙pression for the rec-
tangula⸙   ⸙ould also be made to Data
S⸙   ⸙regular wedges of stress. It must
    ⸙ resultant is at the center of gravity of
    ⸙ is a proportional variation in stress (triangular

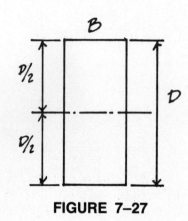

FIGURE 7–27

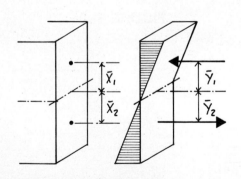

FIGURE 7–28

when looked at from a side view), this variation takes place over irregular areas, and the center of gravity of the wedge will vary accordingly.

IRREGULAR SHAPES

In the preceding discussion the procedure was shown to find the moment of inertia of a rectangular section. Reference was made to Data Sheet D–13, which gives the general expressions for the moment of inertia of variety of shapes. We will now consider using this data for finding the moment of inertia of the infinite variety of cross-sectional shapes that may be encountered.

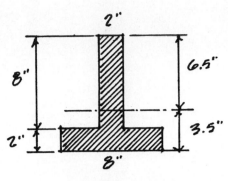

FIGURE 7–29

Example 7–1 (Fig. 7–29)

In this case we wish to determine the moment of inertia of the inverted T section. The moment of inertia must be found with respect to the neutral axis. Therefore, the first step is to determine the location of the centroid of the section.

Using the top of the section as a reference:

$\underline{\Sigma Ax}$

$$(2'' \times 8'')(4'') = \quad 64 \text{ in.}^3$$
$$(2'' \times 8'')(9'') = \underline{144 \text{ in.}^3}$$
$$208 \text{ in.}^3$$

$$\therefore \bar{x} = \frac{208 \text{ in.}^3}{32 \text{ in.}^2} = 6.5'' \text{ from the top}$$

Now, there are essentially two ways to determine the moment of inertia with respect to this axis. The first makes use of the information from Data Sheet D–13 and breaking the section into two parts, with each part having the neutral axis as a base line, as shown in Figs. 7–30 and 7–31. On Data Sheet D–13 we find that the general expression for the moment of inertia of a rectangle with respect to an axis at its base is $BD^3/3$. This is precisely the condition that we have in Fig. 7–30, and evaluating this part

$$I = \frac{BD^3}{3} = \frac{(2)(6.5)^3}{3} = 183.1 \text{ in.}^4$$

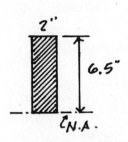

FIGURE 7–30

If it has not yet been discovered, it might be well to point out that the units of moment of inertia must always be "in.[4]"

In order to evaluate the bottom portion of the section, we can use the same expression as above and employ a method of subtraction by computing the moment of inertia with respect to the base line of a solid rectangle 8" wide and then subtracting the void, which is 6" wide and is also resting on the base line. Referring to Fig. 7–31 and remembering that the base line is the neutral axis

$$I = \frac{(8)(3.5)^3}{3} - \frac{(6)(1.5)^3}{3} = 107.6 \text{ in.}^4$$

The total moment of inertia of this section is simply the algebraic addition of the parts and

$$I = 183.1 + 107.6 = 290.7 \text{ in.}^4$$

Another way to determine the moment of inertia of this section involves a new idea which we will refer to in this text as the *transfer equation*.* The transfer equa-

---

*The derivation of the *transfer equation* is beyond the intent or purpose of this text. For those interested, however, the derivation may be found in any text on structural mechanics. While referred to here as the *transfer equation*, it may be listed in other texts as the "Transfer of Axis Theorem" or "Parallel Axis Theorem."

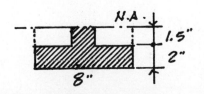

FIGURE 7–31

tion is applicable to irregular sections made up of regular pieces, such as the section under discussion. Indeed, in some cases the use of the transfer equation may be the only feasible way to determine the moment of inertia. We'll now look at this idea in general terms and then apply it to the section of Fig. 7–29. The transfer equation is

$$I_a = I_g + A\bar{x}^2$$

where I_a = the moment of inertia with respect to the reference axis (in our case the neutral axis)

I_g = the moment of inertia of an individual element of the section, with respect to its own centroidal axis

A = the area of the individual element

$\bar{x}$ = the distance from the centroid of the element to the reference axis (the neutral axis) of the section

Let's immediately apply these ideas to the determination of the moment of inertia of the section of Fig. 7–29, which is redrawn with additional data in Fig. 7–32. This additional data shows the distance from the centroid of each piece to the neutral axis, which we will need in using the transfer equation. Let's first determine the moment of inertia of the stem of the inverted T with respect to the neutral axis, following very carefully what the transfer equation tells us,

$$\boxed{\;\rvert\rvert\;}\quad I_a = I_g + A\bar{x}^2$$

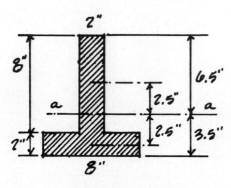

FIGURE 7–32

In order to determine the moment of inertia with respect to the neutral axis (I_a), we must first find the moment of inertia of the element with respect to its own centroidal axis (I_g). This is simply a rectangle, and Data Sheet D–13 tells us that

$$I_g = \frac{BD^3}{12} = \frac{(2)(8)^3}{12} = 85.3 \text{ in.}^4$$

We must now add the product of $A\bar{x}^2$ for this element of the section:

$$A = (2'' \times 8'') = 16 \text{ in.}^2$$

$$\bar{x} = 2.5''$$

$$\therefore A\bar{x}^2 = (16)(2.5)^2 = 100 \text{ in.}^4$$

Therefore, the moment of inertia of the stem of the T with respect to the neutral axis is

$$\boxed{\;\rvert\rvert\;}\quad I_a = 85.3 + 100 = 185.3 \text{ in.}^4$$

Following the same procedure for the flange of the inverted T,

$$\boxed{\;\rule{0pt}{0pt}\;}\quad I_a = I_g + A\bar{x}^2$$

$$I_a = \frac{(8)(2)^3}{12} + (16)(2.5)^2 = 105.3 \text{ in.}^4$$

The total moment of inertia for this section is the algebraic addition of the parts

$$I = 185.3 + 105.3 = 290.6 \text{ in.}^4$$

which, of course, is the same as we found previously.

Example 7–2 (Fig. 7–33)

We'll determine the moment of inertia of this section by using the transfer equation. The first step is to locate the neutral axis of the section:

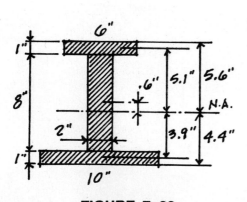

FIGURE 7–33

ΣAx (using the top as a reference)

$(6'' \times 1'')(.5'')$	$=$	3 in.^3
$(8'' \times 2'')(5'')$	$=$	80 in.^3
$(10'' \times 1'')(9.5'')$	$=$	$\underline{95 \text{ in.}^3}$
		178 in.^3

$$\therefore \bar{x} = \frac{178 \text{ in.}^3}{32 \text{ in.}^2} = 5.6'' \text{ from the top}$$

Using the transfer equation,

I

$$\frac{(6)(1)^3}{12} + (6 \times 1)(5.1)^2 = 156.6 \text{ in.}^4$$

$$\frac{(2)(8)^3}{12} + (8 \times 2)(.6)^2 = 91.1 \text{ in.}^4$$

$$\frac{(10)(1)^3}{12} + (10 \times 1)(3.9)^2 = \underline{152.9 \text{ in.}^4}$$

$$I_{\text{total}} = 400.6 \text{ in.}^4$$

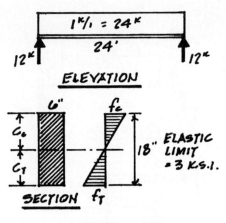

FIGURE 7–34

APPLICATION OF THE BENDING STRESS EQUATION

Now that we have a method for determining the moment of inertia of a cross section we'll put this idea in context and look at several examples using the bending stress equation.

Example 7–3 (Fig. 7–34)

Consider the beam shown, rectangular in cross section and carrying a total uniformly distributed load of 24^K. It is desired to determine the maximum bending stress in the section due to the load. In order to determine the magnitude of the stress we will use the bending stress equation,

$$M = \frac{fI}{c} \quad \text{or} \quad f = \frac{Mc}{I}$$

Now, given that the cross section will be constant for the entire span of the beam, c and I, which are based on the cross-sectional geometry, will be constant. Therefore, the stress f due to bending will be at a maximum (and we are primarily concerned with maximum stresses) where M is a maximum. If a slice at the end of the member, of infinitesimal thickness, is taken as a free body, as shown in Fig. 7–35, then the bending moment (and, therefore, the resisting moment) about point x is zero, since the only force on this free body passes through this point. As we move into the span and take free bodies as shown in Fig. 7–36, there will be a bending moment and, therefore, stresses due to bending. If the deformed shape of the beam (Fig. 7–37) is studied, it should be intuitively obvious that the maximum stresses due to bending will occur at the center of the span. It may be said that

> In a simply supported and symmetrically loaded beam the maximum bending moment, and, therefore, maximum stress due to bending, will occur at the midspan.

In order to evaluate the maximum moment in this example, a free body diagram is taken, cutting the member at the midspan. Referring to Fig. 7–38; the bending moment at x is

$$M_x = (12^K)(12') - (12^K)(6') = 72'^K$$

Now that we have the maximum bending moment and we know the c distances, which are the same in this example for the tension and compression side, it remains only to find I, which is a simple matter for this rectangular section:

$$I = \frac{BD^3}{12} = \frac{(6)(18)^3}{12} = 2916 \text{ in.}^4$$

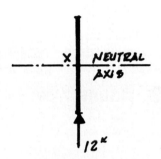

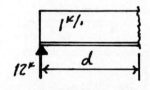

FIGURE 7–35

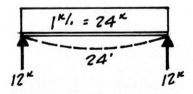

FIGURE 7–36

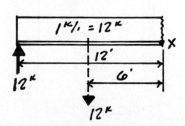

FIGURE 7–37

FIGURE 7–38

98   Structural Principles

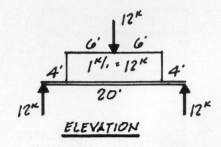

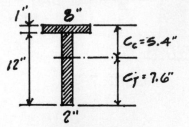

SECTION

ALLOW. STRESS = 15 K.S.I.
ELASTIC LIMIT = 20 K.S.I.

FIGURE 7–39

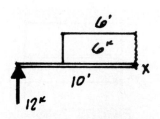

FIGURE 7–40

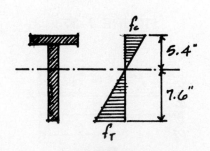

FIGURE 7–41

and the bending stress equation yields

$$f = \frac{Mc}{I} = \frac{(72'^K \times 12''/')(9'')}{2916 \text{ in.}^4} = 2.67 \text{ k.s.i.}$$

which is below the elastic limit of the material (given on Fig. 7–34); therefore, the equation is applicable. It should be noted that this single computation yields the maximum tensile and compressive stress, since the c distance is the same for both sides. When dealing with an unsymmetrical cross section with two different c distances, two computations must be made to determine the tensile and compressive stresses.

Example 7–4 (Fig. 7–39)

In order to determine the maximum bending stresses in the section due to the given loading, it is first necessary to determine the distances from the neutral axis to the outermost fibers of the cross section.

ΣAx (using the top as a reference)

$$(1'' \times 8'' \times 1/2'') = 4 \text{ in.}^3$$
$$(2'' \times 12'' \times 7'') = \underline{168 \text{ in.}^3}$$
$$\Sigma A\bar{x} = 172 \text{ in.}^3$$

$$\therefore \bar{x} = \frac{172 \text{ in.}^3}{32 \text{ in}^2} = 5.4''$$

and $c_c = 5.4''$, $c_t = 7.6''$, as shown.

The next step is to determine the moment of inertia of the section. Using the transfer equation:

$\underline{I}$

$$\frac{(8)(1)^3}{12} + (8 \times 1)(4.9)^2 = 192.8 \text{ in.}^4$$
$$\frac{(2)(12)^3}{12} + (12 \times 2)(1.6)^2 = \underline{349.4 \text{ in.}^4}$$
$$I_{total} = 542.2 \text{ in.}^4$$

Since the loading is symmetrical and the member is simply supported at the ends, the maximum moment will occur at the center of the span. Using the free body diagram of Fig. 7–40

$$M_x = (12^K)(10') - (6^K)(3') = 102'^K$$

The important thing to note in this problem is that there are two different c distances involved. If the stress diagram is sketched, as shown in Fig. 7–41, with the idea in mind that the proportional variation in stress is constant for a homogenus material, then it is easily seen that the critical stress due to bending occurs on the tension side of the beam.

$$f_t = \frac{Mc_t}{I} = \frac{(102'^K \times 12''/')(7.6'')}{542.2 \text{ in.}^4}$$
$$= 17.2 \text{ k.s.i.}$$

This is in excess of the given *allowable* stress, but it is valid since it is below the elastic limit. On the compression side

$$f_c = \frac{Mc_c}{I} = \frac{(102'^K \times 12''/')(5.4'')}{542.2 \text{ in.}^4}$$
$$= 12.2 \text{ k.s.i.}$$

which is less than the *allowable* stress. When making an investigation of this type, care must be taken to incorporate the critical values. In an unsymmetrical section with an allowable stress which is the same for tension and compression, the longest c distance will be critical.

PHYSICAL SIGNIFICANCE OF *E* AND *I*

Earlier in this chapter the expression *modulus of elasticity* was presented, with the idea that this is an indicator of the strength of a material. If a structural member were made of a material for which E is known, or can be determined by testing, then this would be only one indicator of the strength of the structural member. Certainly, the cross-sectional geometry, which is something apart from the material of which it is made, will play a very important role in the load-carrying capacity of the member. The moment of inertia of a cross section may be properly thought of as an indicator of strength based on geometry. For example, it should be intuitively obvious that a wood two-by-four spanning across a space will have a greater load-carrying capacity if it is oriented vertically rather than flat (see Fig. 7–42). The vertical orientation provides greater depth in the direction of bending, which means that the moment of inertia is greatest in this direction. The I of a rectangle will increase as the cube of the depth (that is, D^3), and linearly with the width of the rectangle. Therefore, the two-by-four oriented vertically is *significantly* stronger than that oriented horizontally.

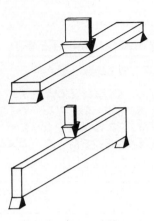

FIGURE 7–42

PROBLEM 7-1

STRESS / STRAIN

TENSION TEST

A STEEL BAR, ½" DIAMETER → AREA = 0.196 in², LENGTH = 6"

AS AXIAL LOAD IS APPLIED THE <u>INCREASE IN LENGTH</u> (DEFORMATION: LISTED AS "D" IN THE DATA) IS ACCURATELY MEASURED AT VARIOUS INTERVALS, WITH THE <u>TOTAL LOAD</u> (LISTED AS "P" IN THE DATA) RECORDED AT EACH INTERVAL.

TEST DATA			
ϵ "/"	D"	P #	f KSI
	.0021	1960	
	.0042	3920	
	.0063	5880	
	.0084	7840	
	.0105	9800	
	.0126	11,760	
	.0150	12,050	
	.0180	12,030	
	.0300	12,040	
	.0480	12,060	
	.0600	12,050	
	.0900	13,100	
	.1200	14,100	
	.1500	15,050	
	.1800	16,000	
	.2100	16,900	

TESTING STOPPED - LONG BEFORE FAILURE WHICH WOULD OCCUR AT f = 105 KSI ± AND ϵ = .050 "/"

REQUIRED

1. CONVERT DEFORMATION DATA (D) INTO <u>UNIT STRAIN</u> ϵ "/" = D/L.

2. CONVERT LOAD DATA (P #) INTO UNIT STRESS f KSI = P/A.

3. PLOT STRESS/STRAIN DIAGRAM AT SCALE SHOWN BELOW.

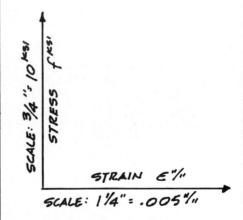

SCALE: ¾" = 10 KSI

STRESS f KSI

STRAIN ϵ "/"

SCALE: 1¼" = .005 "/"

PROBLEM 7–2

STRESS/STRAIN APPLICATIONS

Based on the stress/strain diagrams (Data Sheets D–4 to D–10). Use the Modulus of Elasticity within the elastic limit and the diagrams when beyond the elastic limit.

NOTE: The Modulus of Elasticity is the same for tension and compression in the steels and the aluminums.

---

1. The steel columns in a tall building are shop fabricated to equal a total height (unstressed) of 75 stories at 13′–4″ per story. Determine the *actual* height when the average unit stress in the columns = 14,500 p.s.i.

---

2. The steel towers of a large suspension bridge are planned to equal a total height of 750′–0″ (unstressed). They will be designed to have an average compressive unit stress of 50 percent of the yield strength of the steel. Determine the total loss of height if the towers are built of:

    a. S1 grade of steel
    b. S2 grade of steel
    c. S3 grade of steel

---

3. A "7-wire strand" cable is installed in a vertical line between anchorages at points A and B (30′–0″ apart) as shown.
       Force F is then applied horizontally at midpoint C, pulling it x out of line, as shown.

       Cable cross-sectional *area* = .0799 sq. in.

       Determine the total tension, P kips, in the cable and the force, F kips, necessary to displace point C by:

    a. x = 19 inches
    b. x = 33 inches

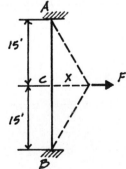

PROBLEM 7–3

STRESS/STRAIN APPLICATIONS

Stress Caused by Resistance to Temperature Change [D = cΔtL]

1. A steel wide flange beam with a cross-sectional area of 20.0 in.^{2} is anchored in heavy concrete abutments at each end of a 50′–0″ effective length when the temperature is 40° F. Coeff. of linear expansion for steel = .00000667 in./in./°F.

   a. Determine the compressive unit stress (f k.s.i.) in the beam, and the total force (P kips) against the abutments when the temperature rises to 100°F. if the abutments:
      i. Are immovable
      ii. Move .150″ apart

   b. If one end of the beam is detailed to allow free (unrestrained) movement, how much change of length would occur between a low temp. of 25° F. *below zero* and a high temp. of 125°F.?

NOTE: Use information from stress/strain data sheets (D4 − D10).

---

2. a. A steel wire 0.1″ in diameter is stretched at 70°F. between fixed supports 6 ft apart with a total tensile force of 40 lbs. What will the tension be in the wire at 30°F.? Use $E = 30 \times 10^6$ p.s.i..

   b. At what temperature will the stress be zero?

PROBLEM 7–4

<u>MOMENT OF INERTIA</u>

● DETERMINE THE MOMENT OF INERTIA OF EACH SECTION WITH RESPECT TO THE CENTROIDAL AXIS.
<u>NOTE</u>: THESE SECTIONS ARE THE SAME AS THOSE OF PROBLEM 4-4.

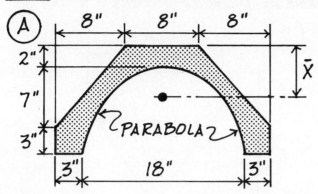

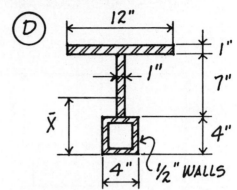

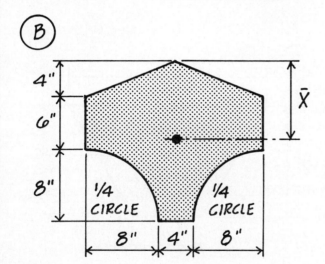

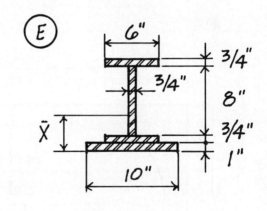

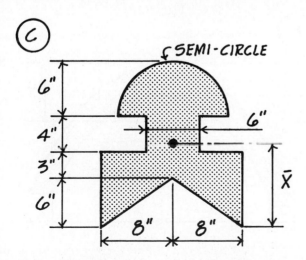

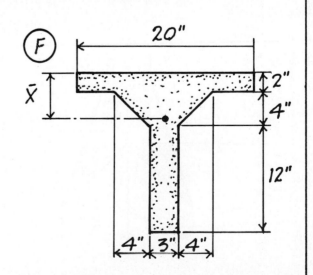

PROBLEM 7–5

MOMENT OF INERTIA

- DETERMINE THE MOMENT OF INERTIA OF EACH SECTION, WITH RESPECT TO THE NEUTRAL AXIS.

NOTE: THESE SECTIONS ARE THE SAME AS THOSE OF PROBLEM 4-3.

(A)
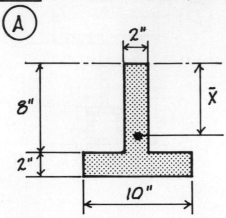

(B)
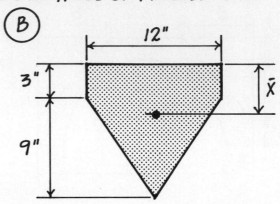

(C)
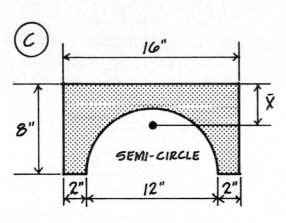

SEMI-CIRCLE

(D)
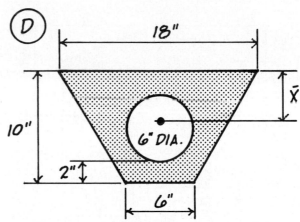

6" DIA.

(E)
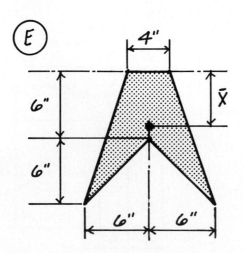

(F)
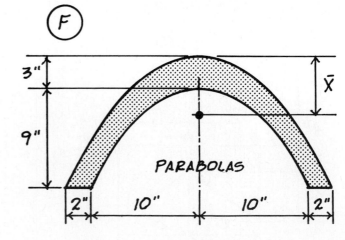

PARABOLAS

PROBLEM 7–6

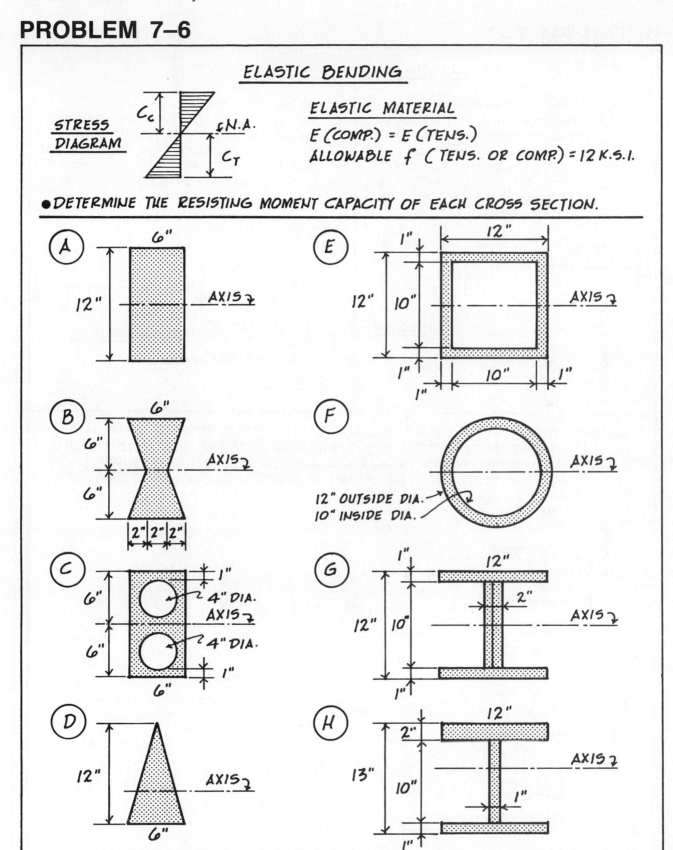

ELASTIC BENDING

STRESS DIAGRAM

C_C | $N.A.$ | C_T

ELASTIC MATERIAL
E (COMP.) = E (TENS.)
ALLOWABLE f (TENS. OR COMP.) = 12 K.S.I.

● DETERMINE THE RESISTING MOMENT CAPACITY OF EACH CROSS SECTION.

A
6"
12"
AXIS

B
6"
6"
6"
AXIS
2" 2" 2"

C
6"
6"
1"
4" DIA.
AXIS
4" DIA.
1"
6"

D
12"
AXIS
6"

E
1"
12"
10"
12"
AXIS
1"
1"
10"
1"

F
AXIS
12" OUTSIDE DIA.
10" INSIDE DIA.

G
1"
12"
12"
10"
2"
AXIS
1"

H
2"
12"
13"
10"
AXIS
1"
1"

PROBLEM 7-7

ELASTIC FLEXURE THEORY

(I) THE CROSS SECTION SHOWN IS MADE OF A MATERIAL WITH AN ALLOWABLE STRESS IN BENDING (TENS. & COMP) OF 2000 P.S.I. — THIS IS WITHIN THE ELASTIC RANGE OF THE MATERIAL. WHAT MUST THE DIMENSION "b" BE FOR THE BEAM TO BE SAFE FOR THE GIVEN LOADING?

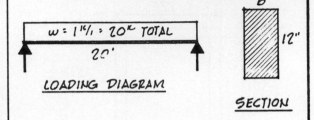

LOADING DIAGRAM

SECTION

(II) THE SECTION SHOWN IS MADE OF A MATERIAL WITH AN ALLOWABLE BENDING STRESS OF 18 K.S.I. IN TENSION AND 16 K.S.I. IN COMPRESSION. WHAT IS THE MAXIMUM SAFE LOAD CARRYING CAPACITY FOR THE LOADING CONDITION SHOWN?

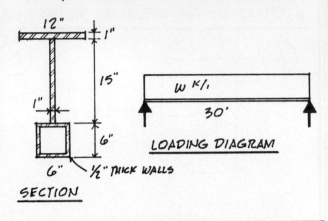

SECTION

LOADING DIAGRAM

(III) THE CROSS SECTION SHOWN IS USED AS A SIMPLY SUPPORTED BEAM WITH THE LOADING SHOWN BELOW. CONSIDERING THAT THE ALLOWABLE BENDING STRESS IN COMPRESSION IS 16 K.S.I. AND 20 K.S.I. TENSION, WHAT LOADS "P" MAY BE PLACED ON THE BEAM, IN ADDITION TO THE LOADS SHOWN?

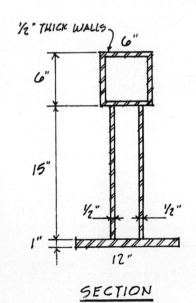

SECTION

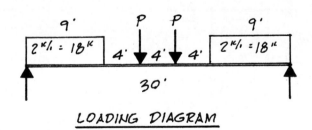

LOADING DIAGRAM

PROBLEM 7–8

ELASTIC BENDING

(I) CONSIDER 4 - 2 × 6's (FULL DIMENSION) ARRANGED AS SHOWN:

    (A) DETERMINE THE MOMENT OF INERTIA FOR EACH CROSS SECTION.
    (B) FOR AN ALLOWABLE STRESS OF 1500 P.S.I. IN BENDING, DETERMINE
    THE LOAD CARRYING CAPACITY FOR EACH SECTION BASED ON THE
    FOLLOWING LOAD AND SPAN CONDITION:

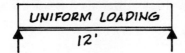

UNIFORM LOADING   SIMPLY SUPPORTED BEAM
12'

THE ABOVE LOADING IS USED ON THE FOLLOWING SECTIONS:

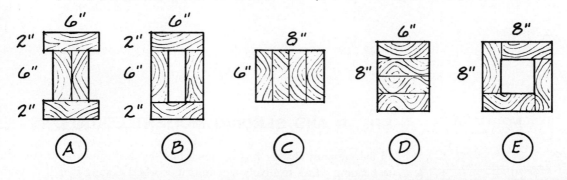

    Ⓐ   Ⓑ   Ⓒ   Ⓓ   Ⓔ

(II) FOR THE TIMBER "T" SECTION SHOWN, DETERMINE THE MAXIMUM POINT LOAD "P" THAT MAY BE PLACED ON THE MEMBER IN ADDITION TO THE GIVEN UNIFORM LOADING. THE MATERIAL HAS AN ALLOWABLE STRESS IN BENDING OF 1800 P.S.I.

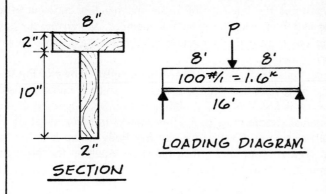

SECTION   LOADING DIAGRAM

(III) FOR EACH OF THE LOADING PATTERNS SHOWN, DETERMINE THE MOMENT OF INERTIA (I) REQUIRED AT THE QUARTER POINTS.
ALLOWABLE BENDING STRESS = 20 K.S.I.

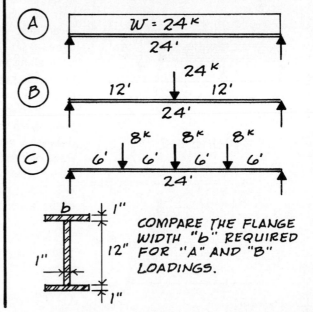

COMPARE THE FLANGE WIDTH "b" REQUIRED FOR "A" AND "B" LOADINGS.

8

Shear and Bending Moment

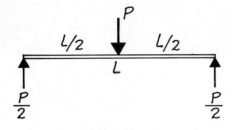

FIGURE 8-1

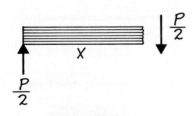

FIGURE 8-2

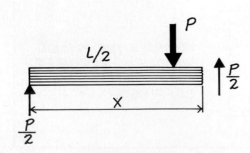

FIGURE 8-3

SHEAR AND BENDING MOMENT EQUATIONS

In the previous chapter it was shown that the determination of stresses due to bending was a relatively simple matter when dealing with *symmetrically loaded and simply supported beams*. In cases such as this, the maximum bending moment, and consequently the maximum stresses due to bending, occur at the center of the span; no bending moment exists at the end of the span. Obviously, there are numerous cases where the loading is not symmetrical and intuition cannot be relied on in the determination of the location of the maximum bending moment. Also, it will be important to understand how the bending moment and the shear vary throughout the span. In order to do this, we will develop equations for various loading cases that will allow us to determine the intensity of shear and bending moment at any point along the length of a loaded structural member. Moreover, the equations will point out some general relationships that always exist between shear and bending moment.

For example, consider the simply supported member shown in Fig. 8–1. If we use a free body diagram of the left end (an arbitrary choice) at some distance X where X is less than $L/2$, as shown in Fig. 8–2, then the shearing force on the internal section must be $P/2$ (as shown) in order to satisfy $\Sigma F_y = 0$.* This will be true in this case for any free body where X is less than $L/2$. Using the symbol V for shear, then the shear equation for the portion of the beam from $X = 0$ to $X = L/2$ is

$$V(\text{from 0 to } L/2) = P/2$$

where X is greater than $L/2$, as shown in Fig. 8–3, the equation becomes

$$V(\text{from } L/2 \text{ to } L) = P/2 - P = -P/2$$

This negative shear is physically more significant than it is algebraically

*The weight of the beam itself is considered to be negligible.

significant. It merely indicates that the tendency of adjacent particles to translate relative to each other is in the opposite direction from that of the left side of the beam. Had we started with free bodies from the right side, the direction of the arrows would have been reversed.

We'll now evaluate the bending moment in the beam of Fig. 8–1. Working from the left-hand side again, as shown in Fig. 8–4, where X is less than $L/2$, the bending moment on the internal section (and, consequently, the resisting moment) is simply equal to the product of the forces involved and their lever arms. In this case,

$$M(\text{from } 0 \text{ to } L/2) = \frac{P}{2}(X)$$

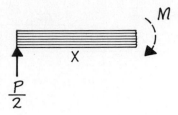

FIGURE 8–4

If a free body is taken where X is greater than $L/2$, then (see Fig. 8–5)

$$M(\text{from } L/2 \text{ to } L) = \frac{P}{2}(X) - P(X - L/2)$$

$$= \frac{P}{2}(X) - PX + \frac{PL}{2}$$

$$= -\frac{P}{2}(X) + \frac{PL}{2}$$

Now let's consider the case of a uniformly loaded member, as shown in Fig. 8–6. In order to evaluate the shear on the internal section, consider the free body taken at some distance X from the left end, as shown in Fig. 8–7. The total upward force on this free body diagram is $\frac{wL}{2}$.

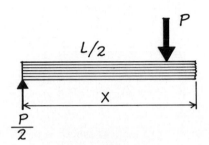

FIGURE 8–5

The total externally applied downward load is $(w^K/')(X')$. The resistance against translation in the vertical direction which must be furnished by the section in order to satisfy $\Sigma F_y = 0$ must be the difference in these external forces. That is,

$$V = \frac{wL}{2} - wX$$

Considering the bending moment at any distance X along the length of the member, as shown in Fig. 8–8, then

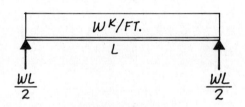

FIGURE 8–6

$$M = \frac{wLX}{2} - wX\left(\frac{X}{2}\right) = \frac{wLX}{2} - \frac{wX^2}{2}$$

In looking at the equation for shear and bending moment, it will be seen that shear (V) is maximum when $X = 0$. Also, the shear (V) is a minimum when $X = L/2$; that is, at the center of this symmetrically loaded member.

A review of the bending moment equation will show that the moment (M) is minimum (zero) in this *simply supported member* when $X = 0$. The bending moment is maximum, in this case, when $X = L/2$.

In looking back at the shear and bending moment equations developed from Figs. 8–1 through 8–5, the same relationships between shear and bending moment may also be noticed, although not so directly. In reviewing these figures and the corresponding equations, it will be seen that the shear (V) is maximum where $X = 0$, and remains constant until $X = L/2$. When X is greater than $L/2$, the direction of the shearing force on the internal section becomes reversed and remains constant until $X = L$. Coincidentally, the bending moment (M) is minimum in this *simply supported member* when $X = 0$ and maximum at $X = L/2$, which is where the shearing force on the internal section changes direction.

Let's now consider a case where the loading is unsymmetrical, such as

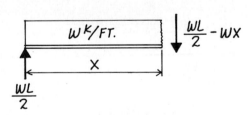

FIGURE 8–7

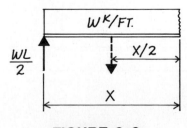

FIGURE 8–8

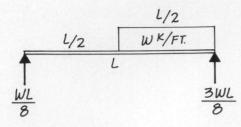

FIGURE 8–9

that shown in Fig. 8–9. In order to evaluate the shear, consider any free body diagram taken where X is less than $L/2$ (measuring X from the left). In this case, shown in the free body diagram of Fig. 8-10,

$$V(\text{from } 0 \text{ to } L/2) = \frac{wL}{8}$$

Where X is greater than $L/2$, measured as shown in the free body diagram of Fig. 8–11, then

$$V(\text{from } L/2 \text{ to } L) = \frac{wL}{8} - w(X - L/2)$$

From this equation it can be determined that $V = 0$ where $X = \frac{5}{8}L$.

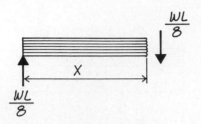

FIGURE 8–10

Now, considering the bending moment on this member, it can be seen from Fig. 8–12, where X is less than $L/2$, that

$$M(\text{from } 0 \text{ to } L/2) = \frac{wLX}{8}$$

Considering a free body diagram where X is greater than $L/2$, as shown in Fig. 8–13, then

$$M(\text{from } L/2 \text{ to } L) = \frac{wLX}{8} - w(X - L/2)\left(\frac{X - L/2}{2}\right)$$

$$= \frac{wLX}{8} - \frac{w}{2}\left(X^2 - XL + \frac{L^2}{4}\right)$$

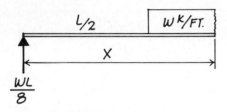

FIGURE 8–11

If numerous values were substituted in these equations for X, it would be found that the bending moment is *maximum* where $X = \frac{5}{8}L$, *which coincides with the point at which the shear (V) is equal to zero.* Also, the bending moment is zero in this *simply supported* member where $X = 0$ and where $X = L$.

In looking back at the two previous examples (Figs. 8–1 through 8–8), it can be seen that the same relationships exist between the shear and bending moment equations. That is, where the shear is minimum or changes direction, the bending moment is at a maximum. It should also be recognized that in every case evaluated, the moment equation was always one degree higher than the corresponding shear equation. Where the shear equation indicated that the shear was constant between two points, the moment *varied linearly* between the same points. Where the shear equation indicated a *linear variation* in the shear between two points, the moment equation was a second degree equation, indicating a *parabolic* variation in the bending moment between the same two points.

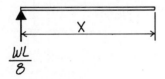

FIGURE 8–12

To summarize the two important principles that come from the preceding discussion:

1. Where the shear is zero, or changes direction, the bending moment will be a maximum.
2. The bending moment equation will always be one degree higher than the shear equation.

SHEAR AND BENDING MOMENT DIAGRAMS

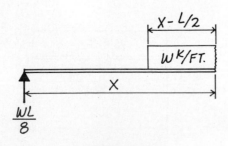

FIGURE 8–13

For our purposes, it will be important to determine the magnitude of shear and bending moment at any point in a structural member. Considering that we may be confronted with a great variety of possible loading patterns, it would clearly be a nuisance to write equations for every condition and then substitute values of X to determine the shear and bending moment at any point along the length of a structural member. In order to do this in a somewhat swifter and more spontaneous manner, the relationships pointed out earlier between shear and bending moment will now be used to present a method for sketching diagrams which will be indicative of the shear and bending moment intensity at any point along the length of a member.

Consider the simply supported member shown in Fig. 8–14. Using the shape of the member as a base line on which to construct the shear (V) and bending moment ($B.M.$) diagrams (in this case the base line is a horizontal line), it should be recognized that the shear at the left end is 5^K and remains constant on any free body taken for the first $10'$. At the point of application of the concentrated load, the direction of the internal shear reverses and remains constant until the 5^K reaction on the right-hand end is encountered. A diagrammatic representation of the shear intensity is shown in Fig. 8–15. Referring to the bending moment diagram of Fig. 8–15, we will now make use of the two important principles established in the discussion of shear and bending moment equations. That is:

1. Where the shear is zero, or changes direction (crosses the base line), the bending moment will be a maximum.
2. The bending moment equation (and, consequently, the bending moment diagram) will always be one degree higher than the shear equation (or, in this discussion, the shear diagram).

In Fig. 8–15 the bending moment is shown as zero at the ends, which is always true of a simply supported member. Therefore, the values of bending moment at the ends of the beam are known, and the location of the "peak" is known (the shear diagram tells where this occurs). The variation between the ends (zero) and the maximum is a linear variation, as shown on the diagram, because the shear is a constant.

All that remains now is to determine the value of the maximum bending moment. Since we know the location of the maximum moment, take a free body diagram at that point, as shown in Fig. 8–16, and evaluate the moment

$$M_x = (5^K)(10') = 50'^K$$

Now consider the shear and bending moment diagrams for the uniformly loaded member of Fig. 8–17. First recognize that since the loading is a constant, the variation in the shear will be of the first degree (linear variation). Referring to the shear diagram shown in Fig. 8–18, the shear at the end section is 20^K, as shown. Visualizing a series of free bodies (going from left to right), this 20^K end shear is being negated by the uniform rate of loading. Finally, at the center of the span, the free body of the left-hand side of the span includes enough downward load to completely negate the 20^K end shear. In this case, therefore, the shear is zero at the center of the span. Free bodies taken to the right of where the shear is zero will indicate that the direction of shear on the internal section is opposite to that where free bodies are taken to the left of the zero point. Therefore, the shear diagram to the right of zero is drawn below the base line. This is drawn continuously to the right end of the member where the 20^K end reaction is encountered, which brings the diagram back to the base line.

In order to draw the bending moment diagram properly, it is important to recognize that since the shear diagram was of the first degree, then the bending moment diagram will be a second degree curve, indicating a para-

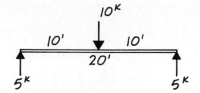

FIGURE 8–14

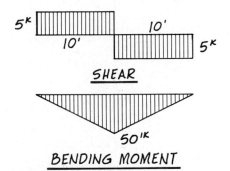

SHEAR

BENDING MOMENT

FIGURE 8–15

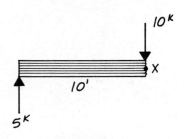

FIGURE 8–16

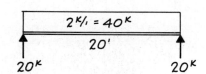

FIGURE 8–17

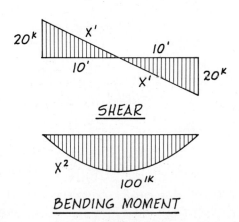

SHEAR

BENDING MOMENT

FIGURE 8–18

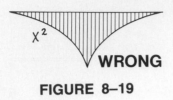

WRONG

FIGURE 8–19

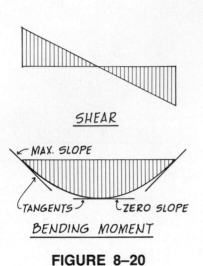

SHEAR

BENDING MOMENT

FIGURE 8–20

bolic variation in the moment intensity. Also, since this is a simply supported member, the moment is zero at the ends. The "peak" in the moment diagram is indicated by the shear diagram crossing the axis. In order to connect these points, a second degree curve is sketched, as shown in Fig. 8–18. There may be a question at this point regarding the manner in which the parabolic variation of Fig. 8–18 was drawn. For example, why was the parabolic variation in the moment not drawn as shown in Fig. 8–19? In this diagram the moment is also zero at the ends and has a "peak" in the middle! Actually, the variation shown in Fig. 8–19 is *incorrect*. In order to determine the proper way in which to draw curves such as this, we must use the shear diagram and the following relationship between the shear and bending moment diagram:

> Values at any position on the shear diagram indicate the value of a slope of a tangent to the bending moment diagram at the same position.

For example, consider this relationship and the diagrams shown in Fig. 8–20. In the shear diagram, the maximum value of shear is at the end; this value diminishes until it reaches zero. This means that the slope of a tangent line to the moment curve has the greatest slope at the end and diminishes until the slope of the tangent is zero (horizontal). As we move to the right of the zero value on the shear diagram there are values that are opposite in sign from those encountered to the left of zero. This indicates that the tangency to the moment curve has a slope, but it is now opposite in direction from the slopes on the left side of the diagram. Using this procedure will allow the drawing of correctly shaped diagrams. It might be well to note at this point that we are not interested in the numerical values of the slopes of the tangents to the moment curve. We are only interested in the relative differences in the intensities of the slopes.

Now let's consider several examples of unsymmetrically loaded members, using the relationships between shear and bending moment already established.

Example 8–1 (Fig. 8–21)

The shear diagram. Working from the left, there is a 6^K reaction, which produces 6^K of internal shear. This remains *constant* for the first 12' of the beam. Then a uniform rate of loading is encountered, which produces a linear variation in the shear diagram. Since the shear diagram is 6^K above the base line, it will take 3' of this uniform load before the 6^K value is negated. Generally, the point at which the shear will become zero due to a *uniform load* may be found by dividing the shear value above the base line by the *rate of loading*. Expressed mathematically,

$$d = \frac{V}{w}$$

where d = the distance from the beginning of the uniform load at which the shear diagram will cross the axis

V = the value of shear at the beginning of the uniform load

w = the rate of loading (on a "per linear foot" basis)

Referring again to Fig. 8–21, beyond the point at which the shear is zero, the direction of the internal shear changes. This is expressed by continuing to draw the linear variation below the base line. At the right-hand end of the member the value of shear is 18^K. The end reaction of 18^K brings the diagram back to the base line.

If it has not been discovered yet, it might be well to point out that the shear diagram must always "close out." It must start at the base line, and the forces encountered at the opposite end must return the diagram back to the base line. If this does not happen, then the requirement for equilibrium ($\Sigma F_y = 0$) has not been satisfied.

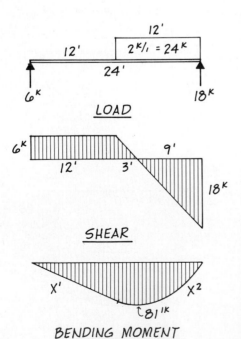

LOAD

SHEAR

BENDING MOMENT

FIGURE 8–21

The moment diagram: In order to sketch the moment diagram for this case, we must first recognize the obvious physical conditions involved. Since this is a simply supported beam, the moments at the supports are zero. Also, the moment diagram will be at a maximum where the shear diagram crossed the base line. Now that we have these three points, it only remains to connect them properly in order to show the correct variation in the moment diagram. As pointed out previously, the shear diagram will instruct us in the correct construction of the moment diagram. Working from the left, the shear diagram has a constant value for the first 12′. Therefore, there will be a linear variation (first degree) in the moment diagram for the first 12′. At this point the shear diagram changes to a linear variation. This means that the moment diagram, at the corresponding location, becomes a second degree variation (parabolic). From 12′ to 15′ into the span the values on the shear diagram are diminishing, indicating that the slopes of the tangents to corresponding points on the moment diagram are also diminishing. At 15′ the value of shear is zero, indicating that the tangent to the corresponding point on the moment diagram is zero (horizontal). Beyond 15′, going to the right, there are again values on the shear diagram, but they are below the base line, indicating that the slope of tangents to corresponding points on the moment diagram are opposite in direction to the slopes indicated by values of shear above the base line. As we continue going to the right, the values on the shear diagram are increasing, indicating that the slopes of the tangents to the moment curve at corresponding points are increasing, with the greatest slope at the end of the diagram.

In order to determine the value of the maximum moment, a free body diagram can be taken of the member to either the left or right of the point where the maximum moment exists. Using a free body taken to the left, as shown in Fig. 8–22,

$$M_x = (6^K)(15′) - (6^K)(1.5′) = 81^{′K}$$

Before going on to further examples, there is another important relationship that exists between the shear and bending moment diagrams. In order to evaluate the magnitude of the maximum moment, after its location was determined by the shear diagram, a free body diagram was taken, and the moments were taken at the point of maximum. However, if the examples of Figs. 8–14, 8–17, and 8–21 are studied carefully, it will be seen that the area under the shear diagram, to either side of where the shear diagram crosses the base line, is equal to the maximum bending moment. The applicable principle governing this relationship may be stated in the following manner:

The *difference* in values between *any two points* on the moment diagram is equal to the area under the shear diagram *between the same two points*

Example 8–2 (Fig. 8–23)

The first step, of course, in any of these problems is to determine the end reactions. These have been computed and are shown on the loading diagram.

The shear diagram: Going from left to right, plot 8.7^K above the base line. This value remains constant for the first 4′. At this point a uniform load is encountered that produces a linear variation in the shear diagram, producing a decrease in the shear at the rate of $1^{K′}$. At this rate of decrease the shear will become zero at 8.7′ from the start of the uniform load. Continuing this linear variation beyond this point, the shear becomes opposite in direction, which is shown below the base line. At 16′ from the left (3.3′ beyond the point of zero shear), the value of shear below the base line is 3.3^K. At this location the 12^K concentrated load is encountered, which brings the shear to 15.3^K below the base line. Then the uniform rate of loading is encountered again for 2′, bringing the shear to a value of 17.3^K below the base line. This value remains constant for the final 2′, at which point the 17.3^K end reaction is encountered, which closes out the diagram.

The moment diagram: Since this is a simply supported member, there is no moment at the ends. The maximum moment will occur at the point where the

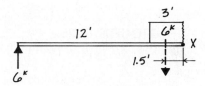

FIGURE 8–22

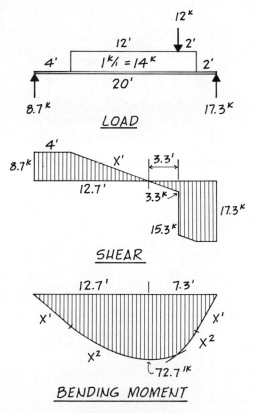

FIGURE 8–23

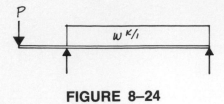

FIGURE 8–24

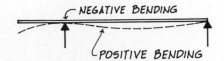

FIGURE 8–25

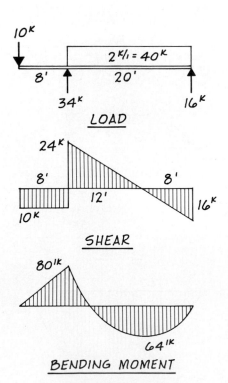

FIGURE 8–26

shear diagram crosses the base line, that is, at 12.7′ from the left end. It remains now to connect these three points to show the correct variation. Again working from the left end of the member, the shear diagram has a constant value for the first 4′. This means there will be a *linear variation* in the moment diagram for the first 4′. At this point the shear diagram changes to a linear variation, indicating that the moment diagram, starting at this point, will vary parabolically (second degree). The transition between the first degree and second degree portions of the moment diagram is smooth because there is only one value on the shear diagram at this point, which is indicative of a common tangent to both portions of the moment diagram.

At 12.7′ from the left end, the shear diagram crosses the base line, indicating that the slope of the tangent to the moment diagram at this point is zero (horizontal). Then, values of shear below the base line are encountered, still varying linearly, which indicates that the slopes of tangents to corresponding points on the moment diagram are opposite in direction to those indicated by values of shear above the base line. At 16′ from the left end of the beam there are two values on the shear diagram, and again, a linear variation beyond this. This indicates that two second degree curves meet at this point, with two different tangents, meaning a non-smooth transition between the two curves. At 18′ the shear values become constant until the end of the member, indicating a linear variation in the moment diagram. Since there is only one value on the shear diagram at the 18′ mark, the transition between the second degree curve and first degree variation on the moment diagram is smooth.

POSITIVE AND NEGATIVE MOMENT

Thus far we have only considered simply supported members. When members such as this are subjected to load, the bending produces tension in the bottom of the member and compression in the top of the member. We shall refer to bending of this type as *positive bending* and the corresponding moment as *positive bending moment*.

In a cantilevered beam, such as that shown in Fig. 8–24, if one visualizes how the member deforms under the loads, it will be recognized that there will be a certain region where bending will produce tension in the top of the member and compression in the bottom of the member. We will refer to this condition as *negative bending* and the corresponding moment as *negative bending moment* (see Fig. 8–25).

In all examples considered to this point only positive bending moment existed, and the moment diagram was drawn below the base line. Conversely, where negative bending moment occurs, the moment diagram will be drawn above the base line. In this book the moment diagrams will always be drawn on the tension side of the member in order to reflect the manner in which the beam bends. It should be noted, for the sake of a student who may be using a supplementary reference source, that many authors use a reverse approach; that is, the moment diagram may be drawn on the compression side of the member. Regardless of the convention used for drawing the diagrams, the mathematical information will be the same in either case. We will now consider examples in which negative as well as positive bending exists.

Example 8–3 (Fig. 8–26)

In the cantilevered beam shown, both negative and positive bending moment will exist. If one visualizes the way this member will deform under the loads, then it will be recognized that negative bending exists over the entire length of the cantilever and for some distance into the main span. In the main span the character of the bending moment changes from negative to positive, and then goes to zero at the unrestrained end.

The shear diagram: Working from left to right, go down 10^K. This remains constant until the 34^K reaction is encountered. At this point go up 34^K, which puts the shear value at 24^K above the base line. The uniform load will then begin to negate this 24^K value at the rate of 2$^{K/\prime}$. At this rate the shear will be zero after 12$'$. The linear variation of 2$^{K/\prime}$ continues below the base line for the remaining 8$'$, which places the shear value at 16^K below the base line. At this point the 16^K end reaction closes the diagram.

The moment diagram: Going from left to right, there is a constant value on the shear diagram, below the base line, for the first 8$'$. This indicates negative moment, which is drawn above the base line on the moment diagram. At this point there is another value of shear above the base line. This indicates that the tangent to the moment diagram changes direction, which means a peak exists on the diagram (the shear diagram crossed the base line at this point). There is a linear variation on the shear diagram, indicating a second degree curve on the moment diagram. At the point of zero shear there will be a horizontal tangent to the moment diagram, indicating another peak. Values of shear below the base line are now encountered, indicating that the tangents to points on the moment diagram have changed direction again. Increasing values on the shear diagram indicate that the slopes of tangents to corresponding points on the moment diagram are increasing.

The values of the maximum moments may be determined from the area under the shear diagram. Since there are two peaks in the moment diagram, both will have to be evaluated. Working from left to right, the difference in value on the moment diagram for the first 8$'$ may be found by evaluating the area under the shear diagram between these two points:

$$M = (10^K)(8') = 80'^K$$

This is a negative moment, meaning that there is tension in the top of the member and compression in the bottom. In order to evaluate the next peak in the moment diagram, evaluate the area under the shear diagram for the next 12$'$. Since this area is above the base line, it must be treated as an algebraic opposite of the area of the shear diagram of the first 8$'$. What must be remembered here is that the area under the shear diagram between two points yields the *difference* in the bending moment between the same two points. Therefore, the maximum positive moment is

$$M = (24^K)(12')\left(\frac{1}{2}\right) - 80'^K = 64'^K$$

In this case the area under the shear diagram indicated an 80$'^K$ negative moment at the first support from the left. Then an area of opposite sign was encountered. This area measured 144$'^K$, but this indicates the *difference* in moment. Since we started with a negative 80$'^K$ moment, this produces a net 64$'^K$ positive moment.

Sometimes, in a cantilevered member, the length of the cantilever in relation to the main span, or the magnitude of load on the cantilever, is such that the negative moment produced is so large that it is never eliminated by the load on the main span. A state of negative bending exists throughout the entire length of the member. A condition of this sort will now be considered.

Example 8–4 (Fig. 8–27)

For this double-cantilevered member the shear diagram is constructed in precisely the same manner as the approach used in previous examples.

In the bending moment diagram, working from left to right, the negative bending moment at the first support is 72$'^K$, which is determined by evaluating the area under the shear diagram for the first 8$'$. For the next 7.5$'$ the area under the shear diagram is 56.25$'^K$, of opposite sign. Starting at the support with a negative moment of 72$'^K$, there was a tendency to go positive by only 56.25$'^K$, leaving a net negative moment of 15.75$'^K$. From this point, going to the right, shear area below the base line is again encountered, indicating that the moment is going further into negative bending. The value at the right-hand support is 36$'^K$.

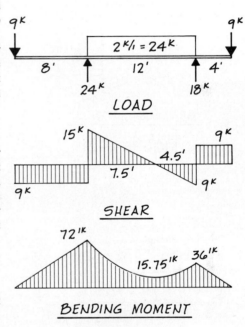

FIGURE 8–27

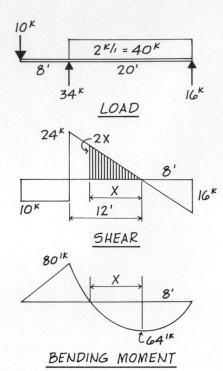

FIGURE 8–28

POINTS OF CONTRAFLEXURE

In the preceding discussion it was shown that in a member where *positive* and *negative* bending moment exists, the bending moment diagram crosses the base line at some point. This point of crossing is known as the *point of contraflexure*. For a variety of purposes that we will encounter in future studies it will be important to determine the location of the point, or points, of contraflexure in a bending moment diagram. The determination of the location of a point of contraflexure is a reasonably simple matter, making use of the idea that the area under the shear diagram between any two points is equal to the difference in values in the moment diagram between the same two points. To demonstrate the procedure, let's look at several examples.

Example 8–5 (Fig. 8–28)

To begin with, let's reconsider Example 8–3, this time concerning ourselves with the location of the point of contraflexure. The diagrams, which have already been determined, are shown in Fig. 8–28. In this example the maximum positive moment was found to be $64'^K$; its location is as shown. What we would like to know is the location of zero bending moment, because this represents the transition between positive and negative moment, which is the point of contraflexure. There are several ways to do this, but in this example the easiest way is to determine the distance noted as X from the point of zero shear. The area under this portion of the shear diagram must be $64'^K$. The slope of the shear diagram is the rate of loading or, in this case, $2^K/'$. Therefore, starting from the point of zero shear and rising at a rate of $2^K/'$ over a distance of X feet, the vertical value on the shear diagram is $2X$ at the point of contraflexure. Remembering that the area under the shear diagram is equal to the *difference* in values on the bending moment diagram, then

$$\left(\frac{1}{2}\right)(2X)(X) = 64'^K$$

and

$$X = 8'$$

It's probably recognized at this point that this portion of the shear diagram is identical to the shear diagram to the right side of zero shear. This is as it should be, since the area under the shear diagram to the right side also represents a change in value on the moment diagram equal to $64'^K$. Since the rate of loading is constant, it follows that the length of the triangular area that will produce $64'^K$ must be the same to either side of the zero point. In a simple case such as this, the point of contraflexure can be determined by inspection. However, the procedure presented is applicable to the infinite variety of possibilities which cannot be determined by inspection. We'll now proceed to a somewhat more complicated case.

Example 8–6 (Fig. 8–29)

In the double-cantilevered beam shown there are two points of contraflexure. If, however, we locate these from the point of zero shear, only one computation will be required. This is so because the rate of loading within the span is constant, indicating that the triangular areas to either side of the zero shear point must be the same.

We are looking for an area under the shear diagram, in this case, equal to $40'^K$, since this is the difference in moment from the maximum positive bending moment to the point of zero bending moment. Starting from the point of zero shear within the span, the shear diagram changes linearly at the rate of $1^K/'$. Therefore, over a horizontal distance of X feet, the vertical value on the shear diagram is $(1)(X)$. Equating the area under this segment of shear diagram to the difference in moment,

$$\left(\frac{1}{2}\right)(1)(X)(X) = 40'^K$$

and $X = 8.94'$ to the left and right of the point of zero shear.

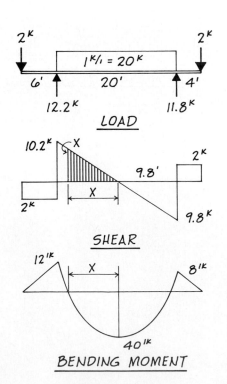

FIGURE 8–29

Example 8–7 (Fig. 8–30)

In this double-cantilevered beam there are two points of contraflexure, and the shear and bending moment diagrams are shown in Fig. 8–30. Unlike the previous example, however, the shear diagram to either side of the point of zero shear is not the same; consequently, two computations will be required to determine the location of the points of contraflexure. We'll define these distances as X and Y, as shown on the shear diagram. The distance X from the point of zero shear is determined in the same manner as in previous examples:

$$\left(\frac{1}{2}\right)(X)(2X) = 120'^K \text{ and } X = 10.95'$$

In order to determine the location of the point of contraflexure on the right-hand side, it would be easiest to work from the right-hand support because we will only have to deal with a rectangular area on the shear diagram. The difference in value that we are looking for on the moment diagram is $24'^K$; therefore,

$$(Y)(12) = 24'^K \text{ and } Y = 2'$$

Sometimes the areas on the shear diagram are not so "clean" as a triangle or a rectangle; the computation of such areas may be a nuisance. Therefore, another way of locating the point of contraflexure would be to use a free body of the beam that is cut loose at the point of contraflexure. Since this is a point of zero bending moment, an equation can be written to locate the point of contraflexure. To demonstrate this technique, let's locate the point of contraflexure on the left-hand side of the beam of Fig. 8–30. In order to do this a free body is taken, as shown in Fig. 8–31, with the distance X from the support to the point of contraflexure. Taking moments about the point where the cut has been made, and equating this to zero, we get

$$\Sigma M = 0 = 4(6 + X) + (2X)\left(\frac{X}{2}\right) - 28X$$

$$0 = X^2 - 24X + 24$$

Solving the quadratic, it should be obvious which root is applicable; we find that

$$X = 1.05'$$

which, of course, locates the point of contraflexure in precisely the same location that we found when using the area of the shear diagram.

SUMMARY OF PRINCIPLES

It will be useful, at this point, to restate the language and the principles used in the construction of shear and bending moment diagrams. This will be done in compact form.

1. Determine reactions at supports.
2. Use obvious physical conditions, such as the positions of zero or maximum amounts, symmetry, etc.
3. Ordinate (amount at any position) of the shear diagram controls the pitch (rate of change in ordinates) at the corresponding position in the bending moment diagram. This enables the construction of properly shaped diagrams.
4. Where the shear diagram is zero, or crosses the axis, this compels a peak in the moment diagram. Location is vital.
5. Between any two positions, the area under the shear diagram is equal to the *difference* in ordinates on the bending moment diagram, between the same two positions (areas above and below the base line on the

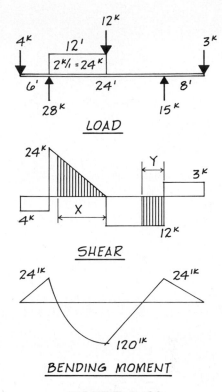

LOAD

SHEAR

BENDING MOMENT

FIGURE 8–30

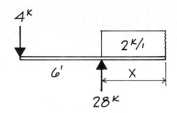

FIGURE 8–31

shear diagram must be treated algebraically). This enables the computation of amounts in the moment diagram from areas in the shear diagram.

CONCLUSION

The importance of the shear and bending moment diagrams, and the information derived from them, can hardly be overemphasized. In the chapters following, we will deal with a variety of issues, including the process of design of structural members. For all of the work to follow, it is absolutely imperative that the student have a great deal of proficiency in the proper construction of the shear and bending moment diagrams. It is, therefore, strongly recommended that every problem at the end of this chapter be done, analyzed, discussed, and, if necessary, redone. If the student fully understands the relationships between the shear and bending moment diagrams, it will make life through the following chapters much more comfortable.

PROBLEM 8–1

SHEAR AND BENDING MOMENT

● DETERMINE THE REACTIONS R_1 & R_2 FOR EACH OF THE FOLLOWING AND DRAW THE SHEAR AND BENDING MOMENT DIAGRAMS. SHOW ALL FEATURES CLEARLY (SUCH AS DEGREE OF CURVE, INTERSECTIONS, ETC.).

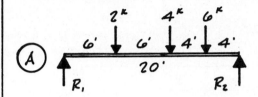

(A) 2^K, 4^K, 6^K ; $6'$, $6'$, $4'$, $4'$; $20'$; R_1, R_2

(B) 2^K, 3^K, 6^K ; $6'$, $6'$, $4'$, $4'$; $20'$; R_1, R_2

(C) 4^K, 8^K ; $4'$, $8'$, $4'$; $16'$; R_1, R_2

(D) 4^K, 8^K ; $4'$, $4'$, $8'$; $16'$; R_1, R_2

(E) 6^K, 3^K, 3^K ; $6'$, $6'$, $12'$, $6'$; $18'$; R_1, R_2

(F) 2^K, 6^K, 6^K, 2^K ; $8'$, $4'$, $8'$; $4'$, $12'$, $4'$; R_1, R_2

(G) $10'$, $1.2^{K/I} = 12^K$; $4'$, $4'$; $18'$; R_1, R_2

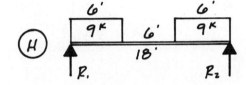

(H) $6'$, 9^K ; $6'$; $6'$, 9^K ; $18'$; R_1, R_2

(J) 4^K, 4^K ; $10'$, $1.2^{K/I} = 12^K$; $4'$, $4'$; $18'$; R_1, R_2

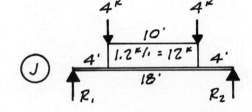

(K) 12^K ; $10'$, $10'$; $1.2^{K/I} = 24^K$; $20'$; R_1, R_2

(L) $1^{K/I} = 24^K$; $8'$, $16'$; R_1, R_2

(M) 4^K, 4^K ; $1.25^{K/I} = 20^K$; $4'$, $16'$, $4'$; R, R_2

PROBLEM 8–2

SHEAR AND BENDING MOMENT

PART I:
DETERMINE THE REACTIONS R_1 & R_2 AND
DRAW THE SHEAR AND MOMENT DIAGRAMS.

PART II:
GIVEN: ONE OF THE DIAGRAMS
DETERMINE: THE OTHER TWO DIAGRAMS

(A)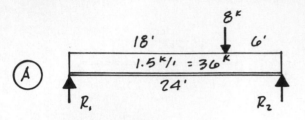

(A)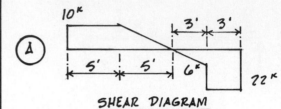
SHEAR DIAGRAM

(B)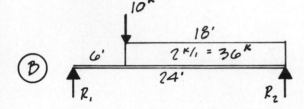

(B)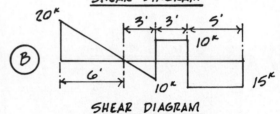
SHEAR DIAGRAM

(C)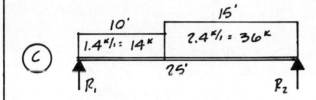

(C)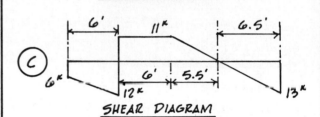
SHEAR DIAGRAM

(D)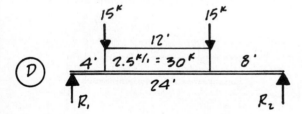

(D)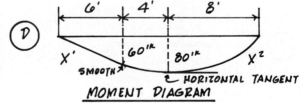
MOMENT DIAGRAM

(E)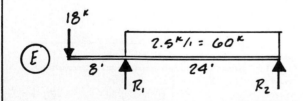

(E)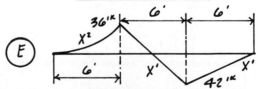
MOMENT DIAGRAM

(F)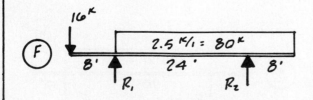

(F)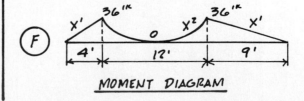
MOMENT DIAGRAM

PROBLEM 8-3

<u>SHEAR AND BENDING MOMENT</u>

DRAW THE SHEAR AND BENDING MOMENT DIAGRAMS FOR EACH OF THE FOLLOWING, AND LOCATE ANY POINTS OF CONTRAFLEXURE - SHOW ALL FEATURES CLEARLY, SUCH AS DEGREE OF CURVE, INTERSECTIONS, ETC.

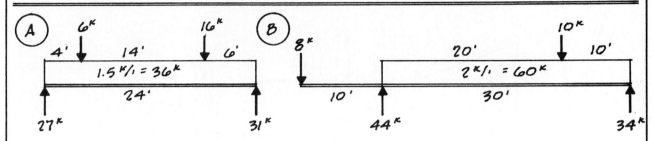

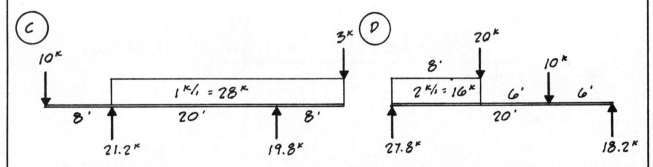

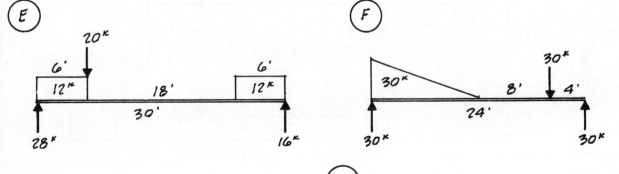

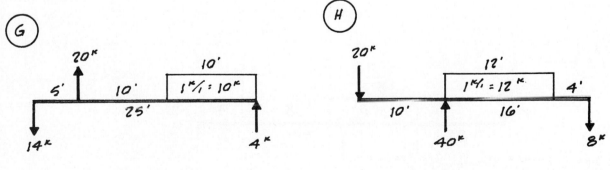

PROBLEM 8–4

SHEAR AND BENDING MOMENT

DRAW THE SHEAR AND BENDING MOMENT DIAGRAMS FOR EACH OF THE FOLLOWING.
SHOW VALUES AT POINTS OF CHANGE AND INDICATE DETAILS CLEARLY (SUCH AS
INTERSECTIONS, DEGREE OF CURVE) AND LOCATE ANY POINTS OF CONTRAFLEXURE

(A)

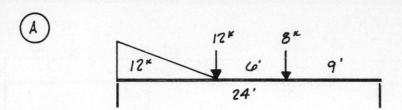

(B)

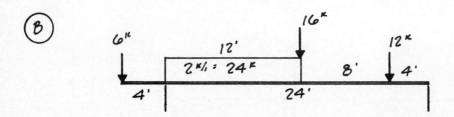

(C)

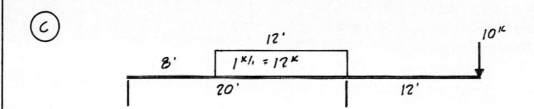

(D)

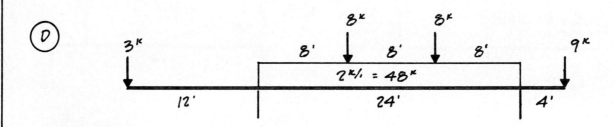

(E)

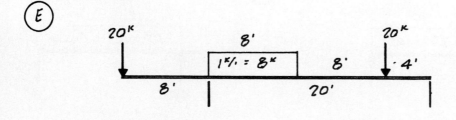

PROBLEM 8–5

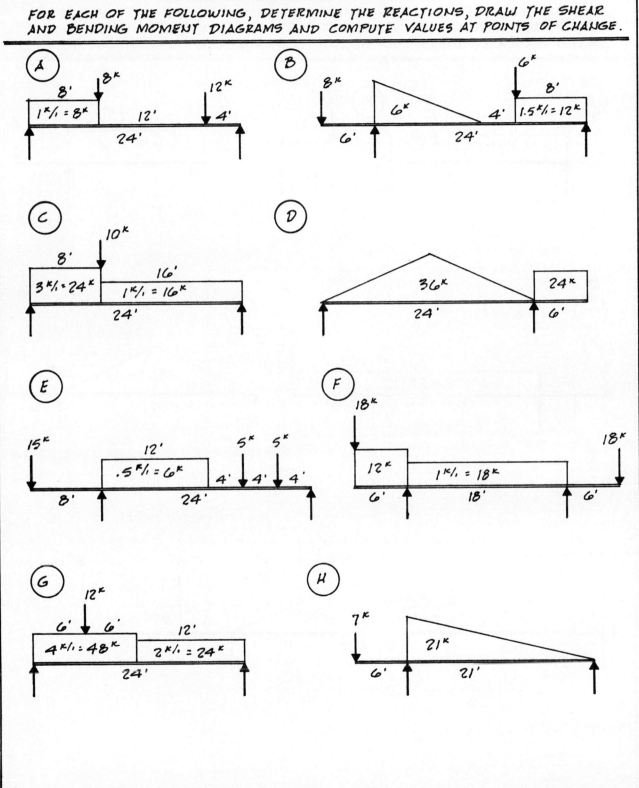

SHEAR AND BENDING MOMENT

FOR EACH OF THE FOLLOWING, DETERMINE THE REACTIONS, DRAW THE SHEAR
AND BENDING MOMENT DIAGRAMS AND COMPUTE VALUES AT POINTS OF CHANGE.

PROBLEM 8–6

<u>SHEAR AND BENDING MOMENT</u>

DRAW THE SHEAR AND BENDING MOMENT DIAGRAMS FOR EACH OF THE FOLLOWING.
SHOW VALUES AT POINTS OF CHANGE AND INDICATE DETAILS CLEARLY (SUCH AS
DEGREE OF CURVE, INTERSECTIONS) AND LOCATE ANY POINTS OF CONTRAFLEXURE.

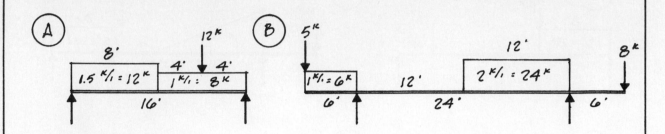

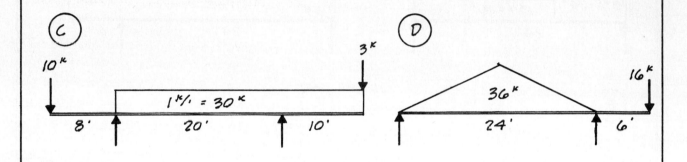

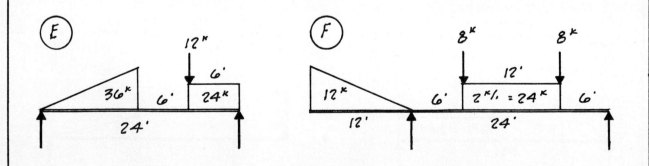

PROBLEM 8–7

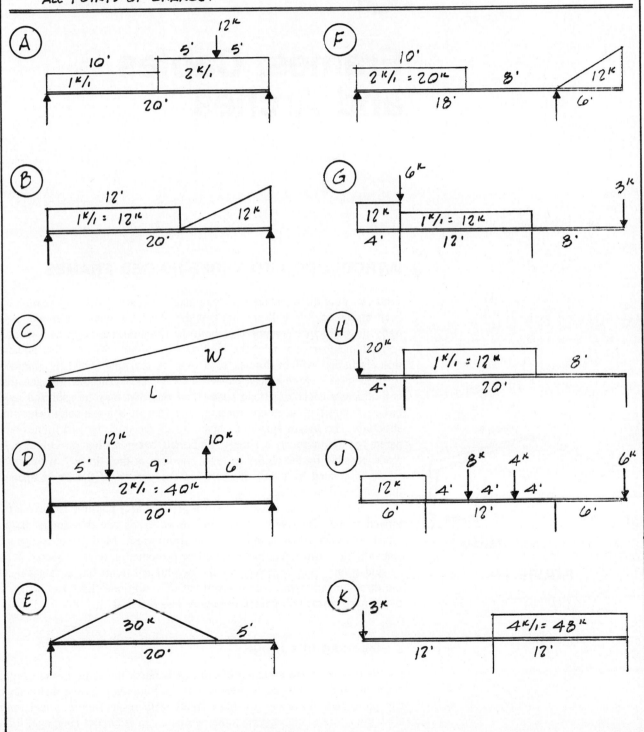

SHEAR AND BENDING MOMENT

FOR EACH OF THE FOLLOWING, DETERMINE THE REACTIONS, DRAW THE SHEAR AND BENDING MOMENT DIAGRAMS AND LOCATE ANY POINTS OF CONTRAFLEXURE – SHOW ALL FEATURES CLEARLY AND COMPUTE VALUES AT ALL POINTS OF CHANGE.

9

Frames, Cables, and Arches

INTRODUCTION TO THREE-HINGED FRAMES

Thus far, only the members that span across a space (primarily beams) have been discussed. We will now look at the idea of a *frame*. A frame, essentially, is a structure composed of both the spanning members as well as the vertical members that take the load to the ground. Specifically, however, this discussion will be limited to a type of structure which is commonly referred to as a *three-hinged frame* (see Fig. 9–1). The three-hinged frame is a structure capable of long spans, and one that may be analyzed by the basic principles of static equilibrium; therefore, it is a statically determinate structure. The longer spanning capability of the three-hinged frame—compared to, for instance, a simply supported beam resting on columns,—is made possible by the ability to resist moment at the "knee" of the frame, which is created by continuity between the portion of the frame spanning the space and the vertical "leg" of the frame.

In addition to the three-hinged frame, it is also possible to have a two-hinged frame. This type of structure, however, is one that cannot be analyzed with only the equations of static equilibrium. Methods of analysis for statically indeterminate structures will be presented in a later chapter. It may be said at this time, however, that the two-hinged frame has advantages over the three-hinged frame. For a given set of conditions (load and span), the two-hinged frame will generally require less material.

THREE-HINGED FRAME

FIGURE 9–1

Meaning of a Hinge

In structures, the meaning of a *hinge* is much the same as the meaning it has in everyday language. Very simply, a hinge is a device that connects two parts and allows one to rotate freely with respect to the other, while allowing no translation between the two parts. In structural language, it may be said that a hinge *cannot* resist moment, but *will* resist shear. It is important to remember this in the analysis of the three-hinged frame (as well as any structural configuration) because the points at which a structure is hinged are points of zero bending moment. This fact will be very useful in the analysis of the three-hinged frame, which will be discussed shortly.

To go one step further in the discussion of hinges and their physical significance, it is useful to point out the general conditions that constitute a hinged connection. Actually, in a structure, a hinge such as that which connects a door to the jamb is not used (although in a physical sense it does the same thing; that is, it permits rotation without translation). It is important to recognize that in a structure, the magnitudes of rotation being referred to are very small. Generally, a simple bolted connection in a structure constitutes a hinged connection. In the three-hinged frame, the connection of the legs to their anchorages is made with bolts. Basically, these are hinged connections. However, it should be mentioned that bolted connections may be designed so that the connection is rigid and no rotation takes place. This statement is primarily applicable to a steel frame*: In a steel frame, it is possible to weld the two parts of the frame together at the peak (Fig. 9–2). This will produce a rigid connection, and the principles used to analyze a three-hinged frame will no longer be applicable. The designer must carefully consider the quality of the connections to be used and employ the proper method of analysis.

In an arch made of laminated timber, two possibilities exist†. On the one hand, if the size of the frame is such that it cannot be made, shipped from the factory to the building site, or properly handled as one piece, then it will necessarily be made in two or more pieces and joined in place at the job site. Most often, where the two halves of the frame join at the peak, a simple bolted connection is used which creates a hinged condition. It is possible, however, to develop a moment-carrying splice between the pieces, thereby producing a two-hinged frame, with the two hinges being at the anchorage between the legs of the frame and the foundation‡. On the other hand, if the scale of the frame is such that it can be manufactured and shipped as one piece, it will inherently be a two-hinged frame which, as previously mentioned, generally requires less material than the three-hinged frame.

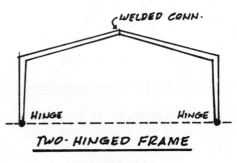

TWO-HINGED FRAME

FIGURE 9–2

Load Collection

The manner in which loads are delivered to three-hinged frames is generally the same as the trussed structure. Secondary members (purlins) are used to support the roof deck, and the purlins deliver concentrated loads to the frames (Fig. 9–3). Unlike the trussed structure, there are no panel points to which these loads must be delivered. The distance between purlins is dependent only on the strength of the roof deck being used. The purlins, of course, must be designed to span between the frames. In very general terms, laminated timber frames may be spaced about 16–22 ft apart and steel frames about 18–30 ft apart.

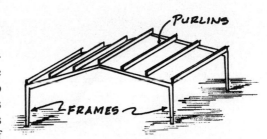

FIGURE 9–3

---

*For those interested in further study of connection details for steel frames, see *Manual of Structural Steel Detailing*, prepared by the American Institute of Steel Construction. Current Edition.

†While reference is made throughout this discussion to three-hinged and two-hinged *frames*, the timber industry generally refers to these kinds of structures as *arches*.

‡For further study of connection details for laminated timber three-hinged and two-hinged frames, see *Timber Construction Manual*, prepared by the American Institute of Timber Construction. John Wiley and Sons, Inc. Current Edition.

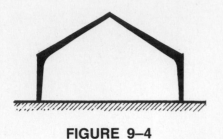

FIGURE 9–4

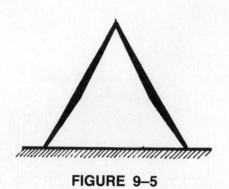

FIGURE 9–5

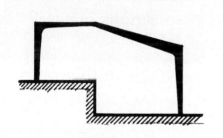

FIGURE 9–6

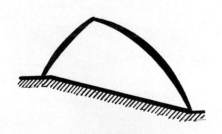

FIGURE 9–7

Configuration of Frames

A variety of configurations may be achieved with three-hinged frames. They need not be symmetrical (although this is mostly what is designed), and the legs need not be anchored at the same elevation. Actually, an infinite number of configurations may be achieved to satisfy the variety of conditions that may be encountered. Several possible configurations are shown in Figs. 9–4 through 9–8, merely as suggestions.

ANALYSIS OF THREE-HINGED FRAMES

As previously mentioned, only the equations of static equilibrium are necessary for the analysis of a three-hinged frame. However, one important phenomenon that occurs in a frame must be recognized before proceeding with the method of analysis. In a frame such as shown in Fig. 9–9, when the loads are applied there is the tendency for the legs of the frame to move outward, as shown in Fig. 9–10. Therefore, in order to provide stability for the frame, the anchorage must provide *horizontal* reactions, as well as vertical reactions, to keep the legs from moving outward. We will now proceed to the method for determining these reactions in a three-hinged frame, which is the first step in the analysis.

Example 9–1 (Fig. 9–11)

To begin, we will find the vertical reactions. In this particular case the loading is not symmetrical; consequently, the vertical reactions cannot be determined by inspection. According to the equations of static equilibrium:

1. $\Sigma F_y = 0$. Therefore, $A_V + B_V = 55^K$. But this involves two unknowns and, at this point, cannot provide a solution.
2. $\Sigma F_x = 0$. This tells us that $A_H = B_H$, but this is no help at this time.
3. $\Sigma M = 0$. This equation of static equilibrium will provide a solution. However, since there are four unknowns involved, we must be very careful in the choice of a point about which to take moments. The idea here is to choose a center of moments through which all but one unknown passes. This will produce an equation with only one unknown, which can easily be solved. In this case, this will happen if either point A or point B is chosen as the center of moments. Taking moments about point A:

$$\Sigma M_A = 0 = (10^K)(15') + (10^K)(30') + \\ (10^K)(45') + (10^K)(60') + \\ (10^K)(75') - B_V (75')$$

NOTE. The lines of action of all other unknowns pass through point A.

Solving this equation:

FIGURE 9–8

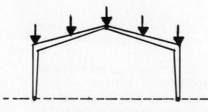

FIGURE 9–9

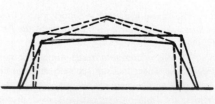

FIGURE 9–10

Frames, Cables, and Arches

129

$$B_V = \frac{2250'^K}{75'} = 30^K$$
$$\therefore A_V = 25^K \ (\Sigma F_y = 0)$$

Now, in order to solve for the magnitude of the horizontal reaction, taking moments at A or B will be of no use since the lines of action of A_H and B_H pass through both of these points. However, if a free body diagram of the left side of the frame is considered, as shown in Fig. 9–12, and moments are taken about the peak hinge (P) then this will provide a solution for the horizontal reaction A_H. Assuming the direction of A_H is to the right, as shown, then

$$\Sigma M_P = 0 = (25^K)(30') - (5^K)(30') - (10^K)(15') - A_H(40')$$

and

$$A_H = \frac{450'^K}{40'} = 11.25^K$$

Since the answer turns out positive, this verifies the assumed direction. It then follows that

$$B_H = 11.25^K \leftarrow (\Sigma F_x = 0)$$

It should be noted at this time that the choice of using the left side or the right side of the frame is purely arbitrary. The left side was chosen in this case simply because fewer numbers were involved. The same conclusion, however, would have been reached had the right side of the frame been chosen as the free body. What is *not arbitrary* in this solution is the choice of a point about which to take moments when solving for the horizontal reactions. As already mentioned, it would be useless to take moments at the anchorage points A or B, since the lines of action of both unknown horizontals pass through these points. Neither of the unknowns would show up in the moment equation. Therefore, the decision was made to take moments about the third, and remaining, hinge in the structure. This choice was not arbitrary because only at the hinges is the moment known to be zero. While it is true that the *summation* of moments ($\Sigma M = 0$) is equal to zero at any point in the structure, if a free body is cut at any point except the hinge, there is an internal bending moment released, and this must be included in the free body. Of course, at the time reactions are being determined, the bending moments in the structure are not known. But, based on the definition of a hinge, the internal bending moment must be zero. It is because of the existence of the third hinge in a three-hinged frame that the structure is statically determinate. If the hinge at the peak of the frame of Fig. 9–11 did not exist, the horizontal reactions could not have been determined by the basic equations of static equilibrium.

Example 9–2 (Fig. 9–13)

In the structure shown, the vertical reactions cannot be determined by inspection despite the group of symmetrical parallel forces. In a situation such as this, where the legs of the frame are anchored at different elevations, the symmetrical group of forces will not divide equally to the reaction points. Therefore, in order to solve for the vertical reactions, take moments about A or B (in this example, point B is chosen as the center of moments)

$$\Sigma M_B = 0 = (5^K)(75') + (10^K)(60') + (10^K)(45') + (10^K)(30') + (10^K)(15') - A_V(75') - A_H(10')$$

Notice that the unknown force A_H has a lever arm to point B and must be included in the moment equation. This presents a dilemma because there are two unknowns in this equation. The other equations ($\Sigma F_x = 0$ and $\Sigma F_y = 0$) will not help at this time either. However, if a free body of the left side of the frame is taken, as shown in Fig. 9–14, another moment equation can be written, using the peak hinge as the center of moments, and the same two unknowns (A_H and A_V) will appear in the

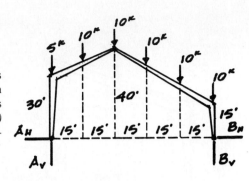

FIGURE 9–11

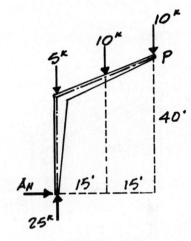

FIGURE 9–12

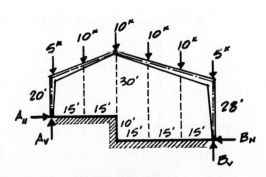

FIGURE 9–13

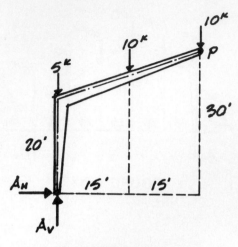

FIGURE 9–14

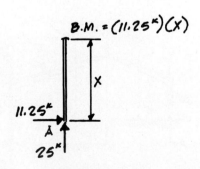

FIGURE 9–15

FIGURE 9–16

equation. This will provide two equations with two unknowns that can be solved simultaneously. Using the free body of Fig. 9–14, the second equation is

$$\Sigma M_P = 0 = (5^K)(30') + (10^K)(15') - A_V(30') + A_H(30')$$

Restating these equations in simplified form,

1. $0 = 1875 - A_V(75') - A_H(10')$
2. $0 = 300 \quad - A_V(30') + A_H(30')$

Solving these equations simultaneously:

$$A_V = 23.2^K \quad \text{and} \quad A_H = 13.2^K$$
$$\therefore B_H = 13.2^K \quad (\Sigma F_x = 0)$$

and

$$B_V = 26.8^K \quad (\Sigma F_x = 0)$$

In the solution for A_V and A_H, both reactions came out with positive answers, indicating that the *assumed* directions shown in Figs. 9–13 and 9–14 and used to write the equations are correct. It is important to emphasize that once an assumption is made regarding the unknown reactions, this assumption must be consistent in writing both equations.

BENDING MOMENT IN FRAMES

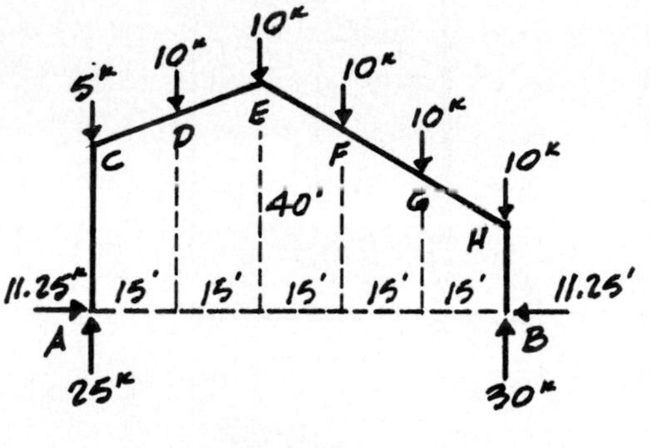

In the analysis of any structure, it is important to understand the bending moment variation and to determine the magnitude of the maximum bending moment. As in Chapter 8, where bending moment diagrams were drawn for beams, we will similarly develop a method for drawing bending moment diagrams for three-hinged frames.

In the same way that bending moment diagrams have been previously drawn, using the beam shape (a horizontal line) as the base line, the three-hinged frame shape will be used as the base line for the diagram. Using the example of Fig. 9–11, a single line (the centerline of the frame) drawing is used, as shown in Fig. 9–15. Starting at point A, where the bending moment is known to be zero, if one visualizes a series of free bodies going up the leg toward point C, the bending moment at any point on the leg is simply the product of the horizontal force (A_H) and the distance from point A, as shown in Fig. 9–16. This will produce a linear variation in the bending moment. It has already been determined that the bending moment will vary linearly between concentrated loads (see Chapter 8). The vertical force (A_V) will not affect the bending moment in the leg of the frame, since the line of action of this force will pass through any point on the leg chosen as a free body. Therefore, the maximum bending moment in the leg of the frame will be at point C:

$$M_C = (11.25^K)(30') = 337.5'^K$$

The convention used to determine whether the diagram is drawn on the outside of the base line or the inside will be as follows:

> If the bending moment produces tension on the outside of the frame, draw the bending moment diagram on the outside. Conversely, if the bending moment produces tension on the inside of the frame, draw the bending moment inside.

In order to determine whether there will be tension outside or inside, it will

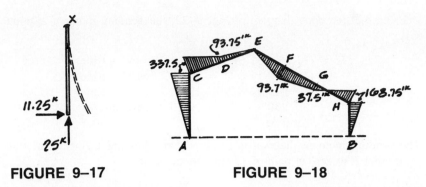

FIGURE 9–17   FIGURE 9–18

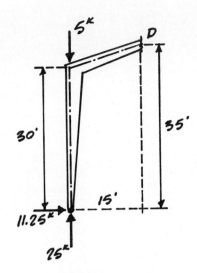

FIGURE 9–19

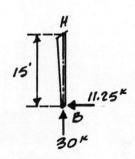

FIGURE 9–20

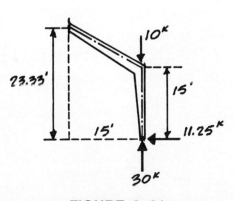

FIGURE 9–21

be necessary to put one's mind's eye to work and do a bit of visualizing. For the segment of the frame under discussion, if a free body is visualized, as shown in Fig. 9–17, and the point at which the free body is cut (point X in the figure) is visualized as being restrained, and the horizontal force is considered to be producing a tendency to push the leg inward, then it should be determined that this bending will produce tension on the outside of the frame leg.

Referring to Fig. 9–18, and continuing to draw the bending moment diagram from point C, another convention that will be used here will be described. The bending moment diagram on any segment of the frame will be drawn perpendicular to the frame shape. At point C, therefore, we will turn the corner and start with the value already determined at this point (point C is common to both the leg and the rising portion of the frame). The next step is to determine the value of the bending moment at point D, and the character of this bending moment (tension inside or outside). If a free body is taken, as shown in Fig. 9–19, then both the magnitude and character of the moment can be determined. Visualizing point D as being restrained:

$$M_D = (11.25^K)(35') = 393.75'^K \text{ (tens. outside)}$$
$$(5^K)(15') = 75.00'^K \text{ (tens. outside)}$$
$$- (25^K)(15') = \underline{375.00'^K \text{ (tens. inside)}}$$
$$\text{Net B.M.} \quad 93.75'^K \text{ (tens. outside)}$$

The net value of the bending moment produces tension on the outside of the frame. Recognizing that there can only be a linear variation in the bending moment between concentrated loads, the bending moment diagram is drawn as shown in Fig. 9–18.

From point D to point E, there is again a linear variation in the bending moment, and the moment at point E must be zero since this is a hinge. Therefore, the diagram is drawn to zero at point E as shown in Fig. 9–18.

Following the procedures already outlined, the bending moments and their characters will now be determined for the right-hand side of the frame. Referring to Fig. 9–20, the moment at point H is

$$M_H = (11.25^K)(15') = 168.75'^K$$

and this moment produced tension on the outside of the frame.

Using the free body diagram of Fig. 9–21, the bending moment at point G is

$$M_G = (11.25^K)(23.33') = 262.5'^K \text{ (tens. outside)}$$
$$(10^K)(15') = 150.0'^K \text{ (tens. outside)}$$
$$- (30^K)(15') = \underline{450.0'^K \text{ (tens. inside)}}$$
$$\text{Net B.M.} \quad 37.5'^K \text{ (tens. inside)}$$

Using the free body diagram of Fig. 9–22, the bending moment at point F is

$$
\begin{aligned}
M_F &= (11.25^K)(31.67') &= 356.3'^K \text{ (tens. outside)} \\
& (10^K)(30') &= 300.0'^K \text{ (tens. outside)} \\
& (10^K)(15') &= 150.0'^K \text{ (tens. outside)} \\
-& (30^K)(30') &= \underline{900.0'^K \text{ (tens. inside)}} \\
&\text{Net B.M.} & 93.7'^K \text{ (tens. inside)}
\end{aligned}
$$

The bending moment diagram is completed by drawing a linear variation from point F to zero at point E where a hinge exists.

Significance of the Bending Moment Diagram

In the example used for drawing the bending moment diagram, it should be noticed that the maximum moments occurred at the knees of the frame. This indicates that more material will be needed at the knees than at other points where the bending moment is less. In general, the bending moment diagram may be thought of as an indicator of the amount of material that will be required to resist the stresses produced by the loading. With this in mind, the designer may begin to visualize the optimum forms that structural elements may take for particular loading situations. It should be added, however, that in spite of the fact that shear diagrams have not been considered in three-hinged frames, shear stresses do exist, and material must be provided to resist shear where necessary. This means that a certain amount of material will have to be provided at the hinge points, where the bending moment is zero, to resist the stresses due to shear, although this may be very small.

INTRODUCTION TO CABLES

Essentially, a cable is a flexible member which, as such, is incapable of resisting compressive stresses. Since a cable cannot resist compressive stresses, it cannot resist a bending moment because, as shown in Chapter 7, in order to resist bending moment a structural member must be capable of providing an internal couple, which is made up of tensile and *compressive* forces. A cable is capable of resisting *only a tensile stress* and, consequently, cannot produce a resisting moment.

Acting in tension, a cable is an extremely efficient structural member because only relatively small cross sections (and, therefore, minimum amounts of material) are required to resist forces of large magnitude. For example, as an indication of strength, cables made of steel wire may be obtained which have breaking strengths of 270,000 p.s.i. This cable makes use of a particularly high grade of steel and is primarily used for prestressing purposes. A more commonly used steel cable for building purposes, generally referred to as *wire rope,* has a breaking strength of about 200,000 p.s.i. While a cable would be designed for an *allowable* stress rather than for its breaking (ultimate) strength, these values should serve as indicators of the incredible strength available in a member acting in a pure state of tension.

In addition to steel cable (or wire rope), other kinds of rope may be used for a cable-supported structure. Many ropes made of synthetic fibers such as nylon, polypropylene, etc., have extremely high breaking strengths, but they suffer from a phenomenon known as "creep." This means that the material will, in time, change dimension under sustained loading. In simple language, ropes made of these synthetic fibers will stretch when subjected

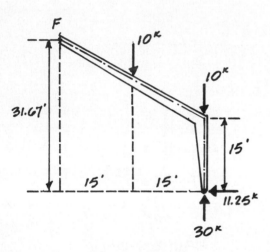

FIGURE 9–22

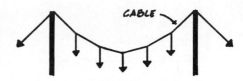

FIGURE 9–23

to tensile forces. The amount of stretch that takes place is generally quite large; consequently, the use of such ropes on large-scale or important structures is not feasible.

ANALYSIS OF FORCES IN CABLES

The primary purpose of this section is to provide a sense of the way cables behave under load and to show how the laws of static equilibrium are applied to determine the magnitude of stresses in cables. Consequently, only two-dimensional configurations will be considered for this purpose (Fig. 9–23). It should be understood, however, that oftentimes, in buildings, cable networks are employed which are multidirectional, such as that shown in Fig. 9–24. The analysis of such cable networks is beyond the scope of intent of this book, but may be found in a more advanced treatise on this topic.

The analysis of a cable subjected to loads makes use of the equations of static equilibrium and the very important idea that a cable can only resist *axial tension*. Because a cable can only operate in a state of tension, it must also be recognized that when a cable is subjected to load, as shown in Fig. 9–25, it will assume a particular geometrical configuration in response to that load. In Fig. 9–25, for example, the cable must deflect from the horizontal position by some amount in order to provide a vertical component that will equilibrate the load ($\Sigma F_y = 0$). These components are resolved at the supports by reactions. It should also be recognized that under the influence of the load, there is also the tendency for the cable to pull inward from the supports. This, of course, must be resisted by horizontal reactions at the supports, as shown in Fig. 9–25. If there are vertical components in a sloping member, which is in a single state of stress such as tension or compression (tension in cables), then there must also be a horizontal component. Also, the resultant of these components will act along the centroid of the member, and it is the resultant that produces the stress in the member.

We will now look at some numerical examples which will serve to clarify some of the above ideas, as well as provide a basis for discovery of some important relationships between loads on cables and the geometrical configurations that must be assumed by the cable in order to resolve these loads in a pure state of tension.

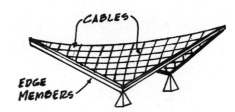

FIGURE 9–24

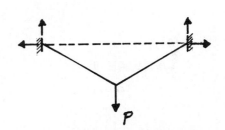

FIGURE 9–25

Example 9–3 (Fig. 9–26)

Because of the symmetry in this particular example, the determination of the vertical reactions is a simple matter. By inspection,

$$V_L = V_R = 6^K$$

In order to determine the horizontal reactions, there are two approaches that may be taken. The first approach makes use of the geometrical configuration and the idea that the resultant force in the cable will act along the centroid of the member. This means that the components *must* be in the same ratio as the slope of the member. Now if a free body is taken at one end of the structure, as shown in Fig. 9–27, it will be seen that the vertical component in the cable must be 6^K to equilibrate the vertical reaction of 6^K ($\Sigma F_y = 0$). The horizontal component must, therefore, be 8^K, since the cable has a slope ratio of 3:4. The resultant force in the cable is 10^K, since we are dealing with geometry of a 3:4:5 ratio.

The free body shown in Fig. 9–27 is taken on the right-hand side of the structure (although the left side could have been used with the same results). Therefore, it is the value of H_R that has been determined. In order to determine the value of H_L:

$$\Sigma F_x = 0, \qquad \therefore H_L = H_R = 8^K$$

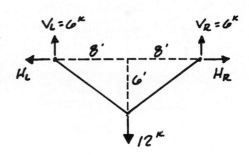

FIGURE 9–26

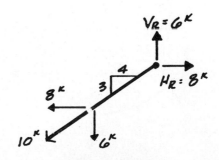

FIGURE 9–27

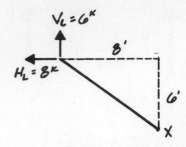

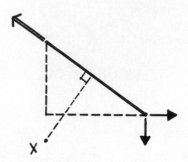

FIGURE 9–28

Another way to determine the value of the horizontal reaction is to make use of the fact that there cannot be a bending moment in a cable. With this in mind, a free body diagram may be taken, as shown in Fig. 9–28, and moments taken about point X, as follows:

$$\Sigma M_x = 0 = (6^K)(8') - H_L(6')$$

$$\therefore H_L = \frac{48'^K}{6'} = 8^K$$

Two important points need to be made about this approach. First, it is important that moments be taken about some point on the cable (such as point X in Fig. 9–28). When the free body diagram is "cut," the internal tensile force in the cable is released. The value of this force is unknown at the time reactions are being determined. However, this is of no consequence if the point about which moments are being taken lies on the cable, since the line of action of this unknown force will pass through the point. If a point is arbitrarily chosen that lies off the cable, such as shown in Fig. 9–29, the line of action of the unknown internal force that was released will have a perpendicular distance to the point, and would have to appear in the moment equation. There would then be two unknowns in a single equation which, of course, cannot be solved.

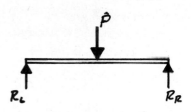

FIGURE 9–29

The second important point is the difference between a member that cannot resist bending moment, such as a cable, and a rigid member that can resist bending moment, such as a simply supported beam. In the example under discussion, it was shown that a free body may be taken and the only internal force released is a tensile force. This allows us to take moments about the point where the free body cut was made without concern for including the unknown tensile force in the moment equation, since its line of action will pass through the point. However, in a rigid member which can resist bending moment, such as the simply supported beam of Fig. 9–30, moments cannot be taken about an arbitrary point. For example, if a free body is taken of the simply supported beam, as shown in Fig. 9–31, it will be recalled from Chapter 7 that there will be a tendency for this free body to rotate; this is the *bending moment*. In order to resist this bending moment there must be an equal and opposite *resisting moment*. This resisting moment is produced by the *internal couple*, which is made up of tensile and compressive forces. Now, if moments are taken about point X, shown in Fig. 9–31, the forces that constitute the internal couple will have lever arms to this point and would, consequently, have to appear in the moment equation. But, of course, the values of the forces that make up the internal couple are unknown and cannot be determined until the reactions are found. This is why, in a beam, reactions can only be found by taking moments about a point where the bending moment is known to be zero, such as the end of the simply supported beam. This dilemma does not arise in a cable, since the bending moment *at any point* is known to be zero.

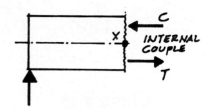

FIGURE 9–30

Example 9–4 (Fig. 9–32)

The analysis of the cable structure of the previous example was particularly simple because it was symmetrical and the vertical reactions could be determined by inspection. When the structure is not symmetrical, such as that of Fig. 9–32, the analysis is not much more difficult. It is only necessary to write a moment equation to determine the vertical reactions, rather than determine them by inspection. In order to determine the vertical reactions

$$\Sigma M_L = 0 = (6^K)(16') + (18^K)(32') - V_R(48')$$

$$V_R = \frac{672'^K}{48'} = 14^K, \quad \therefore V_L = 10^K \,(\Sigma F_y = 0)$$

In order to determine the horizontal reactions, a free body is taken of the left-hand end of the structure, as shown in Fig. 9–33, and moments are taken about point X on the cable:

$$\Sigma M_x = 0 = (10^K)(16') - H_L(10')$$

$$H_L = 16^K, \quad \therefore H_R = 16^K \,(\Sigma F_x = 0)$$

FIGURE 9–31

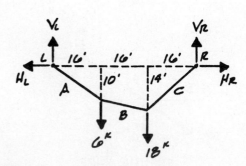

FIGURE 9–32

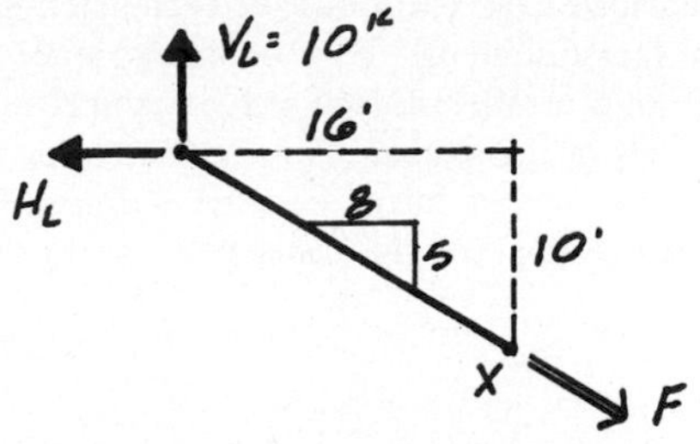

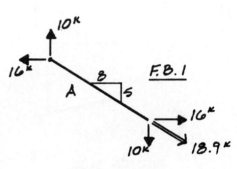

FIGURE 9–33

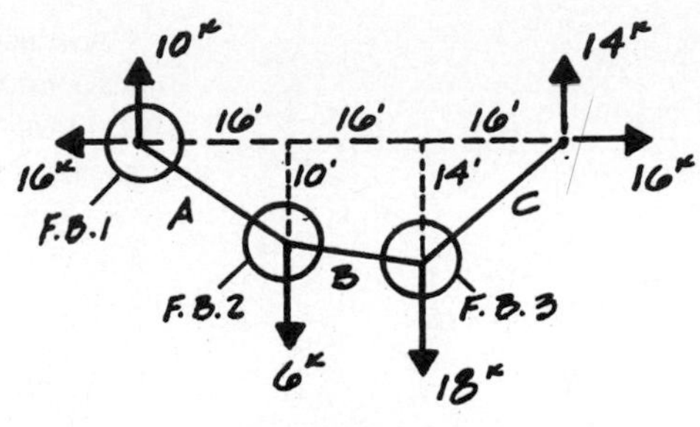

FIGURE 9–34

The horizontal reaction H_L could also have been determined by the geometry of the cable. The reactions must be equal to the vertical and horizontal components of the force in the cable, and the components must be in the same ratio as the slope of the cable. Therefore, still referring to Fig. 9–33, knowing the vertical reaction, the horizontal reaction must be

$$H_L = \frac{8}{5}(10^K) = 16^K$$

The next step in the analysis is to determine the forces acting in the cable. In order to do this, a series of free bodies will be employed, as shown in Figs. 9–34 through 9–37.

In Fig. 9–35, the force in the cable is detemined simply by finding the resultant of the vertical and horizontal components. Therefore the force in segment A is 18.9^K.

In Fig. 9–36, the vertical component in segment B is 4^K and the horizontal component remains 16^K (because there are no horizontal forces being introduced to the system). The resultant force is determined by the solution of the 1:4 right triangle and the value of the resultant, which is the force in segment B, is 16.5^K.

In Fig. 9–37, the vertical component in segment C must be 14^K to satisfy $\Sigma F_y = 0$, and the horizontal component remains 16^K. The resultant stress in segment C is 21.3^K.

FIGURE 9–35

Example 9–5 (Fig. 9–38)

Most often, in an architectural situation, a particular dimension is critical and must be set, such as the $12'$ vertical dimension shown in Fig. 9–38. It should be understood, in a case such as this, that the other vertical dimensions (X and Y) are not arbitrary. There is a unique set of dimensions that will satisfy all the given factors, such as load, span, and the one critical dimension that is set. Therefore, in a practical problem such as this, the complete analysis requires not only the determination of the reactions, but also the determination of unknown vertical dimensions that will conform to the geometry necessary for a pure state of tension to exist.

In order to determine the reactions, take moments about the left-hand support (the right-hand support may be chosen):

$$\Sigma M_L = 0 = (24^K)(12') + (16^K)(24') +$$
$$(8^K)(36') - V_R(48')$$
$$V_R = \frac{960'^K}{48'} = 20^K, \qquad \therefore V_L = 28^K(\Sigma F_y = 0)$$

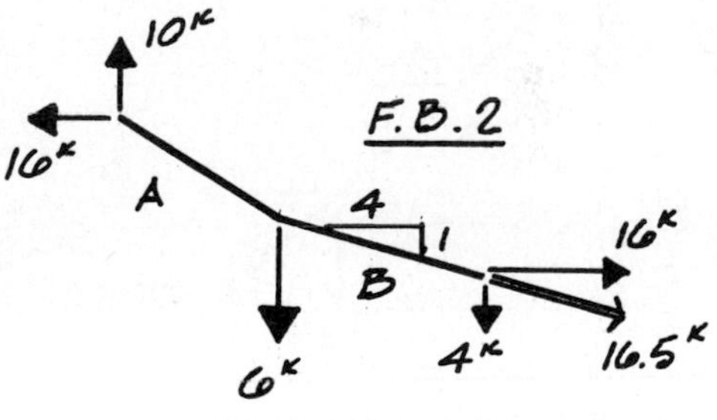

FIGURE 9–36

In order to determine the horizontal reactions, a geometrical solution cannot be used, as in the previous example, because the dimensions X and Y are not yet known; therefore, the slopes of end segments of the cable are not known. To determine the end reactions, the known information must be used. In this case a free body is taken, as shown in Fig. 9–39, and moments are taken about point P:

$$\Sigma M_p = 0 = (28^K)(24') - (24^K)(12') -$$
$$H_L(12')$$
$$H_L = \frac{384'^K}{12'} = 32^K, \quad \therefore H_R = 32^K(\Sigma F_x = 0)$$

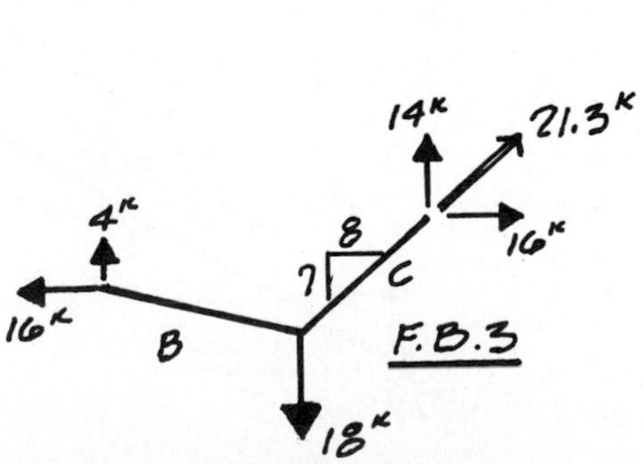

FIGURE 9–37

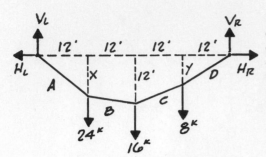

FIGURE 9–38

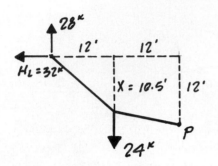

FIGURE 9–39

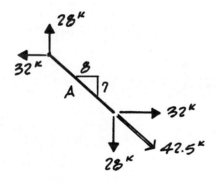

FIGURE 9–40

Now that the vertical and horizontal reactions have been determined, the vertical dimensions X and Y may be found quite easily by geometry, since the slope of the cable must be in the same ratio as the vertical and horizontal components. The components in the end segments of the structure are equal and opposite to the reactions. Considering the free body diagram at the left side of the structure, as shown in Fig. 9–40, it is seen that the slope of the cable must be in the ratio of 7:8. Therefore,

$$X = \frac{7}{8}(12') = 10.5'$$

Investigating a free body diagram on the right-hand side of the structure, it can be determined that the slope of the cable must be in the ratio of 5:8. Therefore,

$$Y = \frac{5}{8}(12') = 7.5'$$

To complete the analysis, it is necessary to determine the forces in the various segments of the cable. In order to determine the stress in segment A, the free body diagram of Fig. 9–40 may be used to determine the resultant of the components. By solving for the proportion of the hypotenuse of the 7:8 right triangle, it is determined that the force in segment A is 42.5^K. Similarly, the force in segment D is determined using the free body of Fig. 9–41. By solving for the hypotenuse of the 5:8 right triangle, it is determined that the resultant force in segment D is 37.7^K.

In order to solve for the resultant force in segment B, the free body shown in Fig. 9–42 will be used. Based on the values of the components (and on the dimensions previously established), it can be determined that the slope of segment B must be in the ratio of 1:8. Based on this geometry, the force in segment B is 32.3^K.

In a similar manner, the stress in segment C is determined by using the free body diagram of Fig. 9–43. Based on the components in segment C, the slope must be in the ratio of 3:8. Therefore, the resultant force in segment C is 34.2^K. The analysis of the cable structure of Fig. 9–38 is now complete.

CABLE SHAPE AND BENDING MOMENT DIAGRAM

In the examples presented it can be seen that in each case, the cable had to assume a particular geometrical configuration in order to respond to the loads. Actually, the cable configuration is closely related to the bending moment diagram that would be produced by a simply supported beam under the same load and span conditions as the cable being considered. This is shown in Figs. 9–44 through 9–46. It may be said that the cable must as-

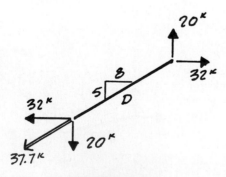

FIGURE 9–41

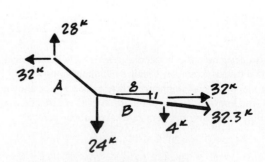

FIGURE 9–42

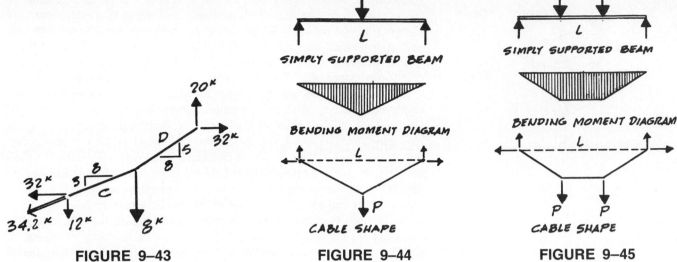

FIGURE 9-43   FIGURE 9-44   FIGURE 9-45

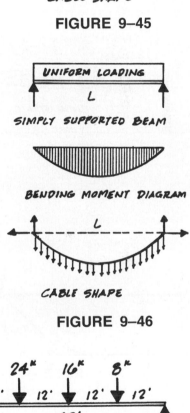

FIGURE 9-46

sume these configurations simply because it cannot resist bending moments. Therefore, the geometry is necessary in order to produce forces that will produce a net zero bending moment at any point along the cable.

The relationship between the cable shape and the bending moment diagram can be particularly useful to the architectural designer when trying to determine the geometry that will be assumed, particularly when dealing with a complex loading pattern. In order to determine the general configuration, it is only necessary to draw the bending moment diagram for a simply supported beam of the same span and the same loads. The cable configuration will be identical to the bending moment diagram.

This relationship can also be used in determining specific information about the cable configuration. For example, let's reconsider the problem of Fig. 9-38. In order to evaluate the geometry, a simply supported member of the same span and loading is used and the bending moment diagram is determined, as shown in Fig. 9-47. Not only will the cable take on the same configuration as the bending moment diagram, but the vertical dimensions of the cable "sag" at various points must be in the same ratio as the bending moments in the simply supported beam at the various points. Therefore, still referring to Fig. 9-38 and setting the 12' vertical dimension at the midspan, the dimensions X and Y may be determined by simple proportions of the bending moment diagram:

$$\frac{X}{336'^K} = \frac{12'}{384'^K} \quad \text{and} \quad X = 10.5'$$

and

$$\frac{Y}{240'^K} = \frac{12'}{384'^K} \quad \text{and} \quad Y = 7.5'$$

THE PROBLEM WITH CABLES

In spite of the fact that a cable acting in a pure state of tension is an extremely efficient structural member, from the standpoint of the amount of

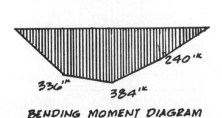

FIGURE 9-47

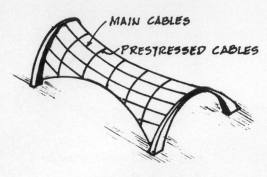

FIGURE 9–48

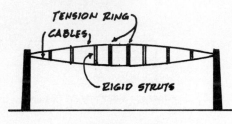

FIGURE 9–49

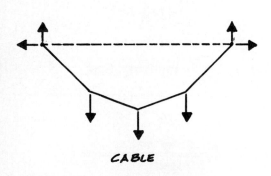

CABLE

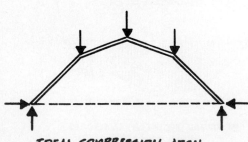

IDEAL COMPRESSION ARCH

FIGURE 9–50

material needed to carry sizable loads, there are problems involved that must be considered when dealing with a cable structure. The primary problem is one of stability or *flutter*, as it is commonly referred to in a cable structure. Flutter is caused primarily by wind forces, which can cause uplift forces as well as downward pressures. Because of the efficiency of a cable structure, there is an inherent lack of weight to resist movement set up by wind forces or, for that matter, any kind of force that might produce vibration.

In order to counteract the possibility of flutter, several techniques are employed. One way to reduce or eliminate flutter is simply to place as much dead weight on the cables as possible. The dead weight acts to dampen movement, thereby stabilizing the structure. Another method of preventing flutter is to use a surface of double curvature, employing two sets of cables, with one set initially prestressed to hold the other cables in a stable configuration, as shown in Fig. 9–48. It is also possible to use a double set of cables (upper and lower), as shown in Fig. 9–49, where each set of cables is prestressed. In a system such as this, the cables are literally "tuned," much like the strings of a guitar. This is done so the tendency of one set of cables to move is counteracted by the other set of cables, which, being tuned differently, acts to dampen the movement.

INTRODUCTION TO IDEAL COMPRESSION ARCHES

The principles involved in the development of an "ideal" compression arch are precisely the same as those already discussed in the section on cables. Where a cable must assume a particular geometrical configuration in order to respond to loads and be in a state of pure tension, an ideal compression arch must conform to a required geometry so that the cross section of each member is subjected to a uniform compressive stress.

The word *ideal* in the name given to this kind of structure comes from the idea that a structure is more efficient when subjected to a pure state of stress (tension or compression), thus eliminating bending moments. In a cable structure, the flexibility of the cable naturally causes it to take on the ideal configuration since it cannot resist bending moments. An arch, on the other hand, is made of rigid rather than flexible material, and can be subjected to bending moments. *Ideally* it is desirable, where possible, to avoid bending moments; this can be achieved if the geometry of the arch is developed in response to the loads.

In order to determine the correct geometrical configuration for the development of an ideal arch, one needs only to remember the principles required to determine cable geometry (zero bending moment configuration) and consider that these principles also apply in the determination of the ideal arch configuration. It is important to recognize at this point than an ideal compression configuration will simply be the mirror image of the pure tensile configuration (see Fig. 9–50). Also, as in a cable structure, an ideal arch must be the same shape (but reversed) as the bending moment diagram of a simple beam of the same span and the same loads.

METHOD OF COMPUTATION

The following examples will serve to show that the principles employed in the analysis of an ideal arch are the same as those already discussed in the section on cables.

Example 9–6 (Fig. 9–51)

In order to analyze this structure, it is first necessary to recognize the forces that will be acting. There will, of course, be vertical reactions at the support points to equilibrate the applied 20^K load. In addition to this, there will be, as there is in any arch structure, horizontal reactions which are necessary to keep the legs of the structure from moving outward. In order to determine the vertical forces, take moments about either the left-hand support or the right-hand support. In this case, the left-hand support will be used:

$$\Sigma M_L = 0 = (20^K)(15') - V_R(50')$$

$$\therefore V_R = \frac{300'^K}{50'} = 6^K \text{ and } V_L = 14^K (\Sigma F_y = 0)$$

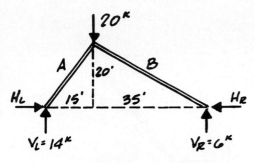

FIGURE 9–51

In order to determine the horizontal reaction, either a free body diagram may be used, knowing that there will be no bending moment (and therefore no internal couple) wherever the cut is made, or the given geometry may be employed. Using the geometry, it must be recognized that the reactions must be equal and opposite to the components in the leg of the arch. Referring to the free body of Fig. 9–52, it is seen that the slope of segment A is in the ratio of 3:4. The components must be in this same ratio. Therefore,

$$H_L = \frac{3}{4}(14^K) = 10.5^K$$

and

$$H_R = 10.5^K \qquad (\Sigma F_x = 0)$$

The force in segment A is simply the resultant of the components, and this is in the proportion of the hypotenuse of the right triangle (segment A forms a 3:4:5 right triangle) to the other sides. Therefore, by a simple proportion,

$$A = \frac{5}{4}(14^K) = 17.5^K \text{ compression}$$

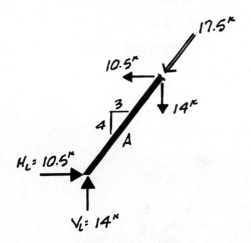

FIGURE 9–52

The compressive force in segment B may be found by the same process.

$$B = 12.1^K \text{ compression}$$

Example 9–7 (Fig. 9–53)

In the discussion on cables, it was mentioned that frequently the designer must set at least one dimension based on some criterion, and other dimensions are determined mathematically. The same is true when it is desired to develop an ideal compression arch. In the example of Fig. 9–53, the 32' vertical dimension has been set, and the dimensions X and Y must be determined for the ideal configuration. It is first necessary to determine the vertical reactions, which will be accomplished by taking moments about the left-hand support:

$$\Sigma M_L = 0 = (20^K)(12') + (30^K)(32') + (24^K)(50') - V_R(60')$$

$$V_R = 40^k \qquad \therefore V_L = 34^K (F_y = 0)$$

In order to determine the horizontal reactions, a free body diagram must be taken using the known 32' dimension. It is not possible to use a geometrical approach, as in the previous example, since the slopes of the various segments are not known due to the unknown dimensions X amd Y. The free body diagram of Fig. 9–54 is used to find the horizontal reactions:

$$\Sigma M_p = 0 = (34^K)(32') - (20^K)(20') - H_L(32')$$

$$H_L = 21.5^K \qquad \therefore H_R = 21.5^K \qquad (\Sigma F_x = 0)$$

Now that the reactions have been determined, the unknown dimension X can be found by determining the slope of segment A, which must be in the same ratio as the components. Referring to the free body diagram of Fig. 9–55, the "rise" of segment A can be determined by a simple proportion, recognizing that the rise takes place over a 12' horizontal run:

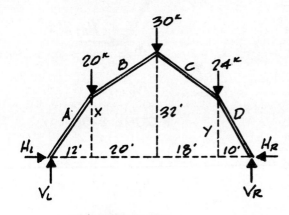

FIGURE 9–53

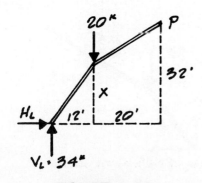

FIGURE 9–54

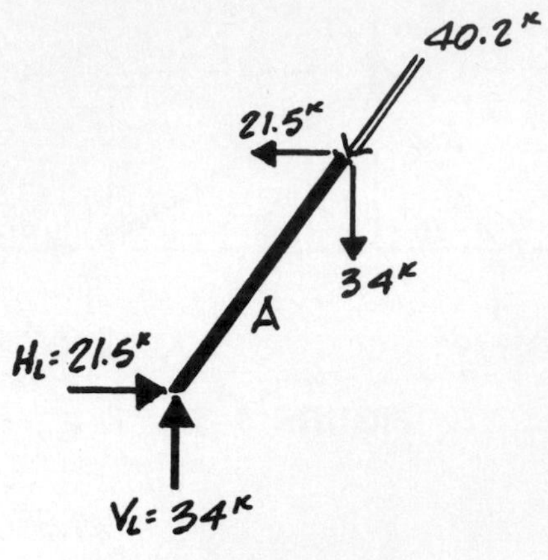

FIGURE 9–55

$$\frac{12'}{21.5} = \frac{X}{34} \text{ and } X = 18.98'$$

By the same process, the dimension Y may be found, using the free body of Fig. 9–56,

$$\frac{10'}{21.5} = \frac{Y}{40} \text{ and } Y = 18.60'$$

Now that the reactions and dimensions have been found, the compressive forces in the various segments may be determined. To find the resultant force in segment A, the free body diagram of Fig. 9–55 may be used. The force in segment A is 40.2^K. The force in segment D is determined by using the free body diagram of Fig. 9–56, and the force is determined to be 45.4^K.

In order to determine the force in segment B, a free body is taken at the junction of segments A and B, as shown in Fig. 9–57. The components are determined so that the free body is in a state of equilibrium. The resultant of these components is the force in segment B, which is determined to be 25.7^K. The force in segment C is found by the same process. Using a free body taken at the junction of segments C and D, as shown in Fig. 9–58, the resultant compressive force in segment C is 26.8^K.

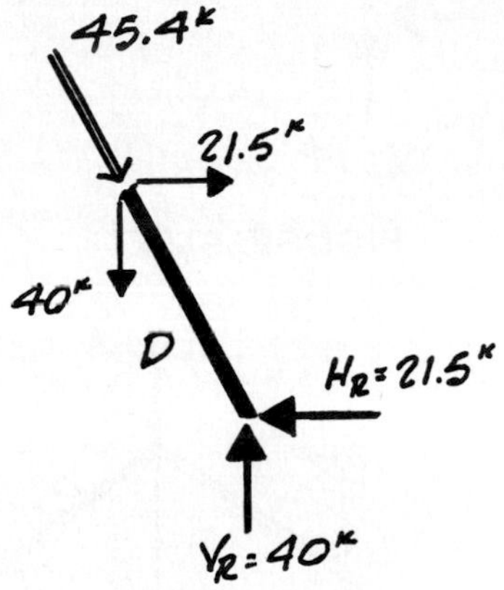

CONCLUDING NOTES ON IDEAL ARCHES

In the previous examples it was shown that the principles involved in the analysis of an ideal compression arch are the same as those for a cable. These ideas are based on the fact that no bending moment exists in the structure.

When designing an ideal compression arch, it is important to remember that it will be made of a rigid material that cannot adjust its configuration when forces other than those for which it was designed are applied, such as wind forces. In a cable, of course, the geometrical configuration will adjust itself when unpredictable forces are applied. This is what causes the cable to move, or flutter, as previously discussed. Being made of a rigid rather than a flexible material, an arch will not readjust itself when unpredictable forces are applied, but will instead develop bending moments. In other words, unpredictable forces will, even though applied for a short period of time, demand an ideal configuration other than that for which the arch was originally designed. The designer of such a structure must be aware of this and must design the structure with supplementary strength, which is required to resist bending moments due to wind forces or other anticipated short-term forces.

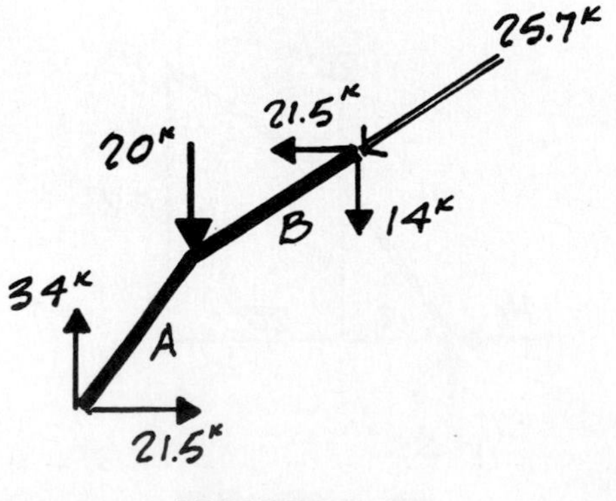

FIGURE 9–56

FIGURE 9–57

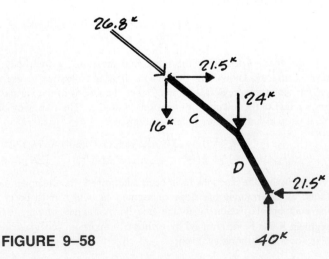

FIGURE 9–58

PROBLEM 9–1

THREE-HINGED FRAMES

FOR EACH OF THE FOLLOWING, DETERMINE THE AMOUNT AND DIRECTION OF REACTIONS H_L, V_L, H_R & V_R (● INDICATES HINGE)

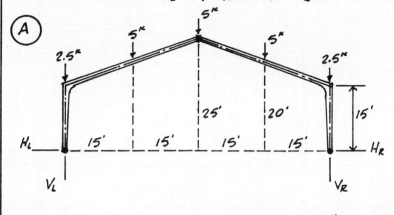

Ⓐ

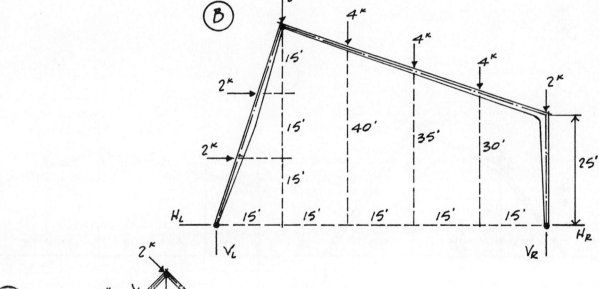

Ⓑ

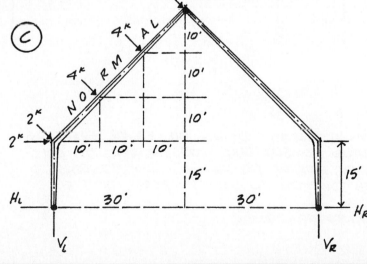

Ⓒ

PROBLEM 9-2

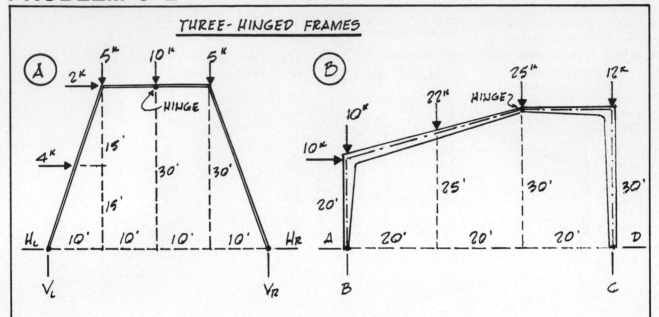

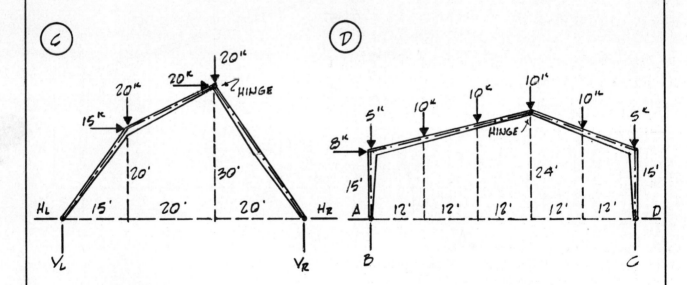

FOR EACH OF THE ABOVE:

(1) DETERMINE THE REACTIONS.

(2) DRAW THE BENDING MOMENT DIAGRAM USING THE FRAME SHAPE AS A BASELINE. DRAW BENDING MOMENT THAT PRODUCES TENSION ON OUTSIDE FIBERS, ON THE OUTSIDE OF THE FRAME, AND BENDING MOMENT THAT PRODUCES TENSION ON INSIDE FIBERS, ON THE INSIDE OF THE FRAME.
SHOW VALUES AT ALL POINTS OF CHANGE.

PROBLEM 9–3

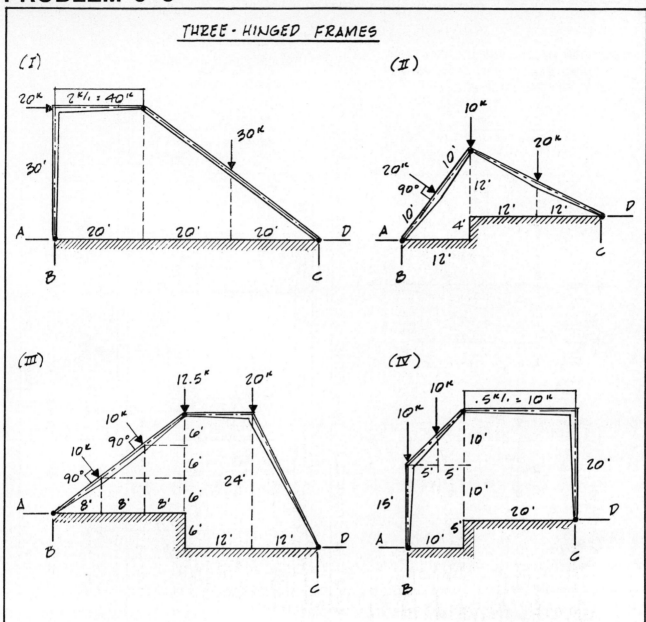

THREE - HINGED FRAMES

● FOR EACH OF THE ABOVE, DETERMINE THE MAGNITUDE AND DIRECTION
OF REACTIONS A,B,C.D. DRAW THE BENDING MOMENT DIAGRAM USING
THE FRAME SHAPE AS THE BASELINE. DRAW THE DIAGRAM ON THE
TENSION SIDE OF THE FRAME. DETERMINE VALUES OF BENDING MOMENT
AT ALL POINTS OF CHANGE.

PROBLEM 9–4

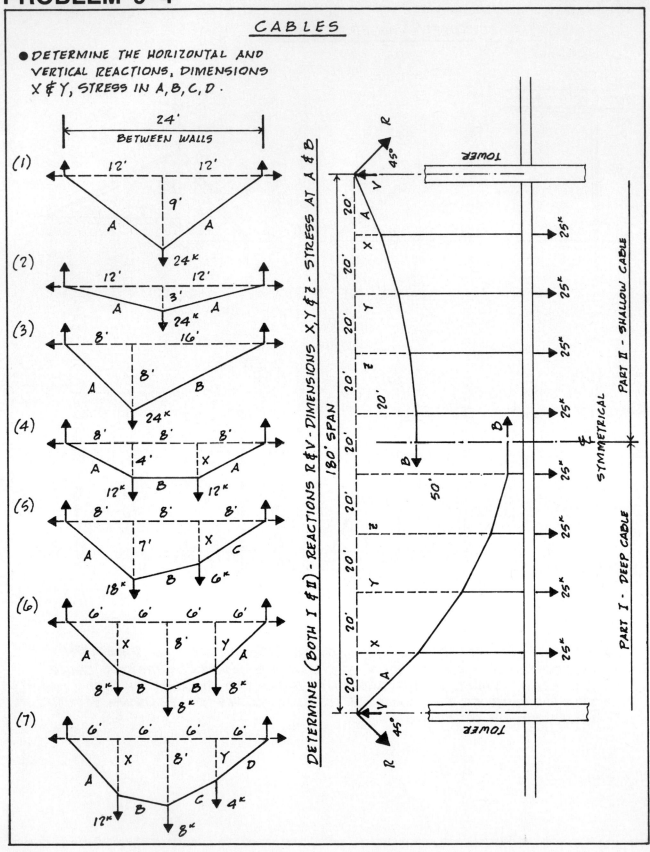

CABLES

● DETERMINE THE HORIZONTAL AND VERTICAL REACTIONS, DIMENSIONS X & Y, STRESS IN A, B, C, D.

DETERMINE (BOTH I & II) - REACTIONS R & V - DIMENSIONS X,Y & Z - STRESS AT A & B

PART I - DEEP CABLE   PART II - SHALLOW CABLE

PROBLEM 9–5

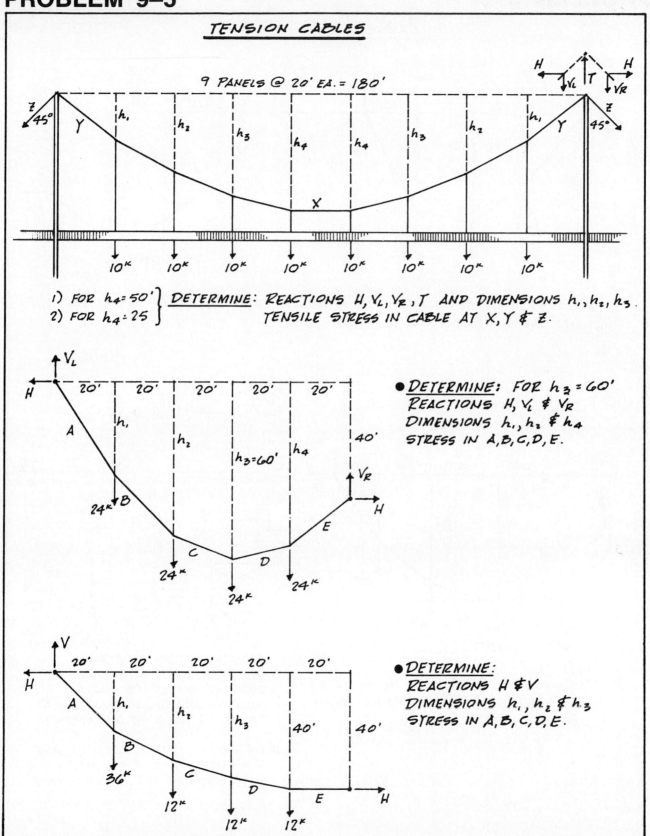

TENSION CABLES

9 PANELS @ 20' EA. = 180'

1) FOR $h_4 = 50'$ ⎫ DETERMINE: REACTIONS H, V_L, V_R, T AND DIMENSIONS h_1, h_2, h_3.
2) FOR $h_4 = 25$ ⎭   TENSILE STRESS IN CABLE AT X, Y & Z.

10ᴷ 10ᴷ 10ᴷ 10ᴷ 10ᴷ 10ᴷ 10ᴷ 10ᴷ

● DETERMINE: FOR $h_3 = 60'$
REACTIONS H, V_L & V_R
DIMENSIONS h_1, h_2 & h_4
STRESS IN A, B, C, D, E.

24ᴷ 24ᴷ 24ᴷ 24ᴷ

● DETERMINE:
REACTIONS H & V
DIMENSIONS h_1, h_2 & h_3
STRESS IN A, B, C, D, E.

36ᴷ 12ᴷ 12ᴷ 12ᴷ

PROBLEM 9–6

<u>CABLES</u>

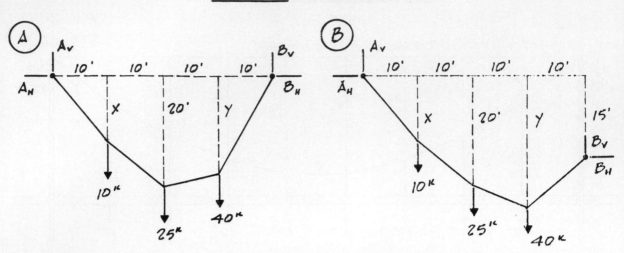

● FOR EACH OF THE ABOVE, DETERMINE THE REACTIONS A_V, A_H, B_V, B_H AND THE DIMENSIONS X & Y

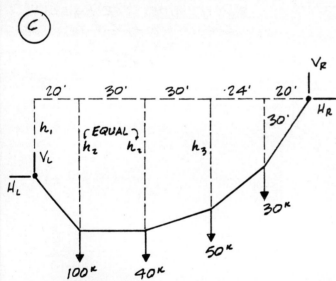

● DETERMINE THE REACTIONS H_L, V_L, H_R, V_R AND THE DIMENSIONS h_1, h_2, h_3.

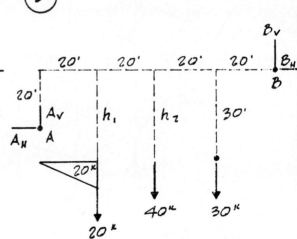

IT IS REQUIRED TO SPAN A CABLE BETWEEN POINTS A & B TO SUPPORT THE LOADS SHOWN. THE GIVEN VERTICAL DIMENSION TO POINT "X" IS A REQUIREMENT.

● DETERMINE THE REACTIONS AND DIMENSIONS h_1 & h_2 AND MAKE A SKETCH OF THE CABLE SHAPE.

PROBLEM 9–7

IDEAL COMPRESSION ARCHES

<u>NOTE</u>: IDENTICAL RELATIONSHIPS (IN REVERSE) AS FOR TENSION CABLES.

● DETERMINE HORIZONTAL & VERTICAL REACTIONS, DIMENSIONS X & Y, AND STRESS IN EACH SEGMENT.

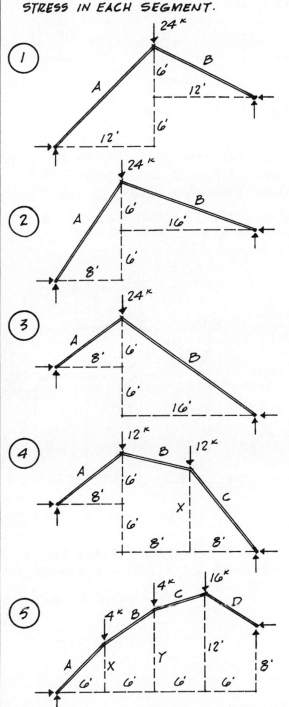

⑥ AN IDEAL COMPRESSION ARCH, TO SUPPORT LOADS FROM BRIDGE DECK (AS SHOWN) HAS REACTIONS AT ABUTMENT POINTS A & B AND IS TO PASS THROUGH POINT C.

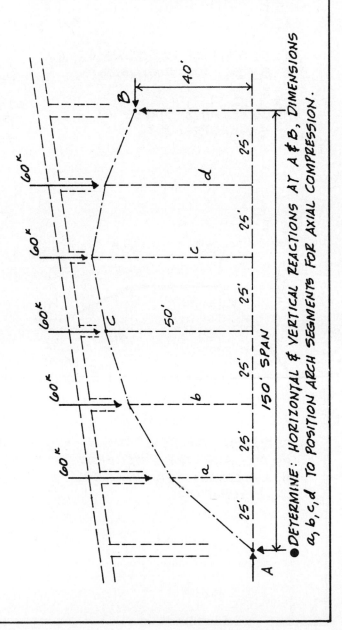

● DETERMINE: HORIZONTAL & VERTICAL REACTIONS AT A & B, DIMENSIONS a, b, c, d TO POSITION ARCH SEGMENTS FOR AXIAL COMPRESSION.

PROBLEM 9–8

IDEAL COMPRESSION ARCHES

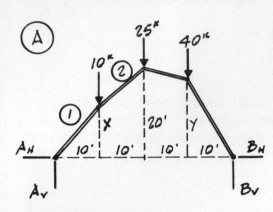

(A) DETERMINE REACTIONS A_H, A_V
B_H, B_V AND DIMENSIONS
X & Y.

(B) DETERMINE STRESS IN
SEGMENTS 1 & 2.

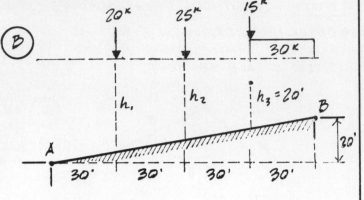

AN IDEAL COMPRESSION ARCH IS TO BE
ANCHORED AT POINTS A & B WITH A 20'
CLEARANCE REQUIRED AT h_3.

● DETERMINE THE REACTIONS AT A & B,
DIMENSIONS h_1 & h_2 AND DRAW A
SKETCH OF THE ARCH.

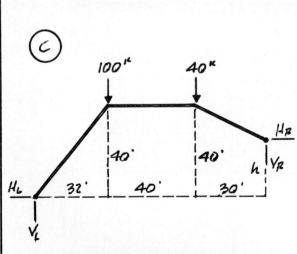

● DETERMINE THE REACTIONS H_L, V_L,
H_R, V_R AND THE DIMENSION
"h" NECESSARY FOR THE IDEAL
CONFIGURATION.

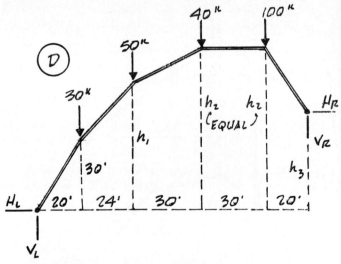

● DETERMINE THE REACTIONS AND THE
DIMENSIONS h_1, h_2, h_3 NECESSARY
FOR THE IDEAL SHAPE.

PROBLEM 9-9

IDEAL COMPRESSION ARCHES

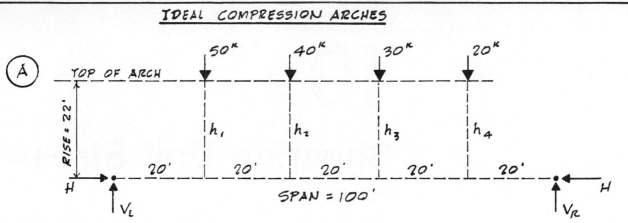

(A)

(1) DETERMINE DIMENSIONS h_1, h_2, h_3, h_4 FOR THE GIVEN LOADS AND RISE.
(2) DETERMINE THE MAXIMUM COMPRESSIVE STRESS (F IN KIPS)
(3) DRAW THE ARCH SHAPE AT SCALE ($1'' = 20'$).

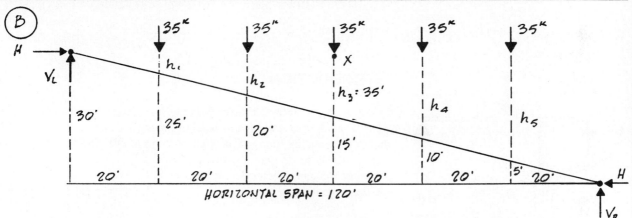

(B)

(1) DETERMINE $h_1, h_2, h_4 \& h_5$ FOR THE GIVEN LOADS, AND h_3 SET AT 35'
(2) DETERMINE THE MAXIMUM COMPRESSIVE STRESS (F IN KIPS)
(3) DRAW THE ARCH SHAPE AT SCALE ($1'' = 20'$)

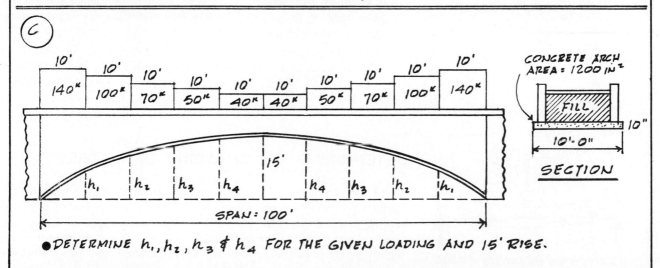

(C)

● DETERMINE $h_1, h_2, h_3 \& h_4$ FOR THE GIVEN LOADING AND 15' RISE.

10

Shearing Unit Stress

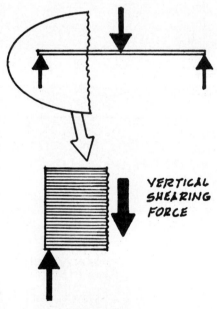

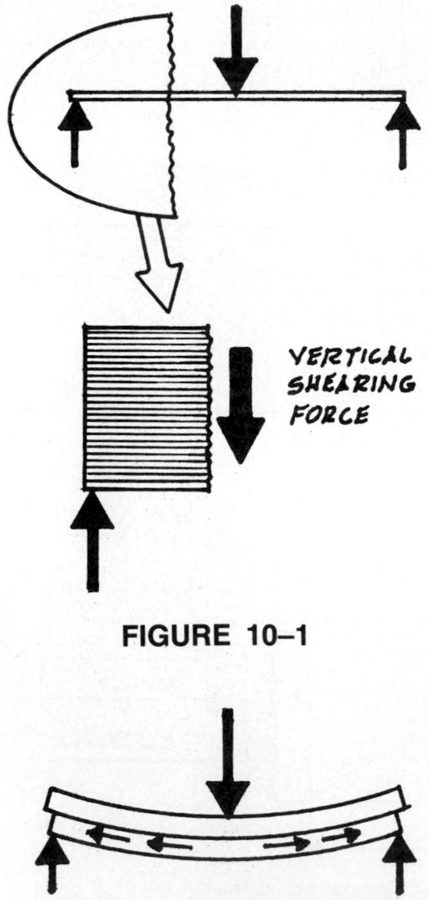

VERTICAL SHEARING FORCE

FIGURE 10–1

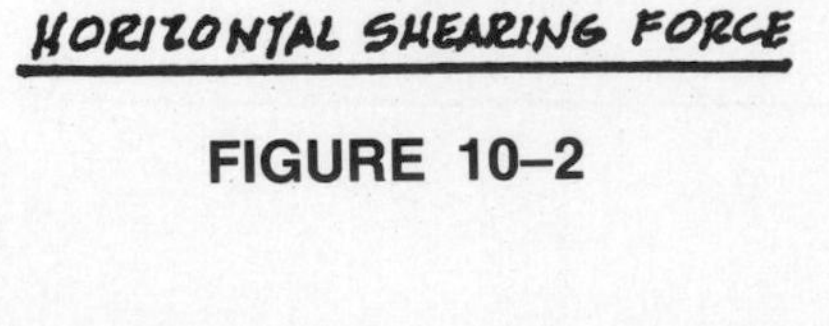

HORIZONTAL SHEARING FORCE

FIGURE 10–2

INTRODUCTION

In the section entitled "Elastic Bending" in Chapter 7, it was shown that when a member is subjected to bending, there are internal stresses produced due to the bending moment, as well as internal shearing stresses, as shown in Fig. 10–1. In Chapter 7, however, we limited ourselves to the determination of the unit stresses due to bending only. We will now concentrate on the determination of the *shearing unit stresses* produced in a member subjected to bending.

Before doing so, it must be recognized that there are two kinds of shearing stresses that must be considered, and they are

1. *Vertical shearing unit stress.* This is produced by the tendency of adjacent parts of the member to translate vertically. In order to satisfy $\Sigma F_y = 0$, an internal vertical force is produced, as shown in Fig. 10–1.
2. *Horizontal shearing unit stress.* This is produced by the tendency of adjacent parts of a member to translate horizontally due to the deformation that takes place in a member subjected to bending, as shown in Fig. 10–2.

DETERMINING THE SHEARING UNIT STRESS

In order to determine the magnitude of the shearing unit stress in a section, consider the beam shown in Fig. 10–3 and the corresponding shear and bending moment diagrams. In Fig. 10–4 an enlarged view of the indicated slice is shown. It should be recognized from the moment diagram that M_2 is slightly higher in magnitude than M_1. Since the free body of Fig. 10–4 must be in equilibrium ($\Sigma M = 0$), then

$$M_2 - M_1 = Vx$$

where V is the total shear on the cross section.

Now, referring to Fig. 10–5, it can be determined that the stress due to bending on the right side of the free body is greater than that on the left side, since M_2 is greater than M_1. Therefore,

$$f_2 > f_1$$

and

$$f_2 = \frac{M_2 c}{I} \quad \text{and} \quad f_1 = \frac{M_1 c}{I}$$

where I is equal to the moment of inertia of the section with respect to the neutral axis.

Considering the force on a horizontal plane (of the slice under consideration) due to the unit stresses, the resultant force on the left side (R_1) is less than the resultant force on the right side (R_2). Therefore, there is a tendency for the piece above the horizontal plane being considered to translate with respect to the piece below. Referring now to Fig. 10–6, the slice being considered is shown three-dimensionally with the resultant forces R_1 and R_2 acting and:

A = cross-sectional area above the horizontal plane being considered
B = the width of the cross section at the horizontal plane being considered

It was shown in Chapter 7 that the value of the resultant force due to proportionally varying unit stress is determined by

$$R = \frac{fA\bar{x}}{c}$$

In Chapter 4 the product of $A\bar{x}$ was given another name, which will be used here; that is

$$Q = A\bar{x}$$

Therefore,

$$R_1 = \frac{f_1 Q}{c} \quad \text{and} \quad R_2 = \frac{f_2 Q}{c}$$

where c is the distance from the neutral axis to the outermost fiber, as shown in Fig. 10–5.

Referring again to Fig. 10–6, let

f_v = the *shearing unit stress* on plane *(XB)* due to the difference of
$\quad R_2 - R_1$

Therefore, the *total shear* on plane *(XB)* is f_v *(XB)*, and

$$f_v(XB) = R_2 - R_1 = \frac{f_2 Q}{c} - \frac{f_1 Q}{c}$$

$$= (f_2 - f_1)\frac{Q}{c}$$

Referring now to Fig. 10–5,

$$(f_2 - f_1)\frac{Q}{c} = \left(\frac{M_2 c}{I} - \frac{M_1 c}{I}\right)\frac{Q}{c}$$

$$\therefore f_v(XB) = \frac{c}{I}(M_2 - M_1)\frac{Q}{c} = (M_2 - M_1)\frac{Q}{I}$$

and referring to Fig. 10–4:

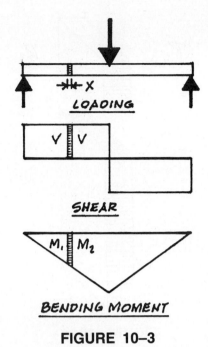

LOADING

SHEAR

BENDING MOMENT

FIGURE 10–3

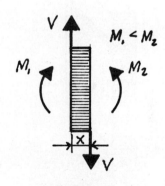

FIGURE 10–4

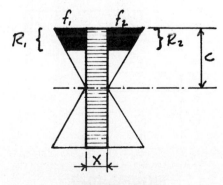

FIGURE 10–5

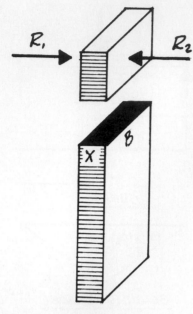

FIGURE 10–6

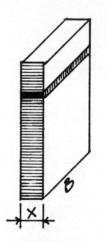

FIGURE 10–7

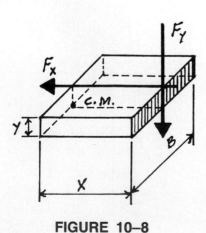

FIGURE 10–8

$$M_2 - M_1 = VX$$

$$\therefore f_v(XB) = (VX)\frac{Q}{I}$$

and

$$\boxed{f_v = \frac{VQ}{IB}}$$

which evaluates the *shearing unit stress* on a horizontal plane.

While we have been defining the various terms in this expression, it would be useful at this point to summarize these definitions:

f_v = *shearing unit stress* on a horizontal plane

V = *total shear* on the section being considered. The magnitude of shear is determined from the shear diagram. Normally we will be concerned with the section being subjected to the maximum shear.

$Q = A\bar{x}$ where

    A = area above the plane where we wish to evaluate the horizontal shearing unit stress. Actually, using the area below this plane would give the same result if we treat areas on opposite sides of the neutral axis as algebraic opposites.

    $\bar{x}$ = the distance from the centroid of the area above (or below) the plane in question to the neutral axis

I = *moment of inertia* of the *total* cross section

B = the width of the cross section *at the plane where the horizontal shearing unit stress is desired*

So far we have developed an expression with which the horizontal shearing unit stress can be determined. As mentioned at the beginning of this discussion, there also exists the tendency for adjacent parts to translate *vertically* with respect to each other, due to the shearing force. Therefore, there is also a *vertical shearing unit stress* which must be evaluated. To do this, we will continue to use the slice of the previous discussion and evaluate the shearing unit stress on a vertical edge of a segment of this slice, as shown in Fig. 10–7. An enlarged view of this segment is shown in Fig. 10–8, with the notations defined as follows:

$$F_x = \textit{total shear on plane } XB$$
$$F_y = \textit{total shear on plane } YB$$

Now, taking moments abut c.m., we get

$$F_yX = F_xY \text{ and } F_y = F_x\frac{Y}{X}$$

Then the horizontal shearing unit stress on plane XB is

$$f_v = \frac{F_x}{XB}$$

and the vertical shearing unit stress on plane YB is

$$f_v = \frac{F_y}{YB}$$

And since $F_y = F_x\frac{Y}{X}$ (from above), then

$$\frac{F_y}{YB} = \frac{F_x Y}{X} \cdot \frac{1}{YB} = \frac{F_x}{XB} = f_v$$

which is precisely the same expression that we got for the horizontal shearing unit stress on plane XB.

Therefore,

$$\boxed{f_v = \frac{VQ}{IB}}$$

evaluates both horizontal and vertical shearing unit stress at any point.

Using the shearing unit stress equation is a relatively simple matter if the definitions of the vaious terms involved are remembered. In order to provide some exercise in the use of the equation, lets look at several examples.

Example 10–1 (Figs. 10–9 and 10–10)

To begin with, we'll consider a very simple cross section, as shown in Fig. 10–9, and we will determine the shearing unit stress due to a given shearing force on the section of 15,000#.

$$\therefore V = 15,000\#$$

We can also evaluate the moment of inertia of this section, with respect to the neutral axis:

$$I = \frac{BD^3}{12} = \frac{(12)(20)^3}{12} = 8000 \text{ in.}^4$$

Also, since this is a section of constant width, the width *(B)* is known at any level within the section

$$B = 12''$$

We now have magnitudes for all the terms involved in the shearing unit stress expression except for Q. All the terms we have evaluated to this point are constant for this example, but Q will vary depending on the point in the cross section under consideration.

In order to determine the variation of the shearing unit stress, let's consider the values at 2″ increments from top to bottom and graphically plot the values, as shown in Fig. 10–10. First, we must determine the value for Q at each level. We need only consider the top half of the section, since it is symmetrical and the bottom half will be the same as the top. Remembering that Q is defined as the product of $A\bar{x}$ of everything above the plane in question, with respect to the neutral axis, then

At the top: $Q = A\bar{x} = 0$ (since $A = 0$ above this plane)

2″ down: $Q = A\bar{x} = (2 \times 12)(9'')$
$\qquad = 216 \text{ in.}^3$

4″ down: $Q = A\bar{x} = (4 \times 12)(8'')$
$\qquad = 384 \text{ in.}^3$

6″ down: $Q = A\bar{x} = (6 \times 12)(7'')$
$\qquad = 504 \text{ in.}^3$

8″ down: $Q = A\bar{x} = (8 \times 12)(6'')$
$\qquad = 576 \text{ in.}^3$

At N.A.: $Q = A\bar{x} = (10 \times 12)(5'')$
$\qquad = 600 \text{ in.}^3$

Now, using the values already established for V, I, and B, we can determine the shearing unit stress at these levels and plot them:

FIGURE 10–9

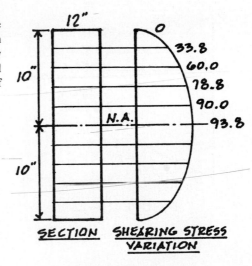

FIGURE 10–10

At the top: $f_v = \dfrac{VQ}{IB} = \dfrac{(15{,}000\#)(0)}{(8000 \text{ in.}^4)(12'')}$

$ = 0$

2" down: $\quad f_v = \dfrac{(15{,}000\#)(216 \text{ in.}^3)}{(8000 \text{ in.}^4)(12'')}$

$ = 33.8 \text{ p.s.i.}$

4" down: $\quad f_v = \dfrac{(15{,}000\#)(384 \text{ in.}^3)}{(8000 \text{ in.}^4)(12'')}$

$ = 60 \text{ p.s.i.}$

6" down: $\quad f_v = \dfrac{(15{,}000\#)(504 \text{ in.}^3)}{(8000 \text{ in.}^4)(12'')}$

$ = 78.8 \text{ p.s.i.}$

8" down: $\quad f_v = \dfrac{(15{,}000\#)(576 \text{ in.}^3)}{(8000 \text{ in.}^4)(12'')}$

$ = 90 \text{ p.s.i.}$

At N.A.: $\quad f_v = \dfrac{(15{,}000\#)(600 \text{ in.}^3)}{(8000 \text{ in.}^4)(12'')}$

$ = 93.8 \text{ p.s.i.}$

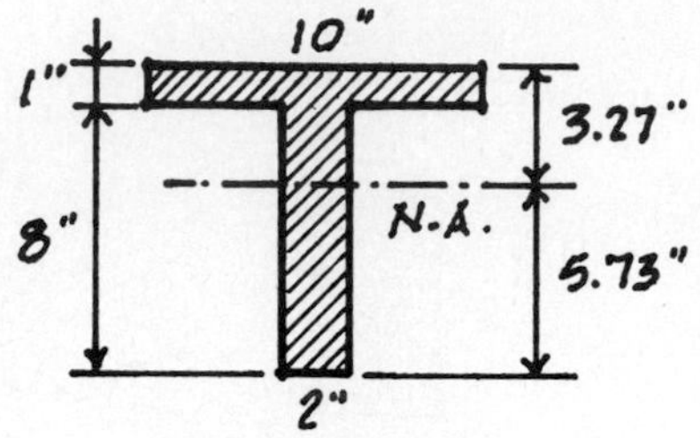

FIGURE 10–11

FIGURE 10–12

Below the neutral axis the shearing unit stress will diminish in intensity. The student can verify this by following the same procedure as above.

Now, if we plot the values determined, as shown in Fig. 10–10, it can be seen that the shearing unit stress is increasing at a decreasing rate, which indicates a parabolic variation. Also, the maximum shearing unit stress within the cross section occurs at the neutral axis.

Several important points should be noted at this time. The section with which we are dealing in this example is symmetrical about the neutral axis, and the maximum shearing unit stress occurs at this point. It can, in fact, be said that in any section with which we would normally be concerned, the maximum shearing unit stress will occur at the neutral axis, whether the section is symmetrical or unsymmetrical. The exceptions to this rule occur in cross-sectional shapes that are not likely to be used for any structural member subjected to bending. The reader can verify this by plotting the shearing unit stress variation for the triangular cross section given on the problem sheet at the end of the chapter.

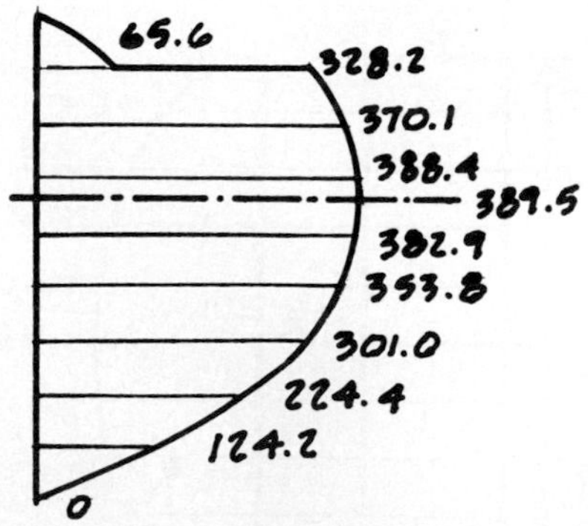

Example 10–2 (Figs. 10–11 and 10–12)

It is desired to make a diagram of the shearing unit stress variation for the cross section of Fig. 10–11, due to a shearing force of 5000#. First, we must determine the location of the neutral axis and the moment of inertia with respect to that axis.

Locating the neutral axis, using the top of the section as the reference:

ΣAx

$\qquad (10'' \times 1'')(.5'') = 5 \text{ in.}^3$

$\qquad (8'' \times 2'')(5'') = 80 \text{ in.}^3$

$\qquad\qquad\qquad \Sigma Ax = 85 \text{ in.}^3$

$\qquad \bar{x} = \dfrac{\Sigma Ax}{A} = \dfrac{85 \text{ in.}^3}{26 \text{ in.}^2} = 3.27 \text{ in.}$

Determining the moment of inertia:

$\qquad \dfrac{(10)(1)^3}{12} + (10 \text{ in.}^2)(2.77'')^2$

$\qquad = 77.6 \text{ in.}^4$

$$\begin{array}{l} \dfrac{(2)(8)^3}{12} + (16 \text{ in.}^2)(1.73'')^2 \\[2mm] = \underline{133.2 \text{ in.}^4} \\[1mm] I = 210.8 \text{ in.}^4 \end{array}$$

Now let's determine the shearing unit stress at 1″ increments from top to bottom, including a computation at the neutral axis. The number of points in the cross section used to plot the variation is somewhat arbitrary. The more computations we make, the more accurate the picture will be.

At the top:

$$f_v = \frac{(5000\#)(0)}{(211 \text{ in.}^4)(10'')}$$
$$= 0$$

At 1″ from the top: At this point there may be some confusion because there are two widths involved. Shall we consider the shearing unit stress at the bottom of the flange, which is 10″ wide? Or shall we compute the shearing unit stress at the top of the stem, which is only 2″ wide? Actually, we must compute both values in order to plot the variation of stress throughout the cross section. Let's make this computation and then discuss this issue further.

For B = 10:

$$f_v = \frac{(5000\#)(10 \text{ in.}^2 \times 2.77'')}{(211 \text{ in.}^4)\,(10'')}$$
$$= 65.6 \text{ p.s.i.}$$

For B = 2″:

$$f_v = \frac{(5000\#)(10 \text{ in.}^2 \times 2.77'')}{(211 \text{ in.}^4)(2'')}$$
$$= 328.2 \text{ p.s.i.}$$

What is happening at this level is that the shearing unit stress, which is the magnitude by which the flange and stem of the cross section tend to slide with respect to each other, is transitioning from a 10″ width to a 2″ width. The total *shearing force* by which these two elements tend to slide is, of course, the same at the plane of contact, whether for the 10″ width or the 2″ width. The *shearing unit stress,* therefore, must in this case be five times as intense when distributed over a 2″ width than when distributed over the 10″ width. The plot of the shearing unit stress distribution, shown in Fig. 10–12, shows this relationship. We'll now proceed with the computations.

At 2″ from the top (B = 2″):

$$f_v = \frac{(5000\#)(10 \times 2.77 + 2 \times 1.77)}{(211)(2'')}$$
$$= 370.1 \text{ p.s.i.}$$

At 3″ from the top (B = 2″):

$$f_v = \frac{(5000\#)(10 \times 2.77 + 4 \times 1.27)}{(211)(2)}$$
$$= 388.4 \text{ p.s.i.}$$

At the neutral axis:

$$f_v = \frac{(5000\#)(10 \times 2.77 + 4.54 \times 1.14)}{(211)(2)}$$
$$= 389.5 \text{ p.s.i.}$$

At 4″ from the top:

$$f_v = \frac{(5000\#)(10 \times 2.77 + 6 \times .77)}{(211)(2)}$$
$$= 382.9 \text{ p.s.i.}$$

At 5" from the top:

$$f_v = \frac{(5000\#)(10 \times 2.77 + 8 \times .27)}{(211)(2)}$$

$$= 353.8 \text{ p.s.i.}$$

At 6" from the top:

$$f_v = \frac{(5000\#)(10 \times 2.77 - 10 \times .23)}{(211)(2)}$$

$$= 301 \text{ p.s.i.}$$

Note: The minus sign in the expression for Q in this computation is necessary because the centroid of this piece of the stem is below the neutral axis.

At 7" from the top:

$$f_v = \frac{(5000\#)(10 \times 2.77 - 12 \times .73)}{(211)(2)}$$

$$= 224.4 \text{ p.s.i.}$$

At 8" from the top:

$$f_v = \frac{(5000\#)(10 \times 2.77 - 14 \times 1.23)}{(211)(2)}$$

$$= 124.2 \text{ p.s.i.}$$

At the bottom:

$$f_v = \frac{(5000\#)(10 \times 2.77 - 16 \times 1.73)}{(211)(2)}$$

$$= 0$$

The shearing unit stress at the outer fiber (top and bottom) must be zero. The plot of the shearing unit stress variation is shown in Fig. 10–12. It can be seen in this diagram that the shearing unit stress increases greatly in intensity from the 10" width to the 2" width, and that it is a maximum at the neutral axis. One of the questions that arises at this point is that of the physical significance of the shearing unit stress and its high intensity between two elements that make up a cross section such as that of this example problem. If the section is made of one piece of material, then the only issue involved is whether or not the material is capable of resisting the maximum shearing unit stress, which occurs at the neutral axis. If, however, the section is made of two rectangular pieces that must be joined, then not only must the material quality be examined for capability of resisting the maximum shearing unit stress, but the method of joining the two (or more) pieces must also be given careful consideration. The horizontal shearing unit stress at the junction indicates that the pieces will have the tendency to slide with respect to each other. For the section to behave as an integral structural unit, this tendency must be resisted and the pieces not allowed to separate from each other. It is this idea that forms the basis for the following discussion.

CONNECTORS TO RESIST HORIZONTAL SHEAR

In order to join two (or more) pieces of a structural section so they will behave compatibly, it is necessary to determine the *total shearing force* that occurs at the junction between them. For example, consider the beam of Fig. 10–13, which has a cross section as shown (same as Fig. 10–11). The location of the neutral axis and the moment of inertia for this section have been established in Example 10–2; this information is given in Fig. 10–13.

Now the shearing unit stress at the junction of the two pieces will be determined:

$$f_v = \frac{VQ}{IB} = \frac{(1000\#)(10" \times 2.77")}{(211 \text{ in.}^4)(2")}$$

$$= 65.6 \text{ p.s.i.}$$

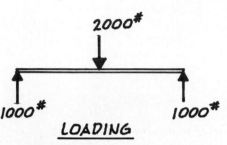

2000#

1000#   1000#

LOADING

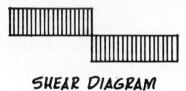

SHEAR DIAGRAM

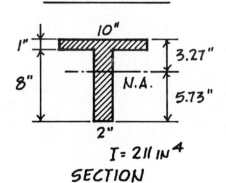

$I = 211 \text{ in}^4$

SECTION

FIGURE 10–13

There are two widths at this junction, but only the 2″ width is used in the computation. It should be obvious that the tendency to slide occurs at the plane of contact between the two pieces, which is 2″ wide, and it is this value of shearing unit stress with which we are concerned.

Let's consider now that this section is made of two pieces of lumber. How shall we join them so that the joint will resist a shearing unit stress of 65.6 p.s.i.? One possible solution would be to glue the two pieces together. This would require a glue of sufficient strength to resist the shearing unit stress of 65.6 p.s.i. The strength of the glue would have to be verified by published data or by testing, which would necessitate access to appropriate testing equipment. Another approach to this sort of problem is to use screws to connect the pieces. Oftentimes a combination of glue and screws (or nails) is used to assure an absolute bond.

When using screws to produce an integral structural unit, the shearing strength of the screw must be known and the spacing of the screws (consequently, the number required) must be determined. The strength of the screw can be determined from actual testing or from published data pertaining to the grade of material of which the connector is made. The material should be a good quality of steel. For the sake of this discussion, the safe shearing strength has been determined, by whatever means, to be 250# each. The problem now is to determine the spacing of these connectors. For the section and loading of Fig. 10–13 this is a reasonably simple matter. The *total* shearing force on the contact plane between the two pieces to be joined is the product of the *shearing unit stress* (f_v) and the area of contact between the pieces involved. The area of contact is the product of the width of the contact plane (B) and the spacing between connectors, which we will call s. Expressed mathematically, the above statement becomes

$$P_{conn} = sBf_v$$

where P_{conn} = the *total* shearing force the connector can resist
  s = spacing between connectors
  B = width of the contact plane
  f_v = shearing unit stress at the plane of contact

Applying the above expression and solving for the required spacing, we get:

$$s = \frac{P_{conn}}{Bf_v} = \frac{250\#}{(2″)(65.6\#/in.^2)} = 1.9″$$

In order for the section to behave as an integral structural unit, this is the maximum spacing that can be used. In this particular example the spacing of the connectors is constant because the value of total shear is constant.

In Fig. 10–14, a simply supported beam with a uniform load is shown, and in this case the total shear varies from a maximum at the ends to zero at the midspan. It should be recognized here that the spacing of the screws used can be increased as we move from the point of maximum shear (the ends) to zero shear. Let's now determine the spacing required:

By inspection, the neutral axis is at the midheight of the section.

Determine I

$$2\left[\frac{(10)(1)^3}{12} + 10(4.5)^2\right] = 406.7 \text{ in.}^4$$

$$\frac{(2)(8)^3}{12} = \quad \frac{85.3 \text{ in.}^4}{492.0 \text{ in.}^4}$$

Now we can determine the shearing unit stress at the joint. There are two joints involved, but because of symmetry, the shearing unit stress will

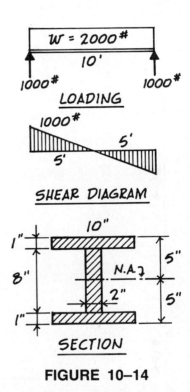

W = 2000#
10′
1000#   1000#

LOADING

1000#
5′
5′

SHEAR DIAGRAM

10″
1″
8″   N.A.?   5″
2″   5″
1″

SECTION

FIGURE 10–14

be the same at each. We'll first evaluate the shearing unit stress at the point of maximum shear, which is at the support:

$$f_v = \frac{VQ}{IB} = \frac{(1000)(10 \times 4.5)}{(492)(2)} = 45.7 \text{ p.s.i.}$$

The spacing at this point can be determined by the expression previously developed. Using screws with a shearing strength of 250# each, the spacing is

$$s = \frac{P_{conn}}{Bf_v} = \frac{250}{(2)(45.7)} = 2.74''$$

This is the spacing required for the amount of shear at the support. However, as we move in from the support, the shear and consequently the shearing unit stress are decreasing linearly. Therefore, the spacing may be increased. The easiest way to decide on the spacing is to compute the spacing required at regular intervals from the support—say, 1′ intervals. The spacing will increase as the reciprocal of the ratio by which the shear is decreasing. For example, the spacing at the support was determined to be 2.74″. At 1′ in from the support the shear is four fifths of the maximum shear that produced the 2.74″ spacing. Therefore, the spacing (s) is

$$\frac{5}{4}(2.74) = 3.43'' \text{ at } 1'$$

At 2′:

$$s = \frac{5}{3}(2.74) = 4.57''$$

At 3′:

$$s = \frac{5}{2}(2.74) = 6.85''$$

At 4′:

$$s = \frac{5}{1}(2.74) = 13.70''$$

At 5′:

$$s = \text{infinity}$$

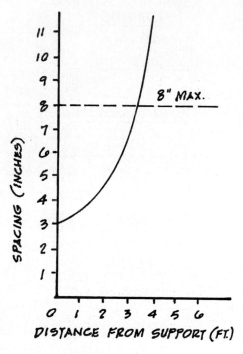

FIGURE 10–15

With the spacing determined at these intervals, one can use this information to make judgments about the spacing that, in fact, should be used. The word *judgment* can hardly be overemphasized at this point. For example, note that the value for the spacing at the 4′ mark is larger than the intervals being considered. Considering this problem in a practical way, it is advisable to set a maximum spacing, based on the judgment of the designer. For this example a maximum spacing for the screws might be set at, say, 8″. A further aid in deciding the spacing between the intervals considered would be the drawing of a *spacing curve*. This consists of a graphic plot of the values established; that is, a plot of distance from the support versus spacing. This is shown in Fig. 10–15. The maximum spacing desired can be plotted on this graph, thereby giving the designer a clear picture as to how the spacing should vary.

Moreover, even with the spacing curve, it is unlikely that any two designers would interpolate the spacing between the intervals considered in precisely the same way. This is not terribly important. What may be of greater importance than the precise spacing, however, is that the total number of connectors used be correct. This can easily be determined by considering the total shearing force over the length of the beam and making sure that the total number of connectors are provided, spaced in reasonable accordance with the information derived from the spacing curve. For the example under consideration, this can be determined for half of the beam because of symmetry. A shearing unit stress diagram showing the values at the joint can be drawn, as shown in Fig. 10–16. This will look just like the shear diagram because the shearing unit stress varies as the total shear varies. The total shearing force along the plane of contact between the pieces, over half the length of the beam, is the product of the area of the shearing unit stress diagram and the width of the contact plane:

$$F = (45.7 \ \#/in.^2)(2'')(60'')\left(\frac{1}{2}\right) = 2742\#$$

The number of connectors to be provided is

$$N = \frac{F}{P_{conn}} = \frac{2742\#}{250\#} = 11$$

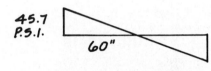

SHEARING UNIT STRESS

FIGURE 10–16

This is the number required for each contact plane for half the length of the beam. Therefore, this beam would require a minimum of 44 connectors, 22 on the top and 22 on the bottom. With this information and that from the spacing curve, it seems that a suitable arrangement might be (from the support toward the center)

$$4 \ @ \ 3'', \ 2 \ @ \ 5'', \ 2 \ @ \ 6'', \ 3 \ @ \ 8''$$

As mentioned previously, this kind of arrangement will vary slightly based on the judgment of the designer. However, note that with all the information determined there isn't much room for great variations.

MATERIALS AND SHEAR

It might be appropriate here to point out the relationship between shearing unit stress and the common structural materials, namely, wood and steel.

Wood is the most critical of the common structural materials for shearing unit stress—specifically, horizontal shearing unit stress. If one can visualize a board used as a beam, as shown in Fig.10–17, it will be recognized that the grain generally runs parallel to the span. The grain, in essence, identifies weak horizontal planes. This produces very low resistance to the tendency for horizontal slices to slide with respect to each other, when the member is subjected to bending. The resistance to horizontal shear varies slightly depending on the species of wood being used, but in general, the *allowable* horizontal shearing unit stress will range from about 60 p.s.i. to 110 p.s.i. While there are a few kinds of wood with an allowable horizontal shearing unit stress over 110 p.s.i., the great majority have allowables lower than this.* To place this value in proper perspective, compare it to an *allowable* shearing unit stress of 14400 p.s.i. in the most commonly used, and lowest grade, of structural steel.

Because of the low allowable shearing stress for wood, this consideration often governs the design of a beam. In any case, an investigation should always be made comparing the actual shearing unit stress, in a wood beam, to the allowable shearing unit stress.

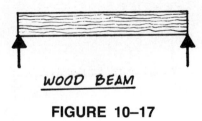

WOOD BEAM

FIGURE 10–17

*For detailed information on allowable stresses, see American Institute of Timber Construction, *Timber Construction Manual*. John Wiley and Sons, Inc. Current edition.

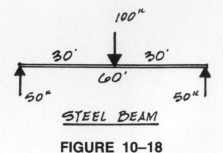

STEEL BEAM

FIGURE 10–18

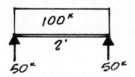

When designing in steel, shear is generally not a problem, when a member is designed based on bending and when load and span conditions are reasonably normal. Problems can arise, however, when selecting steel members for very heavy loads and very short spans, or on normal spans, when a heavy load is placed very close to the support. To illustrate these conditions, consider the span and loading conditions shown in Fig. 10–18. The normal procedure involved in the design of a beam is to size it to resist the maximum bending moment. The maximum bending moment in this case is 1500 ft kips. Without question, the member selected to resist this magnitude of bending moment will easily resist the maximum shear of 50 kips.

Now, to dramatize the problems that may arise when designing a steel section, consider the same load used previously, but now on a very short span, as shown in Fig. 10–19. In this case, the bending moment is 50 ft kips, which would lead to the selection of a considerably smaller member than the 1500 ft kip moment. But the maximum shear is still 50^K. In such a case, it is likely that the small member selected based on the maximum bending moment will not be able to resist the shear. In any case, the selected member should be investigated for shearing stress.

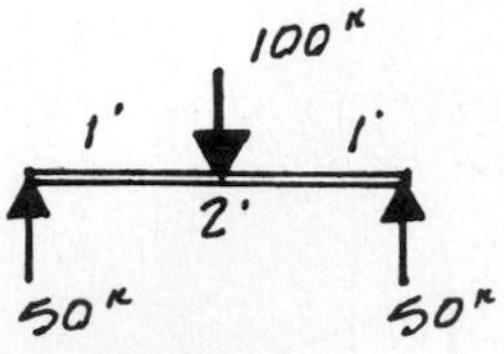

FIGURE 10–19

The loading pattern may also play an important role in design procedures. For example, suppose the point load shown in Fig. 10–19 is now spread out as a uniformly distributed load, as shown in Fig. 10–20. The maximum bending moment now is $25'^K$, but the maximum shear is still 50^K. It is likely that shear will govern the design of the member in this case, and an investigation for both bending and shear should be made.

Another problem concerning shear may arise with normal spans that have an unusual placement of a heavy load, as shown in Fig. 10–21. In this case the maximum bending moment is $190'^K$ and the maximum shear is 95^K. In cases such as this, the design procedure should include consideration of both bending and shear.

The extreme cases shown to illustrate the conditions that may cause shearing problems in steel beams are unusual, though possible. Generally, such problems will not be encountered in buildings such as schools, office buildings, etc., but it is likely that extreme conditions may be encountered in heavy industrial facilities, where extremely heavy equipment is supported. The designer of a steel structure must be wary of these unusual conditions.

To this point, we have discussed two of the major structural materials, steel and wood. Reinforced concrete, of course, is the third major structural material in use today. Shear produces problems that must be given strong consideration in the design of reinforced concrete beams. However, since a reinforced concrete memeber is not homogeneous and the concrete is not really an elastic material, the fundamental principles involved are quite different from those discussed in this chapter; in fact, they are beyond the scope of this book.

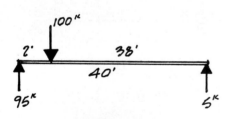

FIGURE 10–20

FIGURE 10–21

PROBLEM 10–1

SHEARING UNIT STRESS DISTRIBUTION

$$f_v = \frac{VQ}{IB} \quad \text{IN WHICH:}$$

f_v = SHEARING UNIT STRESS (HORIZONTAL AND VERTICAL) <u>AT ANY FIBER</u>

V = TOTAL SHEAR ON SECTION

Q = STATIC MOMENT ($A\bar{x}$) WITH RESPECT TO THE NEUTRAL AXIS FOR THE AREA ABOVE OR BELOW THE LEVEL AT WHICH f_v IS DESIRED.

I = MOMENT OF INERTIA OF THE CROSS SECTION WITH RESPECT TO THE NEUTRAL AXIS.

B = WIDTH OF THE SECTION AT POSITION f_v IS DESIRED.

● DETERMINE f_v (P.S.I.) AT 2" INTERVALS (AND AT THE NEUTRAL AXIS) FOR EACH SECTION. PLOT VALUES AND DRAW DISTRIBUTION DIAGRAM.

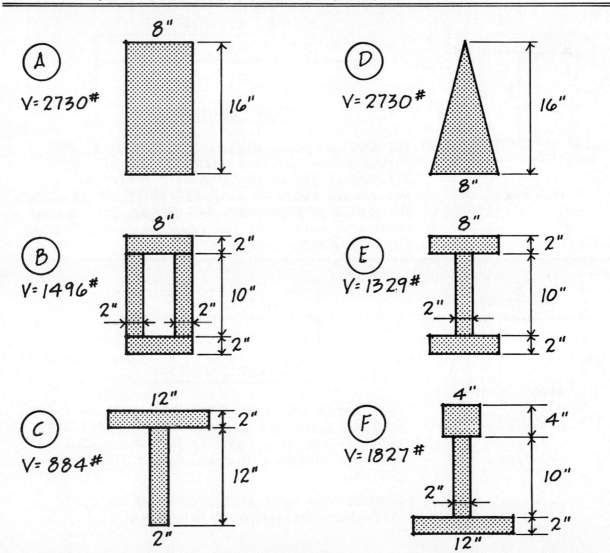

(A)　V= 2730#　　8"　16"

(D)　V= 2730#　　8"　16"

(B)　V= 1496#　　8"　2"　10"　2"　2"　2"

(E)　V= 1329#　　8"　2"　10"　2"　2"

(C)　V= 884#　　12"　2"　12"　2"

(F)　V= 1827#　　4"　4"　2"　10"　2"　12"

PROBLEM 10–2

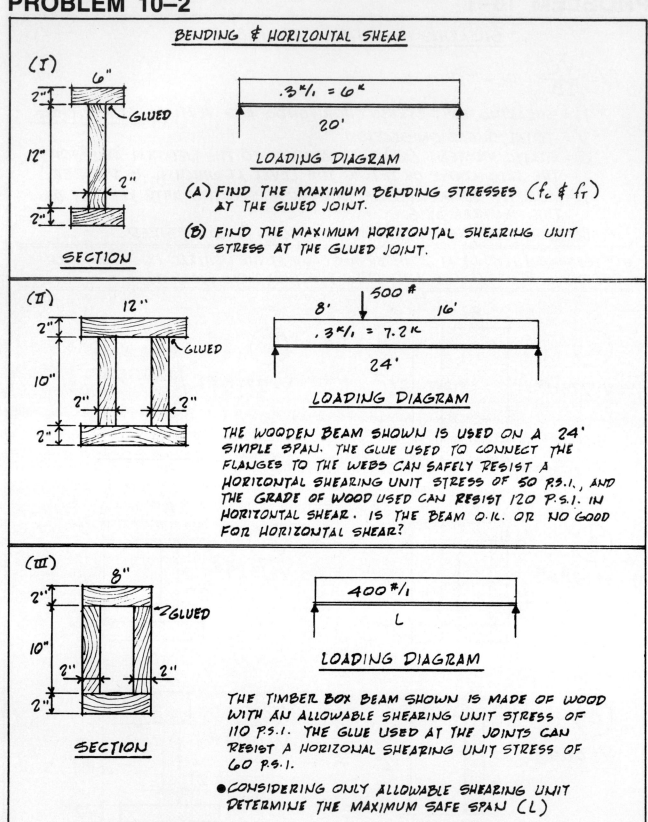

BENDING & HORIZONTAL SHEAR

(I)

SECTION

$.3^k/_1 = 6^k$

20'

LOADING DIAGRAM

(A) FIND THE MAXIMUM BENDING STRESSES $(f_c \& f_T)$ AT THE GLUED JOINT.

(B) FIND THE MAXIMUM HORIZONTAL SHEARING UNIT STRESS AT THE GLUED JOINT.

(II)

500#

8'   16'

$.3^k/_1 = 7.2^k$

24'

LOADING DIAGRAM

THE WOODEN BEAM SHOWN IS USED ON A 24' SIMPLE SPAN. THE GLUE USED TO CONNECT THE FLANGES TO THE WEBS CAN SAFELY RESIST A HORIZONTAL SHEARING UNIT STRESS OF 50 P.S.I., AND THE GRADE OF WOOD USED CAN RESIST 120 P.S.I. IN HORIZONTAL SHEAR. IS THE BEAM O.K. OR NO GOOD FOR HORIZONTAL SHEAR?

(III)

SECTION

400#/1

L

LOADING DIAGRAM

THE TIMBER BOX BEAM SHOWN IS MADE OF WOOD WITH AN ALLOWABLE SHEARING UNIT STRESS OF 110 P.S.I. THE GLUE USED AT THE JOINTS CAN RESIST A HORIZONAL SHEARING UNIT STRESS OF 60 P.S.I.

● CONSIDERING ONLY ALLOWABLE SHEARING UNIT DETERMINE THE MAXIMUM SAFE SPAN (L)

PROBLEM 10–3

<u>BENDING & HORIZONTAL SHEAR</u>

(I)

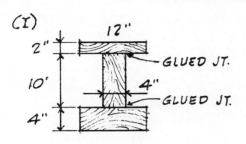

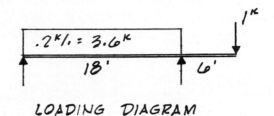

LOADING DIAGRAM

SECTION

●THE WOOD BEAM SHOWN IS USED FOR THE GIVEN LOADING. THE GLUE CAN SAFELY RESIST A HORIZONTAL SHEARING UNIT STRESS OF 75 P.S.I., AND THE WOOD CAN RESIST 90 P.S.I. BASED ON SHEARING UNIT STRESS, IS THE BEAM O.K. OR NO GOOD?

(II)

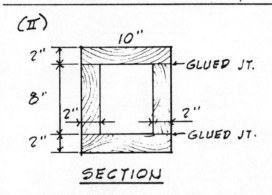

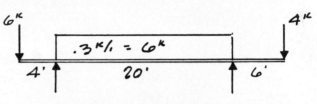

LOADING DIAGRAM

SECTION

ALLOWABLE f_v (GLUE) = 140 P.S.I.
ALLOWABLE f_v (WOOD) = 150 P.S.I.
ALLOWABLE f (BENDING) = 1500 P.S.I.

●THE GIVEN SECTION IS USED AS A DOUBLE CANTILEVER, AS SHOWN. IS THE BEAM O.K. OR NO GOOD FOR THE GIVEN LOADING?

(III)

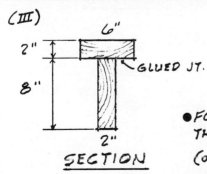

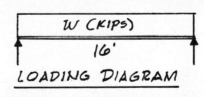

W (KIPS)

16'

LOADING DIAGRAM

SECTION

●FOR THE TIMBER "T" SECTION SHOWN, DETERMINE THE MAXIMUM PERMISSIBLE UNIFORM LOAD (W):

(a) BASED ON ALLOWABLE BENDING STRESS = 1800 P.S.I.

(b) BASED ON AN ALLOWABLE SHEARING UNIT STRESS AT THE GLUED JOINT OF 100 P.S.I.

PROBLEM 10–4

BENDING & HORIZONTAL SHEAR

● FOR EACH OF THE FOLLOWING, DETERMINE THE MAXIMUM STRESS DUE TO BENDING AND THE MAXIMUM STRESS DUE TO SHEAR. CONSIDER ALL STRESSES TO BE BELOW THE ELASTIC LIMIT OF THE MATERIAL.

Ⓐ

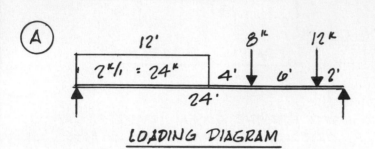

LOADING DIAGRAM

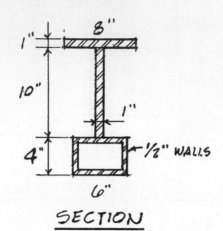

SECTION

Ⓑ

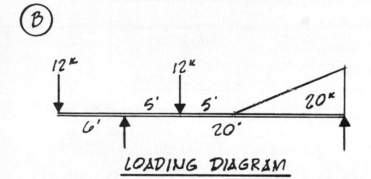

LOADING DIAGRAM

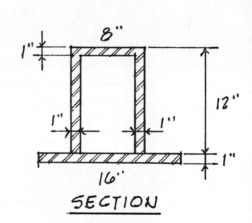

SECTION

Ⓒ

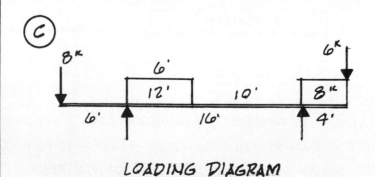

LOADING DIAGRAM

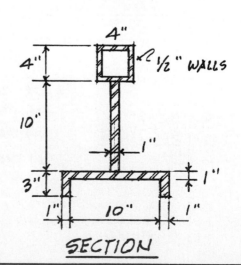

SECTION

PROBLEM 10-5

HORIZONTAL SHEAR - CONNECTORS

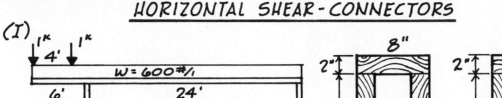

(I)

SECTION THRU BUILDING

SECTION AT CANTILEVER

SECTION AT MAIN SPAN

(A) DETERMINE THE MAXIMUM BENDING STRESS IN THE BEAM SHOWN.

(B) DETERMINE THE SPACING OF CONNECTORS REQUIRED TO SECURE THE FLANGE TO THE WEB. EACH CONNECTOR HAS A SHEARING RESISTANCE OF 250#.

(II) FOR EACH OF THE FOLLOWING LOADINGS, AND THE GIVEN SECTION, DETERMINE THE NUMBER OF CONNECTORS REQUIRED, WITHOUT CONSIDERATION OF SPACING, BETWEEN FLANGE AND WEB. EACH CONNECTOR HAS A SHEARING RESISTANCE OF 200#

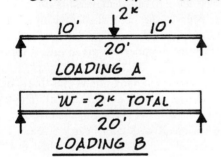

LOADING A

LOADING B

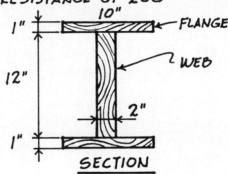

SECTION

(III)

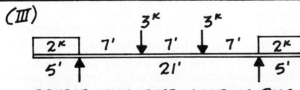

DOUBLE CANTILEVER - LOADING DIAG.

THE GIVEN SECTION IS USED AS A DOUBLE CANTILEVER, AS SHOWN.

● DETERMINE THE NUMBER OF CONNECTORS REQUIRED TO CONNECT THE FLANGES TO THE WEB. EACH CONNECTOR HAS A SHEARING RESISTANCE OF 200#.

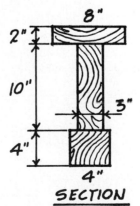

SECTION

11

Design of Wood Beams

INTRODUCTION

Until now we have primarily been concerned with the development of necessary background material and the derivation of various formulas. The application of this material has been in the context of investigation (or analysis) problems. An investigation problem is one in which the cross-sectional dimensions of a structural member are given and a variety of information—such as maximum stresses, load carrying capacity, etc.—may be determined. Now that the necessary background has been developed, we are ready to deal with the *design* of structural members. The typical design problem is one in which the cross-sectional dimensions of a structural member must be determined so that it will safely carry a load. The design process is, of course, the practical and realistic issue when one creates a building.

In this chapter we will limit ourselves to the design of beams made of wood. This will include the design of beams made of solid sawn lumber as well as those made of laminated timber. In subsequent chapters we will deal with the design of beams made of structural steel and reinforced concrete.

It was suggested in the preface, and it must be emphasized at this point, that analysis and design procedures in any of the common structural materials are, to a large degree, affected by requirements of building codes and industry specifications. Such requirements are often empirical in nature, and while the rules presented in codes and specifications are important in the long run, such concerns are considered to be beyond the scope of a book on principles. Consequently, in this chapter and in subsequent chapters when design procedures are presented, the design process should be regarded as rudimentary, with minimal (if any) specific references to building codes or specifications. In other words, emphasis will be placed on the design of structural members, with the application of the fundamental scientific facts that have been presented in this book to this point.

DESIGN CRITERIA

Essentially, allowable bending stresses and shearing stresses are the basis for the design of sawn lumber and laminated timber beams. The usual approach in the design of wood beams is first to determine the size of the member required, based on the allowable bending stress for the grade of wood being used.* However, since wood is a material that is weak in its resistance to horizontal shear, the member selected based on bending stresses must be analyzed to be certain that the maximum shearing unit stress is less than, or equal to, the stipulated allowable.

Aside from our concern for allowable bending stress and shearing stress, another concern of major importance in the design of beams is that of deflection. Deflection is the amount of sag a beam will experience under a given load, as shown (exaggerated) in Fig. 11-1. A beam supporting a load may be stressed within the stipulated allowables for bending and shear, but it may lack sufficient stiffness and hence deflect an intolerable amount to be acceptable for its intended function. The issue of deflection and the techniques employed for predicting the amount of deflection that would occur in a particular situation will be dealt with in some detail in Chapter 16. Let it suffice for now to say that the potential problems created by excessive deflection can be quite serious and, in the final analysis, must be a concern in the design process.

Another issue that's worthy of mention here is that of the potential crushing of the grain in a wood beam where it bears on the support, as shown in Fig. 11-2. The kind of stress produced here is referred to as "compression perpendicular to the grain." The allowable stress for this is quite different from the allowables for bending or shear in a particular grade of wood. As suggested by Fig. 11-2, the important concern is that the contact area between the beam and its support be sufficient so that the beam reaction is distributed and the intensity of stress is thereby kept within the stipulated allowables. Since this is largely a matter of proper detailing rather than design, we will not deal with this issue in the numerical examples. The student, however, should be aware that compression perpendicular to the grain can be a problem and, in the final analysis, details of the support conditions should be carefully examined.

DESIGN OF WOOD BEAMS

In view of our primary concern in this chapter, which is the design of beams based on allowable bending stresses and shearing stresses, there are several bits of information that will now be developed. These will prove to be useful, if not necessary, in the design process.

In Chapter 7, the bending stress equation was developed, which is

$$M = \frac{f}{c}I$$

In the design process, it is the cross-sectional properties that must be determined, as shown in Fig. 11-3. We will limit ourselves to the design of

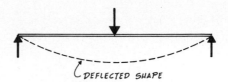

FIGURE 11–1

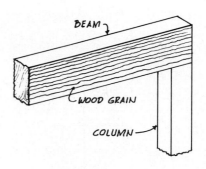

FIGURE 11–2

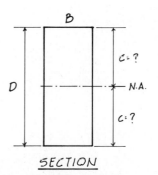

SECTION

FIGURE 11–3

---

* Allowable stresses for all grades of structural lumber and laminated timber may be found in the current edition of the *Timber Construction Manual* prepared by The American Institute of Timber Construction. Allowable stresses may be altered by a "size factor." See this reference for a complete discussion.

rectangular sections since, in wood, this is the most common situation. Consequently, the determination of the width and depth of a section will be our concern. These dimensions must be such that, for a given bending moment (which is a function of the span of the beam and the load being supported), the allowable bending stresses in tension and compression will not be exceeded. This, in essence, means that the moment of inertia (I), and the distance from the neutral axis to the outermost fiber (c) must be determined. Since I and c are dependent upon each other, this presents the problem of determining two unknown factors in the design process. To deal with this dilemma, we will now introduce a factor known as the *section modulus* (S) which is simply equal to the ratio of I/c. Therefore, the bending stress equation may now be rewritten in a more convenient form for the design process, as follows:

$$M = \frac{fI}{c} \text{ or } M = fS$$

The required section modulus may be determined from this form of the equation in order to satisfy the maximum bending moment and the allowable bending stress of the wood being used. Data Sheets D–20, D–21, and D–22, in the Appendix, contain the values of the section moduli for a variety of cross-sectional dimensions in both sawn lumber and laminated timber. The proper use of the Data Sheets will be discussed in some detail as we get ready to look at some example problems.

Another convenient bit of information to have is a general expression for the section modulus (S) of a solid rectangular section, which is the most common section used when designing in wood. Since we know, by definition, that

$$S = I/c$$

and (referring to Fig. 11-4)

$$I = \frac{BD^3}{12} \text{ and } c = D/2$$

then

$$S = \frac{I}{c} = \frac{\dfrac{BD^3}{12}}{D/2} = \frac{BD^2}{6}$$

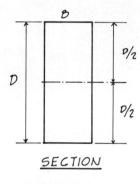

SECTION

FIGURE 11–4

This expression is the basis for the section moduli given on the Data Sheets; its use will prove to be convenient for applications in subsequent chapters. It must be emphasized that the derivation of this expression was based on a *solid rectangular section*. This, however, is not too limiting when dealing with wood. One more piece of information that will prove to be most convenient is a general expression for the maximum shearing unit stress in a *solid rectangular section*. It was shown in Chapter 10 that the general expression for shearing unit stress in any cross section is

$$f_v = \frac{VQ}{IB}$$

Furthermore, it was mentioned, also in Chapter 10, that the maximum shearing unit stress in the sections with which we are normally concerned will occur at the neutral axis. In the design process, we are normally concerned only with the evaluation of the maximum shearing unit stress. In the solid rectangular section shown in Fig. 11-4, we can evaluate Q as follows:

$Q = A\bar{x}$ of everything above (or below) the plane in question with respect to the neutral axis. Since the plane in question *is* the neutral axis, then:

$$Q = \left[(B)\left(\frac{D}{2}\right) \right]\left(\frac{D}{4}\right) = \frac{BD^2}{8} \quad \text{and}$$

$$I = \frac{BD^3}{12}$$

$$\therefore f_v = \frac{VQ}{IB} = \frac{V\left[\dfrac{BD^2}{8}\right]}{\left[\dfrac{BD^3}{12}\right]B} = 1.5\frac{V}{BD}$$

and since the product of BD is simply the total cross-sectional area, then

$$f_v = 1.5\frac{V}{A}$$

This expression provides a very convenient way to determine the maximum shearing unit stresses in wood beams, the majority of which are solid rectangles in cross section. For sections other than solid rectangles, the basic expression

$$f_v = \frac{VQ}{IB}$$

must be used.

We're now ready to look at a few example design problems, which will necessitate the use of information on the properties of sawn lumber and laminated timber sections. Data Sheet D–22 in the Appendix gives the properties for sawn lumber sections. It should be noted that the actual size of a sawn lumber section is not the same as the nominal size. For example, a nominal $2'' \times 4''$ section actually measures $1\frac{1}{2}'' \times 3\frac{1}{2}''$. While the difference between the actual and nominal dimensions seems slight, there is a considerable difference in the moment of inertia and, consequently, the section modulus. All data given on the Data Sheets is based on the actual dimensions of the cross sections.

Data Sheets D–20 and D–21 in the Appendix give the properties for laminated timber sections. Laminated timber sections are made from sawn lumber. Fig. 11-5 shows a laminated timber section that is made up of $2'' \times 6''$ pieces. It should be noted that the $6''$ nominal dimension is actually $5\frac{1}{2}''$, yet in the figure, the width of the section is shown as $5\frac{1}{8}''$. The reason for this is that the sides of the beam must be planed and sanded in order to remove glue that was squeezed out during the laminating process and generally to develop a high-quality surface. The depth of a laminated timber beam will generally be some increment of $1\frac{1}{2}''$, which is the actual thickness of any "2 by." It is possible, when necessary, to have a 3/4'' depth increment. This would be based on the actual thickness of a "1 by."

When designing a laminated timber beam, theoretically, we can develop any depth we choose. However, there are certain practical limitations that must be considered. For example, can you imagine a section $3\frac{1}{8}''$ wide by $90''$ deep, as shown in Fig. 11-6? This would be absurd, since there would be extraordinary problems with the lateral stability of such a tall, thin beam. On the other hand, imagine a very wide and shallow section, as shown in Fig. 11-7. The fundamental concepts of moment of inertia tell us that by no means is this an efficient arrangement of material in a structural section.

LAMINATED TIMBER CROSS SECTION

FIGURE 11–5

FIGURE 11–6

LAM. TIMBER SECTION

FIGURE 11–7

Generally, a well-proportioned cross section for a laminated timber beam will have a width-to-depth ratio of about 1:3 to about 1:5. The data given in the Appendix, on Data Sheets D–20 and D–21, starts and stops at roughly these porportions. Data for depths not shown may be determined simply by adding $1\frac{1}{2}''$ increments.

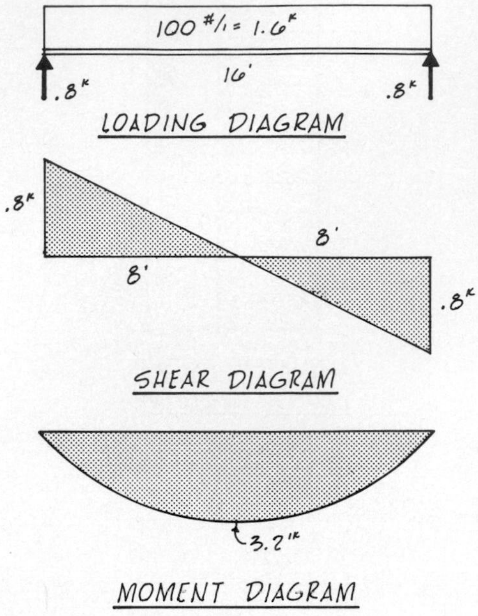

Example 11–1

It is desired to design a sawn lumber section to satisfy the load and span conditions shown in Fig. 11-8. As mentioned earlier in this chapter, we will limit our concerns to the design of beams based on allowable bending stresses and shearing stresses. Therefore, for the problem being discussed, we must first determine the maximum bending moment and select a member that will be satisfactory for the given allowable bending stresses. When this is done, we will use the selected member and investigate it to make sure the allowable shearing stresses are not exceeded. For the purpose of this problem, let's say we're dealing with a grade of wood with the following allowable stresses:

$$\text{Allowable bending } f = 1300 \text{ p.s.i.}$$
$$\text{Allowable shear } f_v = 80 \text{ p.s.i.}$$

In order to size the member for bending, we must first determine the section modulus (S) required to satisfy the maximum bending moment, as discussed earlier in this chapter. The maximum bending moment for this case may be found by drawing the shear and bending moment diagrams. These diagrams, and maximum values, are shown in Fig. 11-8. The maximum bending moment for this case is $3.2'^K$, and the required section modulus based on the given allowable bending stress is

$$S = \frac{M}{f} = \frac{(3.2'^K)(12''/')}{1.3^K/\text{in.}^2} = 29.54 \text{ in.}^3$$

In scanning Data Sheet D–22, it can be determined that a nominal $2'' \times 12''$ sawn lumber section is the most economical (least cross-sectional area) member that will satisfy the bending moment requirement. We will now perform an investigation to determine if this nominal $2'' \times 12''$ will be satisfactory for the given allowable shearing unit stress. It was previously established that the general expression for the maximum shearing unit stress for a solid rectangular section is

$$f_v = 1.5\frac{V}{A}$$

where V = maximum shear in the section (determined from the shear diagram)
A = cross-sectional area

Therefore, considering the $2'' \times 12''$ that was selected based on bending and using the actual cross-sectional area, as given on Data Sheet D–22,

$$f_v = 1.5\frac{V}{A} = 1.5\frac{(800\#)}{16.88 \text{ in.}^2} = 71.1 \text{ p.s.i.}$$

This is less than the given allowable for shearing unit stress; therefore, the nominal $2'' \times 12''$ is okay and the design problem is complete.

Example 11–2

In this example we will design a sawn lumber section to satisfy the load and span conditions shown in Fig. 11–9. The corresponding shear and bending moment diagrams are also shown, with information necessary for the design process. For this problem, we will use a grade of wood with the following allowable stresses:

$$\text{Allowable bending } f = 1400 \text{ p.s.i.}$$
$$\text{Allowable shear } f_v = 90 \text{ p.s.i.}$$

We will first size the member for bending by determining the required section modulus. Referring to the bending moment diagram shown in Fig. 11–9, the maximum bending moment is $2.4'^K$, and

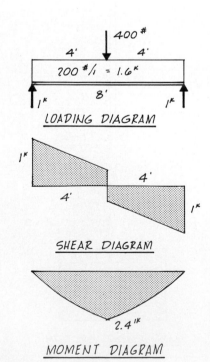

LOADING DIAGRAM

SHEAR DIAGRAM

MOMENT DIAGRAM

FIGURE 11–8

FIGURE 11–9

$$S = \frac{M}{f} = \frac{(2.4'^K)(12''/')}{1.4^K/in.^2} = 20.6 \ in.^3$$

Referring to Data Sheet D–22, a nominal $2'' \times 10''$ is the most economical section that will satisfy the requirement for bending. We must now investigate this section to see if it will be satisfactory for shearing unit stress. The maximum shear is 1000#, as shown on the shear diagram in Fig. 11–9, and the actual cross-sectional area of a $2'' \times 10''$ is 13.88 in.^{2}. Consequently,

$$f_v = 1.5\frac{V}{A} = \frac{1.5(1000\#)}{13.88 \ in.^2} = 108.1 \ p.s.i.$$

This actual shearing stress exceeds the given allowable of 90 p.s.i. Therefore, this beam, which was sized for bending, must be resized to satisfy the stipulated allowable shearing unit stress. The simplest way to do this is to use the shearing stress equation for a solid rectangular section and solve for the required area based on the given allowable shearing unit stress:

$$A = \frac{1.5V}{f_v} = \frac{1.5(1000\#)}{90 \ p.s.i.} = 16.67 \ in.^2$$

In scanning the information on Data Sheet D–22, it can be determined that the most economical section that will satisfy both bending and shearing stress allowables is a nominal $2'' \times 12''$.

Example 11–3

In this example we will design a *laminated timber* beam to safely carry the load on the single cantilever shown in Fig. 11–10. The design will be based on the following allowables:

Allowable bending $f = 2000$ p.s.i.
Allowable shear $f_v = 160$ p.s.i.

Assuming that the cross section will be constant throughout the entire length of the member, we must determine the critical design moment, since there is a negative moment as well as a positive moment. The shear and bending moment diagrams shown in Fig. 11–10 provide us with the information needed for design. The maximum bending moment is the negative moment of 48$'^K$. Based on the given allowable bending stress, the required section modulus is

$$S = \frac{M}{f} = \frac{(48'^K \times 12''/')}{2 \ k.s.i.} = 288 \ in.^3$$

In scanning Data Sheets D–20 and D–21, it will be found that a $5\frac{1}{8}'' \times 19\frac{1}{2}''$ laminated timber section will do the job. We must now check this section to see if it is satisfactory for shear. The maximum shear (V) form the shear diagram is 10^K. Therefore, the shearing unit stress for this solid rectangular section is

$$f_v = 1.5\frac{V}{A} = \frac{1.5(10,000\#)}{99.9 \ in.^2} = 150.2 \ p.s.i.$$

This is less than the given allowable shearing unit stress of 160 p.s.i.; therefore, the design (based on bending and shear) is complete.

THE FRAMING PLAN

The most common floor-framing arrangement in wood construction consists of closely spaced joists supported by either bearing walls, as shown in Fig. 11–11, or beams, as shown in Fig. 11–12. Depending on the magnitude of the loads being delivered and the spans involved, the beams shown in Fig. 11–12 may be of sawn lumber or laminated timber. Actually, when the magnitude of the load and span is such that a sawn lumber member is not

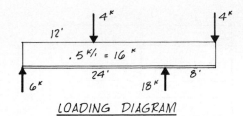

LOADING DIAGRAM

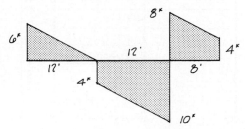

SHEAR DIAGRAM

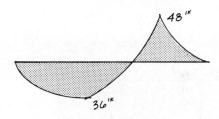

MOMENT DIAGRAM

FIGURE 11–10

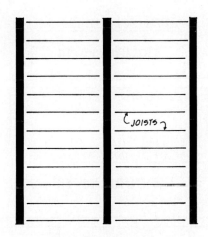

FRAMING PLAN
JOISTS AND BEARING WALLS

FIGURE 11–11

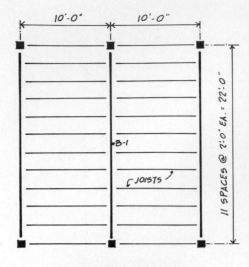

FRAMING PLAN
JOISTS AND BEAMS

FIGURE 11–12

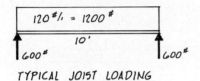

TYPICAL JOIST LOADING

FIGURE 11–13

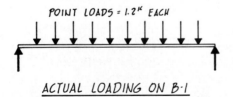

ACTUAL LOADING ON B-1

FIGURE 11–14

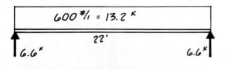

EQUIVALENT UNIFORM LOADING

FIGURE 11–15

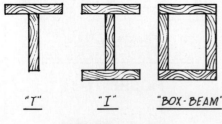

"T"   "I"   "BOX·BEAM"

FIGURE 11–16

feasible, it is probably more common to use a steel beam than a laminated timber beam. This is so because a steel beam of sufficient size would normally be readily available and would only have to be cut to the proper length. The fabrication of a laminated timber beam would require time, and wouldn't be worth the cost or effort unless there was an aesthetic purpose involved. In any case, since this chapter deals with wood, we will assume that all supporting beams are sawn lumber or laminated timber.

Wood joists supporting floors are normally spaced at 16 in. apart. This close spacing produces a great deal of stiffness in the plywood deck that would normally be used for the floor. On a roof system, the members supporting the roof deck may be spaced at 24 in. apart, depending on the live load being used.

When using closely spaced joists, whether the spacing is 16 in. or 24 in., the load pattern on the supporting beams is taken as a uniform load. For example, let's consider the simple framing plan shown in Fig. 11–12 and determine the load on the beam marked B–1. Let's assume that the total floor load (dead load plus live load) is 60#/sq. ft. The plywood deck used for the floor will deliver its load to the joists; the joists, in turn, will deliver their loads to the supporting beams. Therefore, using the lessons learned in the section on "Load Collection" presented in Chapter 5, the load on a typical joist, based on a 24 in. spacing with a 60#/sq. ft. floor load, is 120#/'. A typical joist loading diagram is shown in Fig. 11–13. The loading on B–1 may now be determined. The end reaction of each individual joist may be shown as an action on B–1, as indicated in Fig. 11–14. In the case shown here, the value of the point loads on B–1 would be twice the end reaction of the joists, since there are joists of equal span framing in from each side. The fact is that the maximum shear and bending moment produced by the loading shown in Fig. 11–14 will be virtually identical to the maximums produced by considering the load as uniformly distributed. As a rule, any time a large number of point loads, equal in magnitude, are closely spaced (up to 3 ft apart), the effect approaches that of a uniformly distributed load. It is therefore recommended that the portion of the framing being supported be treated as a one-way slab that delivers an equal amount of load to each linear foot of the supporting member. Therefore, for all practical purposes, the loading on B–1 would be as shown in Fig. 11–15. In both cases the maximum bending moments are practically the same, but this value is much easier to determine from the equivalent uniform loading pattern than it is from the large number of point loads. The maximum shear will be *slightly* higher by using the equivalent uniform load than it would be by using the true loading pattern, but to err slightly on the safe side when considering shear in wood is, in the author's judgment, an acceptable approach.

NON-RECTANGULAR SECTIONS

In the text and problem-solving exercises of Chapter 10, a variety of non-rectangular sections, such as those shown in Fig. 11–16, were used for the purpose of analysis. The *design* of sections such as these is largely a matter of trial and error. In other words, the design process consists of the selection of standard sawn lumber sections, arranging them in the desired cross-sectional configuration, and then investigating that cross section to see that the allowable stresses are not exceeded. When using sections such as those shown in Fig. 11–16, the quality of the joints between the pieces is particularly critical and may be designed using the procedures presented in Chapter 10.

PROBLEM 11-1

TIMBER BEAMS - RECTANGULAR SECTIONS

(1) <u>STANDARD LUMBER SECTIONS</u> - ALLOWABLE BENDING $f = 1600$ P.S.I.

ALLOWABLE SHEAR $f_v = 120$ P.S.I.

(a) DETERMINE THE <u>SAFE UNIFORM LOADING</u> FOR:

2×4, 6' SPAN - 2×6, 9' SPAN - 2×8, 12' SPAN - 2×10, 15' SPAN

(b)

2-3×10 ACTING TOGETHER AS A GIRDER

INVESTIGATE THE SAFETY IN BENDING AND SHEAR (COMPARE MAXIMUM ACTUAL STRESSES WITH ALLOWABLES)

(c)

DETERMINE WHICH 2×? IS NEEDED FOR BENDING UNDER THE GIVEN LOADING.

---

(2) <u>LAMINATED TIMBER SECTIONS</u> - ALLOWABLE BENDING $f = 2500$ P.S.I.

ALLOWABLE SHEAR $f_v = 200$ P.S.I.

(a) DETERMINE THE MOST ECONOMICAL SECTION (LEAST AREA) WHICH WILL BE SAFE FOR THE GIVEN LOADING.

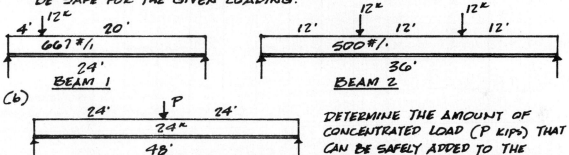

(b)

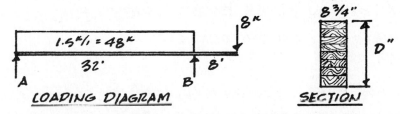

DETERMINE THE AMOUNT OF CONCENTRATED LOAD (P KIPS) THAT CAN BE SAFELY ADDED TO THE GIVEN UNIFORM LOADING.

---

(3) LAMINATED TIMBER - ALLOWABLE BENDING $f = 2400$ P.S.I.

ALLOWABLE SHEAR $f_v = 200$ P.S.I.

LOADING DIAGRAM   SECTION

(a) DETERMINE THE DEPTH (D") REQUIRED FOR BENDING, BASED ON THE GIVEN ALLOWABLE BENDING STRESS.

(b) IS THIS DEPTH O.K. FOR THE GIVEN ALLOWABLE SHEARING STRESS?

PROBLEM 11-2

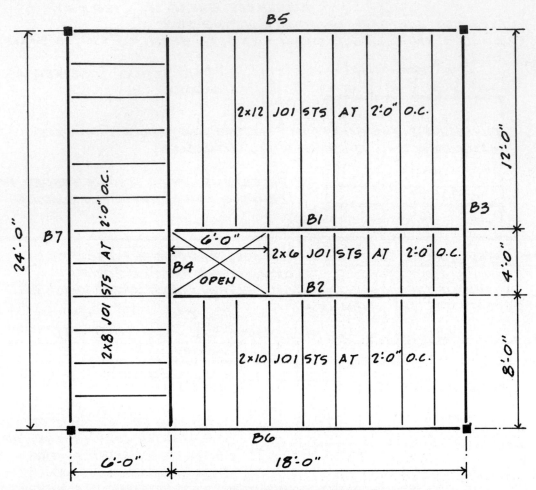

FRAMING PLAN FOR A WEEKEND RESIDENCE

- TOTAL FLOOR LOAD = 100 P.S.F.
- EXTERIOR WALL LOAD ON B3, B5, B6, B7 = 300 P.L.F.

- ASSUME THAT JOISTS DELIVER UNIFORM LOADING TO
  LAMINATED WOOD BEAMS (B1 - B7)

(1) CHECK JOISTS FOR SAFETY - ALLOWABLE BENDING f = 1200 P.S.I.

(2) DESIGN LAMINATED TIMBER BEAMS (B1-B7)
          ALLOWABLE BENDING f = 2000 P.S.I.

PROBLEM 11–3

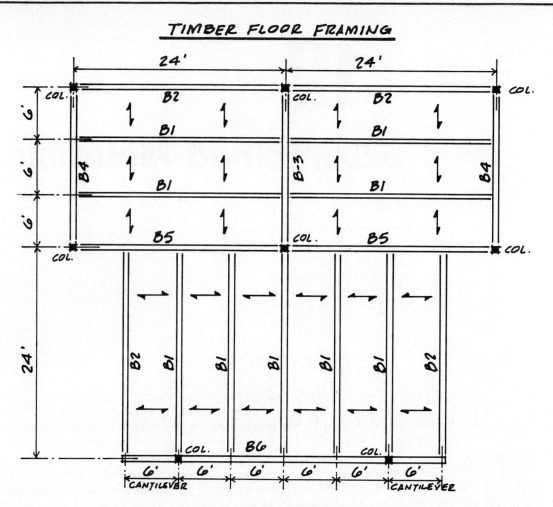

TIMBER FLOOR FRAMING

<u>NOTE</u>: ALL JOISTS ARE 2×6 AT 16" O.C. SPANNING AS INDICATED ⟶
ASSUME JOISTS DELIVER UNIFORM LOADING TO SUPPORTING BEAMS
B1 - B5 ARE LAMINATED TIMBER - SIMPLE SPAN
B6 - LAMINATED TIMBER - SPAN + CANTILEVER AT EACH END.

<u>LOADING</u>: TOTAL FLOOR LOAD = 75 P.S.F. (INCLUDES ALLOWANCE FOR BM. WEIGHT)
EXTERIOR WALLS = 200 #/FT. (ON ENTIRE PERIMETER)

<u>REQUIRED</u>:

(1) CHECK SAFETY OF JOISTS FOR BENDING — ALLOWABLE $f = 1200$ P.S.I.
"   "   "   "   " SHEAR — ALLOWABLE $f_v = 60$ P.S.I.

(2) DESIGN B1-B6 FOR S REQUIRED AT ALLOWABLE $f = 2400$ P.S.I.
AND AREA REQUIRED AT ALLOWABLE $f_v = 135$ P.S.I.
SELECT THE MOST ECONOMICAL SECTIONS (LEAST AREA)

12

Combined Materials

INTRODUCTION

In Chapter 7, a discussion was presented on elastic bending, which led to the development of the expression $f = Mc/I,$ and the concept of "moment of inertia." The development of these concepts, it should be recalled, was based strictly on a *homogeneous* section and an elastic material.

Frequently we will deal with nonhomogeneous sections; that is, sections made of two or more materials working together to provide structural integrity. The most typical member of this sort is one of reinforced concrete, where concrete and reinforcing steel are combined to resist the tensile and compressive forces produced by bending. The concrete resists compression, while the steel resists tension (see Fig. 12–1). Because of the fact that concrete is a brittle material incapable of taking tension, there are certain special considerations that must be taken into account. Consequently, the analysis and design of reinforced concrete members will be treated as a special case of combined materials, and will be covered in the following chapter.

In this chapter we will deal with the basic principles involved in the analysis and design of structural members made of two or more elastic materials, with each material having a modulus of elasticity in compression equal to the modulus of elasticity in tension.

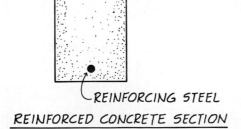

REINFORCING STEEL
REINFORCED CONCRETE SECTION

FIGURE 12–1

COMPATIBLE STRAINS IN MATERIALS

When two or more materials are joined to form a single structural member, it is imperative that they behave compatibly when subjected to a load. This is so whether the load produces bending or axial tension or compression. For the purpose of this discussion we will limit ourselves to bending situations, since they are the more complex.

It was pointed out in Chapter 7, in the section on "Elastic Bending," that experiments have revealed that initially plane sections remain plane during bending, the result being that *strains* vary proportionately with the distance from the neutral axis. Consequently, *stresses* vary proportionately also, when the stresses are below the elastic limit of the material. This idea formed the basis for the development of the bending expression

$$M = \frac{f}{c}I$$

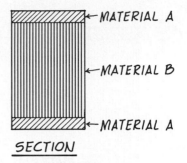

SECTION

FIGURE 12–2

While in Chapter 7 the discussion was based on homogeneous sections, the same ideas, in fact, must hold true of sections made of two or more materials if they are to function as integral structural members. That is, initially plane sections must remain plane during bending. The direct use, however, of the bending stress equation is not possible when dealing with members made of more than one material. Again referring to Chapter 7, it will be recalled that the ideas surrounding the development of the moment of inertia *(I)* were based on the fact that, in a homogeneous section, *stresses* will vary proportionately with the distance from the neutral axis. In a section made of more than one material, this is not true, because if the *strains* vary proportionately, then the *stress* in the stronger of the materials must be far greater than that of the weaker material in order to produce strain compatibility.

To help clarify this, consider the section of Fig. 12–2, made of two different materials bonded together to behave as an integral structural unit. Consider also that this is a cross section of a beam subjected to bending. Let's say that material *A* is stronger than material *B,* meaning that the modulus of elasticity of material *A* is greater than that of material *B.* This relationship is shown graphically in Fig. 12–3. When the beam is subjected to bending, the section shown will (and must) experience strains that are proportional to the distance from the neutral axis. Considering that the member is in a state of positive bending, the strains will be compressive above the neutral axis and tensile below the neutral axis.

Let's now consider what's happening where the section changes from the weaker material *B* to the stronger material *A.* Let's say that at this level, there is a certain amount of strain—call it ϵ_1. Referring to Fig. 12–3, it will be seen that for material *B,* there is a corresponding stress, f_1. It can also be seen from Fig. 12–3 that for strain ϵ_1, material *A* develops a much greater stress, f_2. Consequently, for the proportional variation in strains, a corresponding stress diagram will look like that shown in Fig. 12–4. This, of course, is not a constant proportional variation in stress from the neutral axis to the outermost fiber. This seems to invalidate the use of the bending stress equation because the moment of inertia *(I)* cannot be determined for the section, since the section is not subjected to a constant proportional variation of stress. The constant proportion is the basis for the determination of the moment of inertia. The question then becomes one of how to deal with sections made of two or more materials. The fact is that the bending stress equation, in altered form, will still be used for the analysis and design of members with more than one material involved. The modification necessary is a reasonably simple one, and we will now discuss the fundamental principles involved.

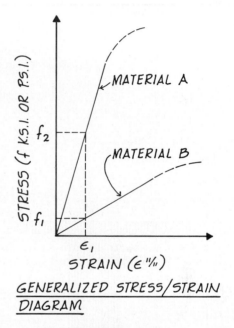

GENERALIZED STRESS/STRAIN DIAGRAM

FIGURE 12–3

THE TRANSFORMED SECTION

In order to deal with sections made of two or more materials using standard elastic bending theory and the bending stress equation, it is necessary to deal with the section as though it were homogeneous. This will allow us to

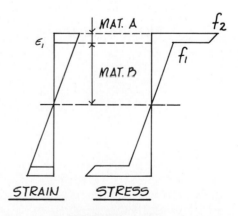

FIGURE 12–4

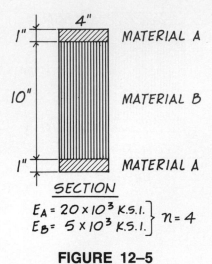

MATERIAL A

MATERIAL B

MATERIAL A

SECTION

$E_A = 20 \times 10^3 \text{ K.S.I.}$
$E_B = 5 \times 10^3 \text{ K.S.I.}$ } $n = 4$

FIGURE 12–5

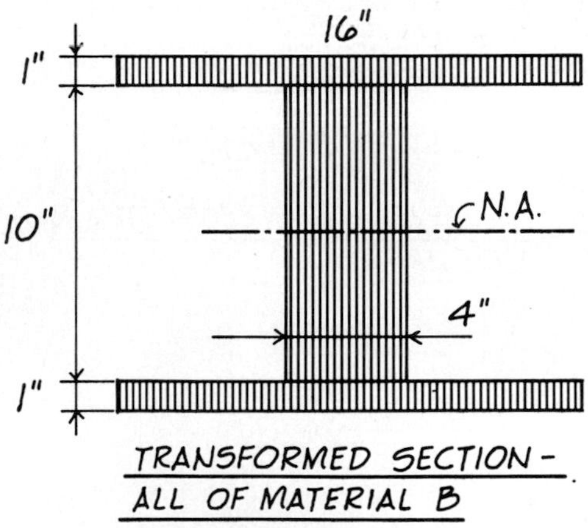

determine a value for the moment of inertia. To do this, a section of combined materials is treated as a *transformed section*. Considering that there are two materials involved, this means that we deal with the section by transforming one material into an equivalent (though hypothetical) area of the other material. This gives us a hypothetical section which, for the purpose of analysis, is homogeneous and to which standard elastic bending theory can be applied.

Now we must examine how the area of one material is treated as the equivalent area of the other material. Referring back to Figs. 12–3 and 12–4, it will be seen that at strain ϵ_1, the stress in material A is n times that of material B. This factor, n, is the relationship between the slopes of the stress/strain diagram for the two materials, in the elastic range of the materials. It should be recalled that the slopes of these lines are determined by the modulus of elasticity, (E), of the materials. Therefore,

$$n = \frac{E_A}{E_B} = \frac{\text{mod. of elast. material } A}{\text{mod. of elast. material } B}$$

where *n is called the modular ratio,* and is the factor by which one material is transformed to an equivalent area of the other material.

In order to clarify the principles involved in the process of transforming a section, consider the section of Fig. 12–5, with the given data. The moduli of elasticity are given; from this it can easily be determined that the modular ratio $n = 4$. In order to create an equivalent homogeneous section, we must increase the area of material A by four times to give us an equivalent area of material B. A picture of the transformed section is shown in Fig. 12–6. It is important to note that the *centroidal position* of the material, which was transformed (material A), is maintained with respect to the neutral axis of the section. *This rule must be followed* in the development of the transformed section. Therefore, when transforming the area of one material to an equivalent area of the other material, the area is altered by changing *the width of the material only,* thereby maintaining its centroidal relationship to the neutral axis.

The next step involved in the process is the determination of the moment of inertia of the *transformed section,* which is now homogeneous. Because the moment of inertia is being determined for this hypothetical section, we give it a special name:

$$I_{tr} = \text{transformed moment of inertia}$$

We can now determine the *transformed moment of inertia* for the section of Fig. 12–6 in precisely the same manner as we would determine the moment of inertia for any section.

16"

c N.A.

4"

TRANSFORMED SECTION –
ALL OF MATERIAL B

FIGURE 12–6

I_{tr}

$$2\left[\frac{(16)(1)^3}{12} + (16)(5.5)^2\right] = 970.7 \text{ in.}^4$$

$$\frac{(4)(10)^3}{12} = \underline{333.3 \text{ in.}^4}$$

$$I_{tr} = 1304.0 \text{ in.}^4$$

This value can then be used in the bending stress equation:

$$M = \frac{f I_{tr}}{c}$$

However, there are several unique considerations in the use of this expression, since we are dealing with a hypothetical section. In order to point these out, we will now proceed to several example problems.

Example 12–1 (Figs. 12–5, 12–6, and 12–7)

Using the section of the previous discussion, we would now like to know the maximum stresses produced in each material, due to bending, when used for the span and loading shown in Fig. 12–7. In order to this, we will use the standard bending stress equation with the value already determined for the transformed moment of inertia:

$$I_{tr} = 1304.0 \text{ in.}^4$$

In order to determine the stresses, the equation will be

$$f = \frac{Mc}{I_{tr}}$$

Since there are two materials involved in the real section and we wish to determine the maximum stress in each, we must make two computations. First, determine the maximum bending moment, since this will produce the maximum stresses:

$$M = \frac{WL}{8} = \frac{(10^K)(20')}{8} = 25'^K$$

(see Data Sheet D–25 for maximum bending moment)

To determine the stress in material B:

$$f = \frac{(25'^K \times 12''/')(5'')}{1304.0} = 1.15 \text{ k.s.i. or } 1150 \text{ p.s.i.}$$

Note that the c distance used is $5''$ because material B does not exist beyond this level. This is shown in the stress diagram of Fig. 12–8. Now let's determine the maximum stress in material A. For this, the appropriate c distance is $6''$. There is something else that must be considered at this point. If we use the bending stress equation in the same form as we did to determine the stress in material B (except for $c = 6''$), we will come up with a stress in the transformed material, which we are considering to be material B. In the real case, however, we only have $1/n$th of the material that was transformed. Therefore, the *actual* unit stress in material A must be n times that of stress determined from the transformed section. The real stress diagram, in fact, looks like that shown in Fig. 12–9. The principle involved would be stated in the following manner:

The *actual* unit stress in *the material which was transformed* is determined by multiplying the bending stress equation by the modular ratio *(n)*.

or expressed mathematically:

$$f = \frac{Mcn}{I_{tr}}$$

Now, to determine the stress in material A:

$$f = \frac{(25'^K \times 12''/')(6'')(4)}{1304.0} = 5.52 \text{ k.s.i. or } 5520 \text{ p.s.i.}$$

In this example, the stronger material was transformed to an equivalent area of the weaker material. However, the transformation could have been made the other way; that is, transforming the weaker material to an equivalent area of the stronger material. In order to do this, the modular ratio would be taken as the *reciprocal* of that previously used, or $n = 1/4$. The original section is shown again in Fig. 12–10 with the appropriate data. Let's go through the example transforming it this way. A very different transformed section will emerge, as shown in Fig. 12–11, and the transformed moment of inertia will be very different. In the end, however, we'll see

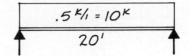

FIGURE 12–7

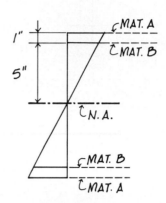

FIGURE 12–8

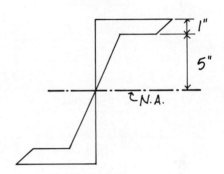

FIGURE 12–9

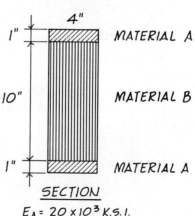

$E_A = 20 \times 10^3$ K.S.I.
$E_B = 5 \times 10^3$ K.S.I.

TRANSFORMING MATERIAL B
TO MATERIAL A: $n = 1/4$

FIGURE 12–10

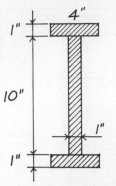

TRANSFORMED SECTION
ALL OF MATERIAL A

FIGURE 12–11

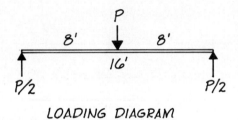

LOADING DIAGRAM

FIGURE 12–12

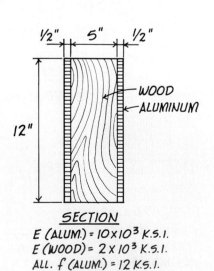

SECTION
$E\ (ALUM.) = 10 \times 10^3\ K.S.I.$
$E\ (WOOD) = 2 \times 10^3\ K.S.I.$
$ALL.\ f\ (ALUM.) = 12\ K.S.I.$
$ALL.\ f\ (WOOD) = 1.6\ K.S.I.$

FIGURE 12–13

that the actual stresses in each material due to the $25'^K$ bending moment will be the same as previously determined, as they must be.

To begin with, we must determine the transformed moment of inertia:

I_{tr}

$$2\left[\frac{(4)(1)^3}{12} + (4)(5.5)^2\right] = 242.7\ \text{in.}^4$$

$$\frac{(1)(10)^3}{12} = \underline{83.3\ \text{in.}^4}$$

$$I_{tr} = 326.0\ \text{in.}^4$$

To determine the stress due to bending in material A (which was not transformed):

$$f = \frac{Mc}{I_{tr}} = \frac{(25'^K \times 12''/')(6)}{326.0\ \text{in.}^4} = 5.52\ \text{k.s.i.}$$

To determine the stress in material B we must, as stated previously, multiply the bending stress equation by the modular ratio which, in this case, is $n = 1/4$:

$$f = \frac{(25'^K \times 12''/')(5'')}{326.0\ \text{in.}^4}\left(\frac{1}{4}\right) = 1.15\ \text{k.s.i.}$$

These, of course, are precisely the same values we found when the section was transformed the other way.

Example 12–2 (Figs. 12–12 and 12–13)

It is desired to know the maximum point load that can be placed on the beam of Fig. 12–12 without exceeding the allowable stresses in either material. The section is shown in Fig. 12–13, along with the necessary data.

In order to determine the desired information, we must first transform the section. For this purpose we'll transform the wood to an equivalent area of aluminum and, therefore, $n = 1/5$. The transformed section is shown in Fig. 12–14.

The next step is to determine the transformed moment of Inertia with respect to the neutral axis:

$$I_{tr} = \frac{BD^3}{12} = \frac{(2)(12)^3}{12} = 288\ \text{in.}^4$$

Now, we know that for the aluminum:

$$f = \frac{Mc}{I_{tr}}$$

and for the wood:

$$f = \frac{Mc(n)}{I_{tr}}$$

Furthermore, the general expression for the maximum bending moment due to the point load at the midspan is

$$M = \frac{PL}{4}\quad \text{(see Data Sheet D–25)}$$

Substituting this value into the bending stress equation, and rearranging to solve for M, we get

Aluminum:

$$\frac{PL}{4} = \frac{fI_{tr}}{c}$$

Wood:

$$\frac{PL}{4} = \frac{fI_{tr}}{cn}$$

Combined Materials   **181**

What must be recognized here is that the allowable stress for either material must not be exceeded. Therefore, it is most likely that, for strain compatibility, one material will be stressed to its full allowable and the other material will not. This means that two computations must be made in order to solve for P. Using the allowable stress for each material, we get

Aluminum:

$$P = \frac{(12 \text{ k.s.i.})(288 \text{ in.}^4)(4)}{(6'')(16' \times 12''/')} = 12^K$$

Wood:

$$P = \frac{(1.6 \text{ k.s.i.})(288 \text{ in.}^4)(4)}{(6'')(16' \times 12''/')(1/5)} = 8^K$$

The least value derived from these computations is the answer to this problem. Therefore, wood governs and the maximum safe load $P = 8^K$. Had we said that 12^K was the answer, then, in fact, we would be saying that the wood was stressed beyond its allowable stress.

Example 12–3 (Figs. 12–15 and 12–16)

In this example we'll deal with an irregular combined-materials section, as shown in Fig. 12–16. We would like to determine the maximum uniformly distributed load (W) that this section can carry when used as shown in Fig. 12–15. The necessary data is given in Fig. 12–16; for this case, we'll transform the steel to an equivalent area of wood. Therefore, $n = 15$; the transformed section is shown in Fig. 12–17.

We must now determine the location of the neutral axis of the transformed section:

ΣAx (using the top as a reference axis)

$$(30 \text{ in.}^2)(.25'') = \quad 7.5 \text{ in.}^3$$
$$(48 \text{ in.}^2)(6.5'') = \underline{312.0 \text{ in.}^3}$$
$$319.5 \text{ in.}^3$$

$$\therefore \bar{x} = \frac{319.5}{78} = 4.1'' \text{ from the top}$$

Now, determine the transformed moment of inertia:

I_{tr}

$$\frac{(60)(.5)^3}{12} + (30)(3.85)^2 = \quad 445.3 \text{ in.}^4$$
$$\frac{(4)(12)^3}{12} + (48)(2.4)^2 = \underline{852.5 \text{ in.}^4}$$
$$I_{tr} = 1297.8 \text{ in.}^4$$

The general expression for the maximum bending moment for a simply supported beam with a uniformly distributed load is

$$M = \frac{WL}{8} \text{ (see Data Sheet D–25)}$$

The bending stress expressions are
Wood:

$$f = \frac{Mc}{I_{tr}}$$

Steel:

$$f = \frac{Mcn}{I_{tr}}$$

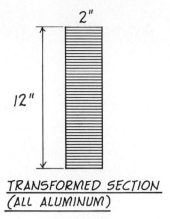

TRANSFORMED SECTION (ALL ALUMINUM)

FIGURE 12–14

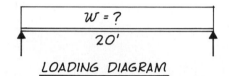

LOADING DIAGRAM

FIGURE 12–15

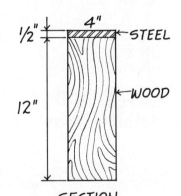

SECTION
E (STEEL) = 30 X 10³ K.S.I.
E (WOOD) = 2 X 10³ K.S.I.
ALL. f (STEEL) = 20 K.S.I.
ALL. f (WOOD) = 2 K.S.I.

FIGURE 12–16

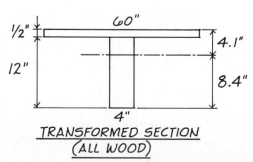

TRANSFORMED SECTION (ALL WOOD)

FIGURE 12–17

Rearranging these and substituting the general expression for the maximum bending moment, we get

Wood:

$$\frac{WL}{8} = \frac{fI_{tr}}{c}$$

Steel:

$$\frac{WL}{8} = \frac{fI_{tr}}{cn}$$

Both materials must be checked to determine W, and it must also be recognized that there are two c distances involved when dealing with the wood. They are

$$c = 8.4'' \text{ to the bottom fiber, and}$$

$$c = 3.6'' \text{ to the top fiber}$$

It should be obvious that the larger of these values will yield the smaller value for W; consequently, this is the one for which the computation will be made.

Wood:

$$W = \frac{(2 \text{ k.s.i.})(1297.8 \text{ in.}^4)(8)}{(8.4'')(20' \times 12''/')}$$
$$= 10.3^K$$

Steel:

$$W = \frac{(20 \text{ k.s.i.})(1297.8 \text{ in.}^4)(8)}{(4.1'')(20' \times 12''/')(15)}$$
$$= 14.1^K$$

Therefore, the maximum value for $W = 10.3^K$.

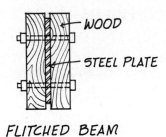

FLITCHED BEAM

FIGURE 12–18

DESIGN OF FLITCHED BEAMS

Thus far, we have discussed the methods for determining stresses or load-carrying capacity for given sections made of more than one material. These are "investigation" problems. The more common problem in a practical sense, however, is not one whereby a given cross section is investigated, but that of design. The design problem is essentially one where the sizes of the various parts of the cross section must be determined. While the principles we have discussed to this point apply, the thought process involved in a design problem is somewhat different than that for an investigation problem.

In practice, one of the most common problems in combined materials (aside from reinforced concrete) is the design of a *flitched beam*. A flitched beam is one made of one or more pieces of lumber combined with one or more steel plates, such as the section shown in Fig. 12–18. Generally, the flitched beam is thought of as wood because it is nailable. Consequently, it would be used in structures that are primarily wood frames, such as residences or small commerical structures.

In spite of the limited use of the flitched beam, the structural principles that underlie its design are of great importance. In the broad sense, we must deal with the idea of strain compatibility, which is fundamental to the concept of various materials combined and working in unison to resist forces. To this end, we will look at several examples dealing with the design of flitched beams.

Example 12–4—Design (Figs. 12–19 and 12–20)

It is desired to design a simply supported beam as shown in Fig. 12–19. In the case of a design problem, everything about the cross section is unknown at the outset; consequently, certain judgments must be made to eliminate some unknowns. In the case of a flitched beam, the wood to be used is chosen based on readily available stock sections. The design problem then becomes one of sizing the steel plate properly so that it works in unison with the wood. The flitched beam section for this example is shown in Fig. 12–20, along with the necessary data. While a $2'' \times 12''$ actually has smaller dimensions, we will use the full $2'' \times 12''$ for the sake of convenience in this example problem.

The first step is to determine the moment-resisting capacity the section must furnish. For the uniformly loaded simple span,

$$\text{Max. } M = \frac{WL}{8} = \frac{(9.6^K)(24')}{8} = 28.8'^K$$

Each material in the section will furnish part of this maximum moment. Since we have already decided on the size of the wood to be used, the next step is to determine that portion of the total moment that the wood can carry, based on its full allowable stress.

$$M(\text{wood}) = \frac{fI}{c} = fS = \frac{(1.5 \text{ k.s.i.})(96 \text{ in.}^3)}{12''/'}$$
$$= 12'^K$$

Note: S is called the *section modulus* and is simply the ratio of I/c. This was first presented in Chapter 11.

Therefore, the steel plate must carry

$$M(\text{steel}) = 28.8'^K - 12'^K = 16.8'^K$$

Now it must be recognized that if the wood is stressed to its full allowable at the outermost fibers, then at this level, the steel must be stressed to n times that of the wood for both materials to be straining by the same amount. This idea forms the basis for determining the depth of the steel plate. If the steel were the full $12''$ depth, the stress at the outer fibers would be

$$n(1.5 \text{ k.s.i.}) = \frac{30}{1.8}(1.5) = 25 \text{ k.s.i.}$$

But this is beyond the allowable stress of 20 k.s.i. given for the steel. In order to determine the depth of the steel plate required so that the allowable stress is not exceeded, a stress diagram as shown in Fig. 12–21 is used. A similar triangle solution will give the desired result:

$$\frac{x}{6} = \frac{20}{25} \quad \text{and} \quad x = 4.8''$$

and the total depth of the plate $= 9.6''$. Now it remains to determine the thickness of the steel plate. Since we know the moment the steel must carry, the allowable stress, and the depth of the plate, this is a relatively simple matter. The thickness can be determined by the bending stress equation (with the thickness being the only unknown):

$$M(\text{steel}) = fS \text{ and } S = \frac{BD^2}{6} \text{ for a solid rectangle}$$

$$(16.8'^K \times 12''/') = \frac{(20 \text{ k.s.i.})(B)(9.6)^2}{6}$$

and

$$B = .66''$$

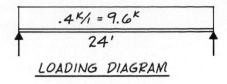

LOADING DIAGRAM

FIGURE 12–19

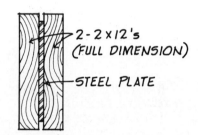

SECTION
E (WOOD) = 1.8 × 10³ K.S.I.
E (STEEL) = 30 × 10³ K.S.I.
ALL. f (WOOD) = 1.5 K.S.I.
ALL. f (STEEL) = 20 K.S.I.

FIGURE 12–20

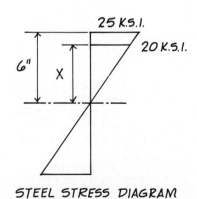

STEEL STRESS DIAGRAM

FIGURE 12–21

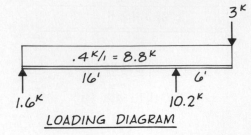

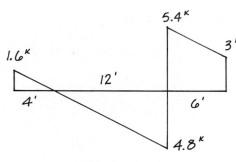

LOADING DIAGRAM

FIGURE 12–22

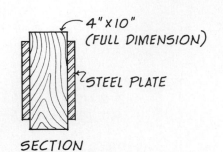

4" × 10"
(FULL DIMENSION)

STEEL PLATE

SECTION

$E\ (WOOD) = 1.7 \times 10^3\ K.S.I.$
$E\ (STEEL) = 30 \times 10^3\ K.S.I.$
$ALL.\ f\ (WOOD) = 1.3\ K.S.I.$
$ALL.\ f\ (STEEL) = 18\ K.S.I.$

FIGURE 12–23

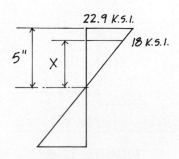

SHEAR DIAGRAM

FIGURE 12–24

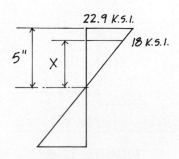

FIGURE 12–25

The theoretical dimensions of the plate, then, are .66″ × 9.6″, and the design of the flitched beam is complete.

Example 12–5—Design (Figs. 12–22 and 12–23)

It is quite possible, in a flitched beam, to eliminate the steel plate where it is no longer needed to carry a portion of the bending moment. This aspect of the design problem will be included in this example. First, however, we must determine the maximum moments in this cantilevered beam and determine the size of the steel plates required. For this purpose we'll use the shear diagram, which is shown in Fig. 12–24:

$$+M = (4')(1.6^K)\left(\frac{1}{2}\right) = 3.2'^K$$

$$-M = (3^K)(6') + (2.4^K)(6')\left(\frac{1}{2}\right) = 25.2'^K$$

For this problem a 4″ × 10″ timber section was chosen; its moment-carrying capacity, based on full dimensions, is

$$M = fS = \frac{(1.3\ k.s.i.)(66.7\ in.^3)}{12''/'} = 7.2^K$$

Therefore, steel plates are required only to help carry the negative moment, which is in excess of what the wood can carry by itself.

$$M(steel) = 25.2'^K - 7.2'^K = 18'^K$$

If the steel plates were the full 10″ depth, the stress at the outermost fibers would be

$$f = n(1.3\ k.s.i.) = \frac{30}{1.7}(1.3) = 22.9\ k.s.i.$$

which exceeds the allowable stress given as 18 k.s.i. Based on similar triangles from the stress diagram, shown in Fig. 12–25, we can determine the depth of the plate

$$\frac{x}{5} = \frac{18}{22.9} \quad \text{and} \quad x = 3.9''$$

and the total depth is 7.8″.

The thickness is determined from the bending stress equation based on the moment the steel must carry at its allowable stress:

$$M = fS = f\left[\frac{BD^2}{6}\right]$$

$$(18'^K \times 12''/') = \frac{(18\ k.s.i.)(B)(7.8)^2}{6}$$

and

$$B = 1.2''$$

Since there are two plates, as shown in the section of Fig. 12–23, each plate must have dimensions of .6″ × 7.8″. These are the minimum theoretical dimensions. However, the thickness would be rounded up to the nearest stock thickness, which in this case would be a 5/8″-thick plate.

It is obvious that there is no need to carry the steel plates throughout the entire length of the beam. The plates, in fact, are only needed in the negative moment region and to points on the right and left side of the support where the moment has diminished to a level the wood can carry by itself. This relationship is shown in the moment diagram of Fig. 12–26. Perhaps the easiest way to solve for the location of the plate cutoff points is to use the shear diagram, remembering that the area under the shear diagram between any two points is equal to the difference in values on the moment diagram between those same two points. Using this principle, we'll now solve for the cutoff point on the right side of the support, using the notations shown on the shear diagram of Fig. 12–27.

Solving for the dimensions x from the right-hand end of the shear diagram, we're looking for an area of $7.2'^K$

$$(3)(x) + (.4x)(x)\left(\frac{1}{2}\right) = 7.2$$

or

$$.2x^2 + 3x - 7.2 = 0$$
$$= x^2 + 15x - 36 = 0$$

and solving the quadratic

$$x = \frac{-15 \pm \sqrt{225 + 144}}{2} = 2.1'$$

and $b = 6' - 2.1' = 3.9'$ from the support.

Now solving for the cutoff point location to the left of the support, we're looking for an area under the shear diagram between the point of zero shear and the cutoff point, equal to $10.4'^K$. This is the total *difference* in value on the moment diagram between these two points.

The equation is

$$(.4y)\,(y)\left(\frac{1}{2}\right) = 10.4 = .2y^2$$

$$y = \sqrt{52} = 7.2' \text{ and } a = 12 - 7.2$$

$$= 4.8'$$

Therefore, the plate required for negative moment must extend 3.9' to the right of the support and 4.8' to the left of the support, or a total length of 8.7'. Beyond these points the wood alone can carry the moment.

The design of flitched beams provides excellent demonstrations and exercise in the theory of sections made of combined materials. It is not possible—in a practical way, based on the usual methods for making a flitched beam—to combine the two materials in such a manner that they will deform *absolutely* compatibly. The usual method for constructing a flitched beam is to use bolts that go all the way through the section to join the materials. In order to get the bolts through, the bolt holes that are drilled must be slightly larger than the diameter of the bolts. This will allow a small amount of slippage between the two materials. The deviation from *absolute* compatibility due to this slippage, however, is slight and may be ignored.

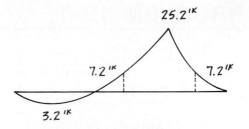

MOMENT DIAGRAM

FIGURE 12–26

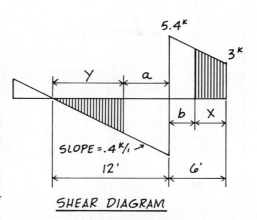

SHEAR DIAGRAM

FIGURE 12–27

PROBLEM 12–1

COMBINED MATERIALS – ELASTIC BENDING

STRENGTH DATA: SAME FOR TENS. & COMP.

MATERIAL A: ALUMINUM $\Big\}$ $E_A = 10000$ K.S.I.
ALLOW. $f = 15$ K.S.I.   E.L. = 25 K.S.I.

MATERIAL B: TIMBER $\Big\}$ $E_B = 2000$ K.S.I.
ALLOW. $f = 2.4$ K.S.I.   E.L. = 6 K.S.I.

BEAM: ALUMINUM SKIN BONDED TO
A TIMBER CORE.

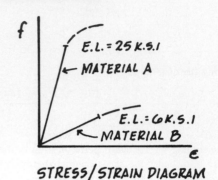

STRESS/STRAIN DIAGRAM

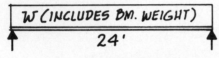

24'

LOADING DIAGRAM

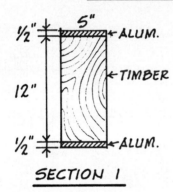

SECTION 1

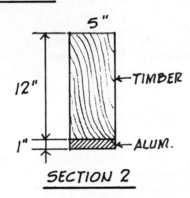

SECTION 2

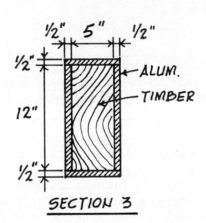

SECTION 3

FOR EACH SECTION:

(A) $W = 14^K$: DETERMINE THE MAXIMUM BENDING STRESS (f K.S.I.) IN
THE TIMBER AND THE MAXIMUM BENDING STRESS IN THE ALUMINUM.

(B) LOAD W IS INCREASED UNTIL ONE MATERIAL REACHES ITS ELASTIC
LIMIT FIRST (E.L. DATA ABOVE) AT MAXIMUM BENDING STRESS (f K.S.I.).
DETERMINE WHICH MATERIAL AND AMOUNT OF W.

(C) DETERMINE THE SAFE LOAD W^K – ALLOWABLE BENDING STRESS MUST
NOT BE EXCEEDED IN EITHER MATERIAL.
  (1) WITH GIVEN ALUMINUM & TIMBER SECTIONS
  (2) WITH AN ALL TIMBER SECTION 5" × 13"

(D) DETERMINE (FOR SECTIONS 1 AND 2) THE STRENGTH OF THE
BONDING MATERIAL REQUIRED FOR BONDING THE ALUMINUM
TO THE TIMBER.

PROBLEM 12-2

COMBINED MATERIALS

(I) INVESTIGATION OF A FLITCHED BEAM

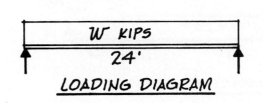

W KIPS
24'

LOADING DIAGRAM

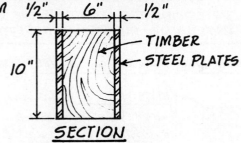

2-2×10
1"×10" STEEL PLATE
E(WOOD) = 2×10⁶ P.S.I.
E(STEEL) = 30×10⁶ P.S.I.
ALL. f(WOOD) = 1.5 K.S.I.
ALL. f(STEEL) = 20 K.S.I.

SECTION

- A SIMPLY SUPPORTED BEAM IS MADE OF 2-2×10's (ACTUAL DIMENSIONS) AND A 1"×10" STEEL PLATE, ALL BONDED TOGETHER TO ACT AS A SINGLE UNIT. DETERMINE THE SAFE UNIFORMLY DISTRIBUTED LOAD (W) BASED ON THE GIVEN ALLOWABLE STRESSES.

(II) INVESTIGATION OF A FLITCHED BEAM

½"   6"   ½"

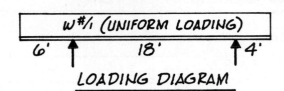

W #/₁ (UNIFORM LOADING)
6'   18'   4'

LOADING DIAGRAM

10"

TIMBER
STEEL PLATES

SECTION

ALLOW. f (WOOD) = 1.8 K.S.I.
ALLOW. f (STEEL) = 22 K.S.I.

E (WOOD) = 2000 K.S.I.
E (STEEL) = 30000 K.S.I.

- FOR THE GIVEN SECTION AND LOADING, DETERMINE THE MAX. W #/₁ THAT CAN BE SAFELY CARRIED BASED ON THE GIVEN ALLOW. STRESSES.

(III) INVESTIGATION OF A FLITCHED BEAM

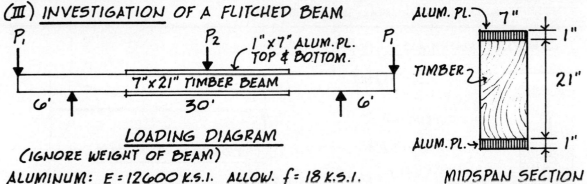

P_1 P_2 P_1

1"×7" ALUM. PL.
TOP & BOTTOM.

7"×21" TIMBER BEAM

6'   30'   6'

LOADING DIAGRAM
(IGNORE WEIGHT OF BEAM)

ALUM. PL.   7"
1"
TIMBER
21"
ALUM. PL.   1"

MIDSPAN SECTION

ALUMINUM: E = 12600 K.S.I.   ALLOW. f = 18 K.S.I.
TIMBER: E = 1800 K.S.I.   ALLOW. f = 2 K.S.I.

(A) DETERMINE THE MAXIMUM VALUES FOR P_1 & P_2 BASED ON THE GIVEN ALLOWABLE BENDING STRESSES.

(B) DETERMINE THE NECESSARY LENGTH OF THE ALUMINUM PLATES SO THAT THE TIMBER WILL NOT BE STRESSED BEYOND ITS ALLOWABLE BENDING STRESS.

PROBLEM 12–3

COMBINED MATERIALS

(I) FLITCHED BEAM DESIGN

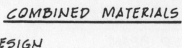
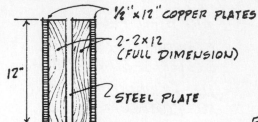
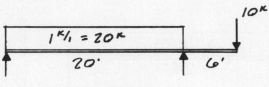

½" x 12" COPPER PLATES

2-2x12
(FULL DIMENSION)

12"

STEEL PLATE

SECTION

1^{K}/₁ = 20^{K}

20'

10^{K}

6'

LOADING DIAGRAM

E (COPPER) = 10 x 10³ K.S.I.   ALL. f (COPPER) = 8 K.S.I.
E (WOOD) = 2 x 10³ K.S.I.   ALL. f (WOOD) = 2 K.S.I.
E (STEEL) = 30 x 10³ K.S.I   ALL. f (STEEL) = 24 K.S.I.

1. DRAW THE SHEAR AND BENDING MOMENT DIAGRAMS FOR THE GIVEN LOADING.
2. THE COPPER PLATES ARE BONDED TO THE WOOD AND ARE CONTINUOUS FOR THE
   FULL LENGTH OF THE BEAM. A STEEL PLATE IS TO BE ADDED WHERE NECESSARY.
   (a) DETERMINE THE LENGTH OF THE STEEL PLATE REQUIRED.
   (b) DETERMINE THE REQUIRED DEPTH AND THICKNESS OF THE PLATE.

(II) COMBINED MATERIALS - INVESTIGATION

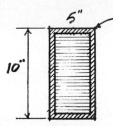

5"

10"

MATERIAL "A": HOLLOW TUBE (5"x10")
WITH ½" THICK WALLS
E = 20x10³ K.S.I.   ALL. f = 15 K.S.I.

MATERIAL "B": SOLID FILLING (4"x9")
BONDED TO THE TUBE.
E = 5 x 10³ K.S.I.   ALL. f = 2 K.S.I.

SECTION

W^{K}

20'

LOADING DIAGRAM

● DETERMINE MAX. W AS SHOWN.

(III) COMBINED MATERIALS - INVESTIGATION

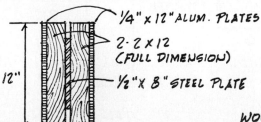
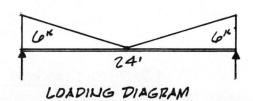

¼" x 12" ALUM. PLATES

2-2x12
(FULL DIMENSION)

12"

½" x 8" STEEL PLATE

SECTION

6^{K} 6^{K}

24'

LOADING DIAGRAM

WOOD: E = 1.5 x 10³ K.S.I.
STEEL: E = 30 x 10³ K.S.I.
ALUM: E = 12 x 10³ K.S.I

● FOR THE GIVEN SECTION AND LOADING, DETERMINE THE MAXIMUM
  BENDING STRESS IN EACH MATERIAL.

PROBLEM 12-4

COMBINED MATERIALS

(I) FLITCHED BEAM DESIGN

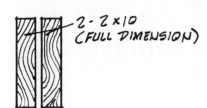

SECTION

2 - 2 × 10
(FULL DIMENSION)

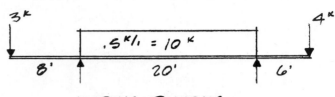

LOADING DIAGRAM

STEEL: E = 30 × 10³ K.S.I.   ALLOW. f = 24 K.S.I.
WOOD: E = 2 × 10³ K.S.I.   ALLOW. f = 1.4 K.S.I.

(1). DETERMINE WHERE STEEL PLATES ARE REQUIRED

(2). DESIGN THE STEEL PLATES SO THAT THEY ARE STRESSED TO THE FULL ALLOWABLE STRESS.

(II) COMBINED MATERIALS - INVESTIGATION

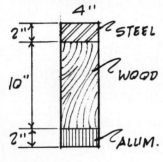

SECTION

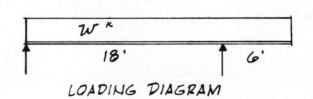

LOADING DIAGRAM

STEEL: E = 30 × 10³ K.S.I.   ALLOW. f = 20 K.S.I.
ALUM: E = 10 × 10³ K.S.I.   ALLOW. f = 10 K.S.I.
WOOD: E = 2 × 10³ K.S.I.   ALLOW. f = 2 K.S.I.

● THE SECTION SHOWN IS MADE OF THREE MATERIALS, AS INDICATED, TO FORM AN INTEGRAL STRUCTURAL UNIT. DETERMINE THE MAXIMUM SAFE LOAD (W ᴷ) BASED ON THE GIVEN LOADING DIAGRAM

(III) COMBINED MATERIALS - INVESTIGATION

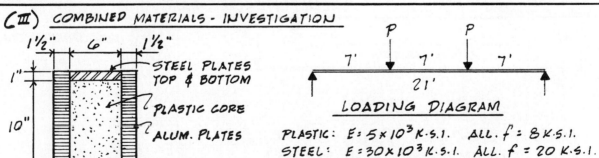

SECTION

STEEL PLATES TOP & BOTTOM
PLASTIC CORE
ALUM. PLATES

LOADING DIAGRAM

PLASTIC: E = 5 × 10³ K.S.I.   ALL. f = 8 K.S.I.
STEEL: E = 30 × 10³ K.S.I.   ALL. f = 20 K.S.I.
ALUM: E = 10 × 10³ K.S.I.   ALL. f = 12 K.S.I.

● IN THE GIVEN SECTION, ALL THE MATERIALS ARE BONDED TOGETHER TO ACT AS A SINGLE STRUCTURAL UNIT. DETERMINE THE SAFE LOAD (P KIPS) AS SHOWN IN THE LOADING DIAGRAM.

PROBLEM 12–5

COMBINED MATERIALS

(I) DESIGN OF A FLITCHED BEAM

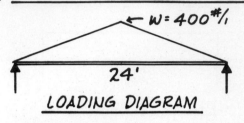

← W = 400 #/₁

24'

LOADING DIAGRAM

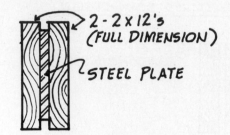

2 - 2 × 12's
(FULL DIMENSION)

STEEL PLATE

DATA:

f ALLOW. (WOOD) = 1.4 K.S.I.   E (WOOD) = 1800 K.S.I.
f ALLOW. (STEEL) = 20 K.S.I.   E (STEEL) = 30000 K.S.I.

• <u>DETERMINE</u> THE SIZE OF THE STEEL PLATE REQUIRED.

(II) DESIGN OF A CANTILEVERED FLITCHED BEAM

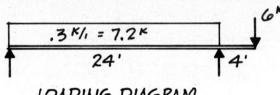

6ᴷ

.3 ᴷ/₁ = 7.2 ᴷ

24'   4'

LOADING DIAGRAM

B   4"   B

STEEL PLATE

2 - 2 × 12's
FULL DIMENSION

12"

SECTION

DATA:

WOOD: E = 2000 K.S.I. f ALL. = 1.5 K.S.I.
STEEL: E = 30000 K.S.I. f ALL. = 22.5 K.S.I.

• IN THE SECTION SHOWN, THE STEEL PLATES ARE BONDED TO THE WOOD
TO FORM AN INTEGRAL STRUCTURAL UNIT. DETERMINE THE LENGTH OF THE
PLATE REQUIRED, AND THE THICKNESS, AT THE CRITICAL SECTION.

(III) DESIGN OF A CANTILEVERED FLITCHED BEAM

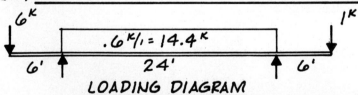

6ᴷ   1ᴷ

.6ᴷ/₁ = 14.4ᴷ

6'   24'   6'

LOADING DIAGRAM

2 - 2 × 12's
FULL DIMENSION

STEEL PLATE

SECTION

WOOD: E = 2000 K.S.I. f ALL. = 1.5 K.S.I.
STEEL: E = 30000 K.S.I. f ALL. = 18 K.S.I.

(A) DETERMINE THE DIMENSIONS OF THE STEEL PLATE REQUIRED
FOR NEGATIVE MOMENT.

(B) DETERMINE THE LENGTH OF THE PLATES REQUIRED FOR
NEGATIVE AND POSITIVE MOMENT.

13

Reinforced Concrete— Working Stress

INTRODUCTION

In this chapter we shall continue with the fundamental principles of combined materials, but specifically applied to reinforced concrete, where, because of the brittle nature of concrete, certain unique considerations must be made.

The methods employed here have been traditionally referred to as *working stress design* methods. The current building code of the American Concrete Institute (ACI) refers to this as the *alternate design method*.* This method is referred to as an "alternate" because a later method, known as *strength design,* is stressed by the ACI Code. Although strength design (formerly called *ultimate strength design*) is the newest method and is winning out as a primary design procedure for reinforced concrete members, the principles employed in the working stress methods are still quite important. In this chapter we shall deal with the principles of working stress, and the methods of strength design will be presented in Chapter 15. It might be worthwhile at this point to explain, briefly, the fundamental differences in the two methods.

Working stress design treats concrete as an elastic material; members are designed much like the combined materials described in the previous chapter. Designs are based on the allowable stresses of the concrete and the reinforcing steel, and the relationship between the moduli of elasticity of the two materials. Also, the design is based on anticipated loads (live load and dead load), which are called *service loads*.

In *strength design,* the design of a structural member is based on the *ultimate* compressive strength of concrete and the yield strength of the rein-

---

*American Concrete Institute, *Building Code Requirements for Reinforced Concrete,* current edition.

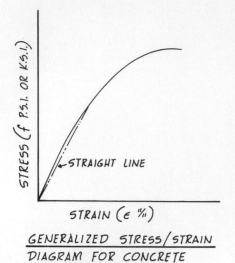

GENERALIZED STRESS/STRAIN
DIAGRAM FOR CONCRETE

FIGURE 13-1

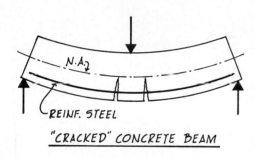

"CRACKED" CONCRETE BEAM

FIGURE 13-2

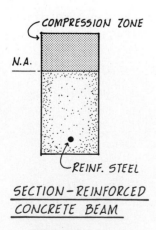

SECTION-REINFORCED
CONCRETE BEAM

FIGURE 13-3

forcing steel. The use of ultimate strength and yield strength invalidates the use of elastic theory, since these values are beyond the elastic limits of the materials. The use of these stresses implies that the member being designed will actually fail. While this is the basis for the philosophy of strength design, members, in fact, will not fail. There are a number of factors included in the design procedure that guard against this. The most noteworthy of these is called the *load factor*. The load factor is a multiplier by which the service loads are inflated to hypothetical design (or ultimate) loads, and it is under these inflated loads, which will not be realized in the building, that the members are designed to "fail." There are other safety factors involved in strength design; these will be discussed in studies dealing with this method.

WORKING STRESSES AND THE "CRACKED" SECTION

It was previously mentioned that in the working stress method, concrete is treated as an elastic material. This idea comes about because up to a certain level of stress in concrete, a plot of a stress/strain diagram, as shown in Fig. 13–1, will indicate that there is nearly a straight-line relationship between stress and strain. It is close enough, in fact, to safely say that for practical reasons, the material obeys Hooke's Law, which indicates an elastic material. Such a stress/strain relationship in concrete is determined from a compression test, since concrete exhibits no significant tensile strength. For purposes of design we say that concrete has zero tensile strength, which, of course, accounts for the development of the reinforced concrete beam where the concrete takes compression and the reinforcing steel takes the tension. It is the lack of tensile strength in concrete that forms the basis for some considerations different than those employed when dealing with sections of combined materials, where each material has compressive strength equal to the tensile strength.

Working stress methods are based on the idea that since concrete resists no tension, the section is cracked below the neutral axis, as shown in Fig. 13–2, with all of the tensile stresses concentrated in the reinforcing steel. Consequently, we consider nothing below the neutral axis, except the steel, to be part of the *effective section* resisting bending. A picture of the effective section for a rectangular reinforced concrete beam is shown in Fig. 13–3. The shaded portion indicates concrete in compression. This, plus the reinforcing steel, constitutes the effective section. Since the unshaded portion of the section does not carry any stress, because it is "cracked," this is not used in the computations for locating the neutral axis or in determining the transformed moment of inertia.

We'll now look at an example that may help to clarify these ideas.

Example 13-1 (Figs. 13-4 and 13-5)

In this example we'll determine the location of the neutral axis and the transformed moment of inertia of the *effective section*. To begin, we must transform the section so that we will be dealing with a homogeneous section. In reinforced concrete problems, this is done by transforming the steel to a hypothetical equivalent area of "tensile resisting concrete." The section and the necessary data are given in Fig. 13–4. It should be mentioned at this point that the modulus of elasticity of the concrete is dependent on the strength of the mix being used. Data Sheet D–23 in the Appendix gives this information, as well as the values for the modular ratio (n) based on steel having a constant modulus of elasticity of 29×10^3 k.s.i. regardless of the grade of steel being used. From this information, it can be seen that we are dealing with concrete whose mix is such that $f_c' = 4000$ p.s.i.* From the Data Sheet we find that $n = 8$. It might be noted here that because of variations that would be found when determining the modulus of elasticity of several specimens of the same mix design, it is permissible to round off the modular ratio to the nearest whole number.

Now, with $n = 8$, we make the transformation by increasing the area of steel (noted as A_s) eight times, observing the rule that only the width can be increased. A point to be made here is that the thickness of the transformed area of steel is negligible, since the bars used for reinforcing are small in diameter compared to the scale of the section. We are, therefore, concerned only with the dimension to the centerline of the transformed area. It should also be noted at this time that the *total* depth of a concrete member is somewhat meaningless for the purposes of structural analysis. We are only concerned with the *effective depth* of the section, which is taken from the extreme fiber in compression to the centerline of the steel, as shown in Fig. 13–4.

Transforming the section of Fig. 13–4 by converting the steel to an equivalent area of concrete, we get the transformed section shown in Fig. 13–5, where A_{tr} is the transformed area of the reinforcing steel.

The next step is to determine the location of the neutral axis of the effective section, which is shown as the shaded portion of the section. In order to do this it must be recognized that by definition, the neutral axis (which, of course, is the centroid of the effective section) is located in a position where the $A\bar{x}$ of everything on the compression side is equal to the $A\bar{x}$ of everything on the tension side.

Expressed in mathematical form,

$$A_c\bar{x}_c = A_t\bar{x}_t$$

where the subscripts indicate the compression side and the tension side and $\bar{x}$ is the distance from the centroids of the areas to the neutral axis.

To find the neutral axis of the transformed section of Fig. 13–5:

$$12(x)\left(\frac{x}{2}\right) = (16 \text{ in.}^2)(20 - x)$$

$$6x^2 = 320 - 16x \text{ or } x^2 + 2.7x - 53.3 = 0$$

Solving the quadratic,

$$x = \frac{-2.7 \pm \sqrt{220.5}}{2} = 6.1''$$

Note that there are two roots that come out of this expression. We are only concerned with the positive value.

The next step is to determine the transformed moment of inertia of the effective section with respect to the neutral axis.

I_{tr}

$$\frac{BD^3}{3} = \frac{(12)(6.1)^3}{3} = 908 \text{ in.}^4$$

$$A\bar{x}^2 = (16 \text{ in.})^2(13.9)^2 = \underline{3091} \text{ in.}^4$$

$$I_{tr} = 3999 \text{ in.}^4$$

Note that in determining the moment of inertia for the transformed steel, the transfer equation is used, which is

$$\frac{BD^3}{12} + A\bar{x}^2$$

However, we are considering that the transformed area of steel is very "thin"; consequently, only the $A\bar{x}^2$ portion of the transfer equation will contribute to the moment of inertia.

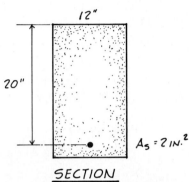

SECTION

E (STEEL) = 29×10^3 K.S.I.
E (CONC.) = 3.65×10^3 K.S.I.

FIGURE 13–4

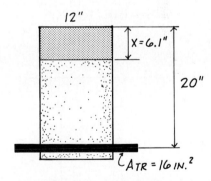

TRANSFORMED SECTION

FIGURE 13–5

---

*f_c' is the notation used to indicate the ultimate compressive strength of the concrete.

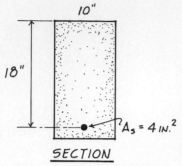

SECTION

E (STEEL) = 29×10^3 K.S.I.
E (CONC.) = 3.2×10^3 K.S.I.
$m = 9$

ALLOW. f (STEEL) = 20 K.S.I.
ALLOW. f (CONC.) = 1.35 K.S.I.

FIGURE 13–6

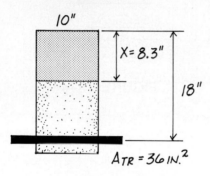

TRANSFORMED SECTION

FIGURE 13–7

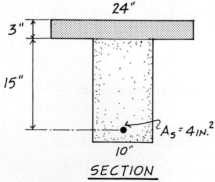

SECTION

E (STEEL) = 29×10^3 K.S.I.
E (CONC.) = 3.65×10^3 K.S.I.
$m = 8$

ALLOW. f (STEEL) = 24 K.S.I.
ALLOW. f (CONC.) = 1.8 K.S.I.

FIGURE 13–8

Example 13–2 (Figs. 13–6 and 13–7)

In this example we'll go a step further in our investigation of reinforced concrete members. For the given section of Fig. 13–6 and the data shown, we'll now determine the moment-carrying capacity. First, however, it is necessary to say a few words regarding the allowable stress shown for the concrete. It was mentioned earlier that up to a certain level of stress, a concrete test specimen will exhibit a nearly straight-line relationship between stress and strain. This level of stress is somewhere close to 50% of the ultimate strength of the concrete (f_c'). The ACI Code, in fact, sets the allowable stress to be used for working stress methods at .45f_c' in order to keep design stresses within the range where concrete behaves nearly elastically. We will use this code requirement; therefore,

$$f_c = .45f_c'$$

where f_c = allowable concrete stress
f_c' = ultimate compressive stress

Transforming the section of Fig. 13–6, we get the transformed section shown in Fig. 13–7, with the shaded portion representing the effective section. In order to determine the moment-carrying capacity of this section based on the given allowable stresses, we must first determine the location of the neutral axis of the effective section and the moment of inertia with respect to this axis.

Locating the neutral axis:

$$A_c \bar{x}_c = A_t \bar{x}_t$$

$$\therefore 10(x)\left(\frac{x}{2}\right) = 36(18 - x)$$

This expression yields a quadratic equation; solving this, we get

$$x = 8.3''$$

Determine the transformed moment of inertia.

I_{tr}

$\dfrac{(10)(8.3)^3}{3} = 1906.0 \text{ in.}^4$

$(36)(9.7)^2 = \underline{3387.2 \text{ in.}^4}$
$ 5293.2 \text{ in.}^4$

In order to determine the moment-carrying capacity of the section, we must consider the fact that we are dealing with two allowable stresses. The moment capacity of the section will be limited by the material that is at its full allowable stress. Therefore, two computations should be made.

Concrete, f = 1.35 k.s.i.

$$M = \frac{fI_{tr}}{c} = \frac{(1.35)(5293.2 \text{ in.}^4)}{(8.3'')(12''/')} = 71.8'^K$$

Steel, f = 20 k.s.i.

$$M = \frac{fI_{tr}}{cn} = \frac{(20)(5293.2)}{(9.7)(9)(12''/')} = 101.1'^K$$

Therefore, the concrete governs and the maximum moment capacity is 71.8$'^K$, with the concrete working at its full allowable stress and the steel under its allowable.

Example 13–3 (Figs. 13–8 and 13–9)

Up to this point, the examples shown have dealt with rectangular sections. In concrete, sections can be made of virtually any shape with relative ease. The analysis of an irregular section poses no real difficulty. As an example, let's determine the moment capacity of the T-shaped section shown in Fig. 13–8. The first step involved here is the location of the neutral axis. A slightly different approach is required from that used for a rectangular section. Because of the irregular shape, we must first determine if the neutral axis falls somewhere within the flange or if it falls below the flange. If the neutral axis falls within the flange, then, as far as we're concerned for the purposes of structural analysis, we are dealing with a rectangular section with the width being that of the flange, since the concrete below the neutral axis is not part of the effective section. If, however, we determine that the neutral axis falls below the flange, then the compression zone will be irregular in shape, thus affecting the setup of the expression used to determine the neutral axis location. The method used for determining whether or not the neutral axis falls in the flange or below it is reasonably simple and makes use of the fact that the $A\bar{x}$ on both sides of the axis are equal to each other.

With this in mind, consider a reference line at the bottom of the flange and compute the $A\bar{x}$ of the tension and compression sides using the transformed section shown in Fig. 13–9.

$$A_c\bar{x}_c = (3'' \times 24'')(1.5'') = 108 \text{ in.}^3$$

$$A_t\bar{x}_t = (32 \text{ in.}^2)(15) = 480 \text{ in.}^3$$

This shows us that with respect to the bottom of the flange, the $A\bar{x}$ of the tension side is much greater than that of the compression side, meaning that the neutral axis is below the flange. With this knowledge, we can properly set up the expression necessary to find the neutral axis. With reference to Fig. 13–9,

$$A_c\bar{x}_c = A_t\bar{x}_t$$

$$(72 \text{ in.}^2)(x + 1.5) + (10x)\left(\frac{x}{2}\right) = (32 \text{ in.}^2)(15 - x)$$

$$72x + 108 + 5x^2 = 480 - 32x$$

or

$$x^2 + 20.8x - 74.4 = 0$$

Solving the quadratic, we get $x = 3.1$; the distance to the neutral axis from the top of the section is 6.1″.

Determining the transformed moment of inertia:

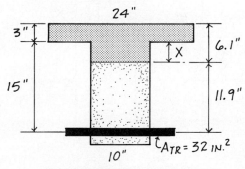

<u>TRANSFORMED SECTION</u>

FIGURE 13–9

$\underline{I_{tr}}$

$$\frac{(24)(3)^3}{12} + (72 \text{ in.}^2)(4.6)^2$$
$$= 1577.5 \text{ in.}^4$$

$$\frac{(10)(3.1)^3}{3} = 99.3 \text{ in.}^4$$

$$(32 \text{ in.}^2)(11.9)^2 = \underline{4531.5 \text{ in.}^4}$$

$$I_{tr} = 6208.3 \text{ in.}^4$$

To determine the moment capacity of the section, we must check both materials, based on allowable stresses.

Concrete (f = 1.8 k.s.i.)

$$M = \frac{fI_{tr}}{c} = \frac{(1.8)(6208.3)}{(6.1)(12''/')} = 152.7'^K$$

Steel (f = 24 k.s.i.)

$$M = \frac{fI_{tr}}{cn} = \frac{(24)(6208.3)}{(11.9)(8)(12''/')} = 130.4'^K$$

The steel governs, and the maximum moment-carrying capacity is 130.4′^K, with the steel stressed to its full allowable and the concrete under its allowable.

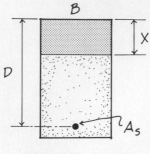

FIGURE 13–10

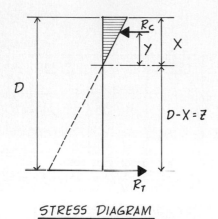

STRESS DIAGRAM

FIGURE 13–11

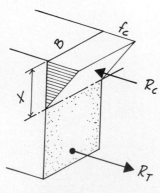

FIGURE 13–12

THE "INTERNAL COUPLE" METHOD

To this point, the discussion and the examples have made use of the bending stress equation. There is another approach that can be taken in analyzing a reinforced concrete member, called the "internal couple" method. While in problems of analysis, the use of the bending stress equation may often be a simpler approach (especially in irregular sections), the internal couple approach is the only way to design a reinforced concrete beam. Also, an understanding of the principles involved in this approach is vital when undertaking studies in strength design methods.

First we'll look at the ideas involved in the internal couple method, and then we'll consider several design problems using this approach.

In order to develop the general expressions used in the internal couple method, consider the generalized section of Fig. 13–10 and the corresponding stress diagrams of Figs. 13–11 and 13–12. The principles of Chapter 7, dealing with elastic bending, are used to develop the expressions. Specifically, the forces that constitute the internal couple, shown as R_c and R_t in the figures, must be equal to each other ($\Sigma F_x = 0$); they produce the resisting moment, which must be equal and opposite to the bending moment ($\Sigma M = 0$). Expressed in equation form,

$$R_c = R_t \text{ and B.M.} = \text{R.M.}$$

In order to expand these expressions, it should be recognized that R_c is the resultant of the compressive stress, and is equal to the volume of the "wedge" of stress shown in Fig. 13–12. R_t is the resultant of the tensile stress and is equal to the product of the area of steel and the stress in the steel (f_s), or

$$R_c = \frac{f_c B x}{2} = R_t = A_s f_s$$

and

$$\frac{f_c B x}{2} = A_s f_s \text{ (since } R_c = R_t)$$

In order to develop an expression for the moment, take moments due to R_c and R_t with respect to the neutral axis.

Referring to Fig. 13–11,

$$\therefore M = R(y + z)$$

which shows that the moment produced by the internal couple is the product of R and the distance between the forces.

At this point it seems that a numerical example, using the internal couple method, might serve to clarify the preceding discussion.

Example 13–4 (Figs. 13–13 through 13–15)

In order to determine the moment capacity of this section of Fig. 13–13 by the internal couple method, we must first locate the neutral axis of the transformed section shown in Fig. 13–14.

$$A_c \bar{x}_c = A_t \bar{x}_t$$

$$(14)(x)\left(\frac{x}{2}\right) = 27(24 - x)$$

or

$$x^2 + 3.9x - 92.6 = 0$$

Solving the quadratic,

$$x = 7.9''$$

Now, referring to the stress diagram for the transformed section, shown in Fig. 13–15, it can easily be determined which material will govern, based on given allowable stresses. This is done by using the similar triangle relationship of the stress diagram. The maximum stress allowed for the concrete is 1.35 k.s.i. The maximum stress on the steel side of the transformed section is $f_s/n = 2.22$ k.s.i. Now, if the concrete is stressed to 1.35 k.s.i. at the outer fiber of the compression zone, the stress in the transformed steel is

$$\frac{16.1}{7.9}(1.35) = 2.75 \text{ k.s.i.} > 2.22 \text{ k.s.i.}$$

This shows that if the concrete is up to its full allowable, the steel will be stressed beyond its allowable, indicating that the steel governs. The actual stress in the concrete can be determined by the similar triangle relationship, now using the full allowable stress in the transformed steel.

$$fc = \frac{7.9}{16.1}(2.22) = 1.09 \text{ k.s.i.}$$

The next step is to determine the value of R. Using the expression previously developed,

$$R = A_s f_s = (3)(20 \text{ k.s.i.}) = 60^K$$

The same value could have been found using the concrete side of the equation, or

$$R = \frac{f_c B x}{2} = \frac{(1.09)(14)(7.9)}{2} = 60^K$$

The moment can now be determined using the moment expression:

$$M = R(y + z)$$

Determining the distance between the forces of the internal couple is a simple matter for this *rectangular section*. The force R_c will act at the center of gravity of the wedge of stress. Therefore,

$$(y + z) = 24 - \frac{x}{3} = 24'' - 2.6'' = 21.4''$$

and

$$M = \frac{60^K(21.4'')}{12} = 107'^K$$

Note that by using the internal couple method we are able to determine the moment capacity directly from the statical relationship involved. The moment of inertia computation was not necessary; consequently, the use of this method seems simpler, but only for a rectangular section where the wedge of stress is regular and its center of gravity is easily determined. When dealing with an irregular section having an irregular wedge of stress, as shown in Fig. 13–16, the center of gravity of the wedge is not easily determined. In a case such as this, it is probably easier to compute I_{tr} and use the bending stress equation.

DESIGN OF REINFORCED CONCRETE BEAMS

Thus far, we have discussed procedures necessary for the investigation of reinforced concrete beams. In an investigation problem the cross section is known, and with this information we can determine the moment capacity or load-carrying capacity and the stresses in the concrete and reinforcing steel.

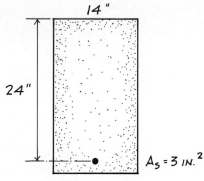

SECTION

$E \text{ (STEEL)} = 29 \times 10^3 \text{ K.S.I.}$
$E \text{ (CONC.)} = 3.2 \times 10^3 \text{ K.S.I.}$
$n = 9$

ALLOW. f (STEEL) $= 20$ K.S.I.
ALLOW. f (CONC.) $= 1.35$ K.S.I.

FIGURE 13–13

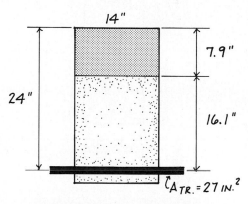

TRANSFORMED SECTION

FIGURE 13–14

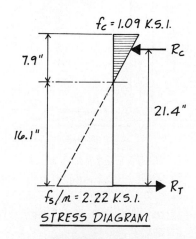

STRESS DIAGRAM

FIGURE 13–15

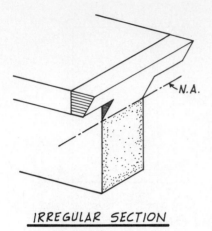

IRREGULAR SECTION

FIGURE 13–16

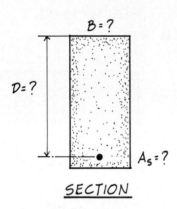

SECTION

FIGURE 13–17

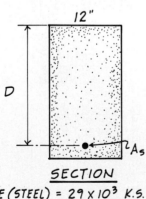

SECTION

$E \text{ (STEEL)} = 29 \times 10^3 \text{ K.S.I.}$
$E \text{ (CONC.)} = 3.2 \times 10^3 \text{ K.S.I.}$
$m = 9$

ALLOW. f (STEEL) = 20 K.S.I.
ALLOW. f (CONC.) = 1.35 K.S.I.

FIGURE 13–18

Investigation problems are useful primarily for academic purposes in that such problems serve as vehicles for presenting the principles involved.

The problem usually encountered by the designer of a reinforced concrete structure is not one of investigation, but one of design. That is, the cross-sectional geometry and the area of steel must be determined in response to the known loads and maximum moments that will be produced by those loads. This means that there are many unknowns regarding the cross section, as indicated in Fig. 13–17. In fact, there are too many unknowns involved to design by mathematical processes. Fortunately, when working with reinforced concrete there are other influences that help us make some judgments regarding the physical dimensions of the cross section, leaving the determination of the area of steel as the design problem. In any case, the section shown in Fig. 13–17 cannot be designed until the three unknowns indicated are reduced to, at most, two unknowns. The width (B) is the least critical dimension; generally, this is set based on standard form board sizes or by some predetermined architectural criterion. This leaves only two unknowns to be designed—the depth of the section and the area of steel required.

In the investigation problems previously done, it was shown that one of the materials involved governed the moment-carrying capacity of the section. If the steel or concrete was at its full allowable stress, then this was the limiting factor, and the other material was working at something less than its full allowable stress. In a design problem, however, we have the opportunity to arrange the geometry so that, for a given moment-carrying requirement, the steel and concrete will be working at their full allowable stresses. Such a section is referred to as a *balanced* section and, in working stress design, this is considered to be ideal, since both materials are being utilized to their fullest extent. We'll now look at an example of a balanced design and discuss other considerations in the design process.

Example 13–5 (Fig. 13–18)

The section shown in Fig. 13–18 is to be designed as a balanced section to carry a moment of $70^{'K}$. The width of the beam has been set at $12''$, as shown. In order to determine the required depth (D) and the area of steel (A_s), we must deal with the internal couple and the geometry of the stress diagram, as shown in Fig. 13–19. Since the goal is to design a balanced section, we know that the stresses in both materials are at their full allowables. Using the similar triangle relationship of the stress diagram,

$$\frac{1.35}{x} = \frac{2.22}{D - x}$$

and

$$1.35D - 1.35x = 2.22x$$

Rearranging this expression and solving for D in terms of x,

$$1.35D - 3.57x = 0$$

and

$$D = 2.64x \qquad (13–1)$$

We know, from the internal couple, that

$$R_c = \frac{f_c Bx}{2}$$

and, for this problem,   (13–2)

$$R_c = \frac{(1.35)(12)(x)}{2} = 8.1x$$

We also know, from the internal couple, that the moment is equal to the force R_c times the distance between R_c and R_t, or

$$M = R_c(D - x/3)$$

Substituting Eqs. (13–1) and (13–2) for D and R_c, we get

$$70'^K \times 12''/' = (8.1x)(2.64x - .33x)$$
$$= 18.7x^2$$

and

$$\underline{x = 6.7''}$$

Substituting this value into Eq. (13–1), we can find the value for D:

$$D = 2.64x = (2.64)(6.7) = 17.7''$$

In order to find the area of steel (A_s), we can use Eq. (13–2) and solve for R_c:

$$R_c = (8.1)(x) = (8.1)(6.7) = 54.3^K$$

We also know that

$$R_c = R_t \quad \text{and} \quad R_t = A_s f_s$$
$$\therefore 54.3^K = A_s(20)$$

and

$$A_s = 2.7 \text{ in.}^2$$

The design of this balanced section is now complete.

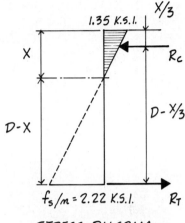

STRESS DIAGRAM

FIGURE 13–19

It must be emphasized, at this point, that given the geometry of the stress diagram with the values of allowable stresses for both materials, there is *only one* value that D can be for the given moment. Any other arbitrarily chosen value for the effective depth will cause this section to be something other than balanced.

It was mentioned earlier that in a building design, often there are certain factors that will force us to set dimensions for a reinforced concrete beam. These factors can set the depth requirement. For example, architectural requirements may call for a beam from the top of an opening to the floor line, as shown in Fig. 13–20. Whatever this total depth may be, the effective depth will be about 2 in. less than this. Except by some remote possibility, this section will be something other than balanced. The design problem here, with the width and depth set, comes down to sizing the area of steel. When dealing with such a condition, where we are restrained from making the depth the unique dimension that will cause the beam to be balanced, there are two other possibilities of stress relationships that must be considered. They are referred to as an *underreinforced* section and an *overreinforced* section.

Summarizing the three possibilities and defining the expressions:

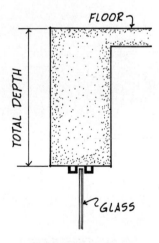

FIGURE 13–20

1. *Balanced section*. Both materials are working at their full allowable stresses. Generally not practical because of architectural requirements or standard form board dimensions which set the width and the depth of the section.

2. *Underreinforced section*. This is a section that is *deeper* than a balanced section would be for a given moment, causing the steel to be at its full allowable stress and the concrete to be below its allowable stress.

3. *Overreinforced section*. This is a section that is *shallower* than a balanced section would be for a given moment, causing the concrete to be at its full allowable stress and the steel to be below its allowable.

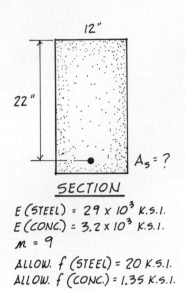

SECTION

$E\ (STEEL) = 29 \times 10^3\ K.S.I.$
$E\ (CONC.) = 3.2 \times 10^3\ K.S.I.$
$m = 9$

$ALLOW.\ f\ (STEEL) = 20\ K.S.I.$
$ALLOW.\ f\ (CONC.) = 1.35\ K.S.I.$

FIGURE 13–21

As mentioned previously, the balanced section is ideal, but because of practical considerations, it can rarely be achieved. Given the other two alternatives, the underreinforced section is more desirable; in fact, practically all members designed by working stress methods are designed as underreinforced members. The reason this is more desirable is because an underreinforced section provides an extra measure of safety. If for some reason a reinforced concrete beam is subjected to a serious overload, the stresses in both materials will increase accordingly. In an underreinforced section, the steel is at its full allowable stress under normal loads, and in an overload situation the stresses will approach the yield strength of the steel with corresponding large strains. Because of its ductility, the steel will not break, but the bottom side of the beam will crack seriously, thereby giving warning of failure. The beam will remain intact although seriously deflected.

In an overreinforced section, where the steel is under its allowable and concrete is at its full allowable, an overload situation may cause the concrete to reach its ultimate strength before the steel yields, thereby causing a failure in the compression zone. A compression failure in concrete is, literally, explosive in nature, and there is little advance warning. This is a dangerous sort of failure; consequently, the design of overreinforced sections is to be avoided.

Since an underreinforced section is rather common, we will now go through an example of the design of such a section.

Example 13–6 (Figs. 13–21 and 13–22)

For this problem we will redesign the section of Fig. 13–18, which was initially designed as a balanced section to carry a moment of $70'^K$. For the balanced design it was determined that a depth of $17.7''$ was required. For the same width and moment we will now make the section $22''$ deep, as shown in Fig. 13–21. This will cause the section to be underreinforced, since it is deeper than the balanced section for the given condition. Since the section is underreinforced, we know that the steel will govern and it will be at its full allowable stress, as shown in the stress diagram of Fig. 13–22. In order to design this section—which, in essence, is to size the area of steel required—we must determine the location of the neutral axis. This is the complex part of the problem. Using the geometry of the stress diagram and the statical relationships involved will provide a solution. To begin, we know that:

$$R_c = R_t\ (\Sigma F_x = 0)$$

and

$$R_c = \frac{f_c(B)(x)}{2} = R_t = A_s f_s \tag{13–3}$$

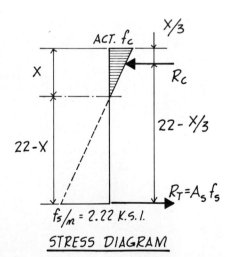

STRESS DIAGRAM

FIGURE 13–22

Using the Internal Couple,

$$M = 70'^K \times 12''/' = A_s f_s\left(22 - \frac{x}{3}\right)$$
$$= R_c\left(22 - \frac{x}{3}\right) \tag{13–4}$$

and, by similar triangles,

$$\frac{f_c}{x} = \frac{2.22}{22 - x},\ \ \therefore f_c = \frac{2.22x}{22 - x} \tag{13–5}$$

Substituting this value into Eq. (13–3),

$$R_c = \frac{2.22x}{22 - x}\left(\frac{1}{2}\right)(12)(x) = \frac{13.32x^2}{22 - x}$$

and substituting this value of R_c into the "concrete" side of Eq. (13–4),

$$70'^K \times 12''/' = \frac{13.32x^2}{22 - x}\left(22 - \frac{x}{3}\right)$$

and

$$840 = \frac{293x^2 - 4.4x^3}{22 - x}$$

which is a cubic equation whose solution is achieved most easily by a trial-and-error method. This is so because we are looking for a positive value for x, and we know it must be somewhere between zero and 22″, which is the depth of the section. The range of possibilities is not even that broad. From the experience of working several problems, we know that the neutral axis is several inches down and certainly nowhere close to the steel. A narrower and more likely range for x is probably somewhere between 5″ and 10″. After two trials the value of x for this problem was found, within acceptable limits of accuracy, to be

$$x = 7''$$

Now we can proceed to determine the area of steel required. Using the "steel" side of Eq. (13–4)

$$70'^K \times 12''/' = A_s(20)(22 - 2.33)$$

and

$$A_s = \frac{840}{440 - 46.6} = 2.14 \text{ in.}^2$$

ONE-WAY SLABS

One-way slabs, as shown in the framing plan of Fig. 13–23, are analyzed and designed in precisely the same manner as beams. For the sake of convenience, we generally deal with a typical 1-ft-wide strip of slab, as shown in the section of Fig. 13–24. This, then, becomes a shallow beam with a width of 12″. The information determined from this typical strip is then repeated for every foot of width. Since the process involved is the same as that of beam design, there is no need to present further examples.

The previous discussions and procedures made no reference to *specific* recommendations made by the ACI Code, which provides a great deal of input into the design process, such as minimum area of steel requirements, minimum steel coverage, and other detail items. In practice, the ACI Code should be adhered to, thereby introducing, into the design process, certain factors that have not been discussed here.

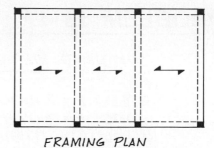

FRAMING PLAN

FIGURE 13–23

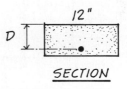

SECTION

FIGURE 13–24

PROBLEM 13-1

<u>COMBINED MATERIALS: REINFORCED CONCRETE: W.S.D.</u>

STEEL: $E = 29000$ K.S.I.   ALLOWABLE $f = 20$ K.S.I.
CONC: $E = 3625$ K.S.I.   ALLOWABLE $f = 1.8$ K.S.I.
ALLOWABLE TENSILE STRESS IN CONCRETE = O

- FOR EACH SECTION, DETERMINE:

(1) LOCATION OF THE NEUTRAL AXIS.
(2) TRANSFORMED MOMENT OF INERTIA.
(3) RESISTING MOMENT CAPACITY (WHICH MATERIAL GOVERNS?)
(4) WHAT IS THE ACTUAL STRESS IN THE OTHER MATERIAL?

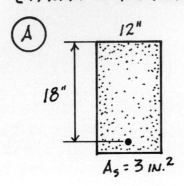

(A) 12" / 18" / $A_s = 3$ IN.^{2}

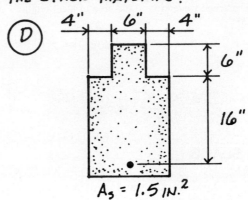

(D) 4" 6" 4" / 6" / 16" / $A_s = 1.5$ IN.^{2}

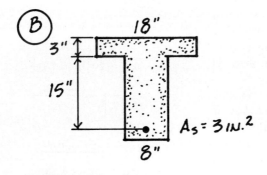

(B) 18" / 3" / 15" / 8" / $A_s = 3$ IN.^{2}

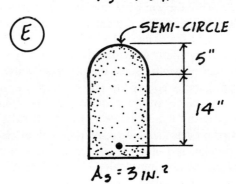

(E) SEMI-CIRCLE / 5" / 14" / $A_s = 3$ IN.^{2}

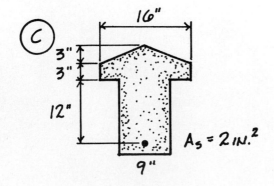

(C) 16" / 3" / 3" / 12" / 9" / $A_s = 2$ IN.^{2}

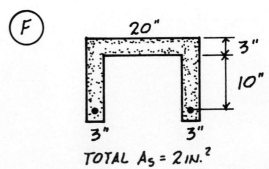

(F) 20" / 3" / 10" / 3" / 3" / TOTAL $A_s = 2$ IN.^{2}

PROBLEM 13-2

COMBINED MATERIALS: REINFORCED CONCRETE: W.S.D.

(I) FOR THE SECTIONS SHOWN IN PROBLEM #13-1, DETERMINE THE NECESSARY LOCATION OF THE NEUTRAL AXIS SO THAT THE ALLOWABLE STRESS FOR STEEL (f_s) AND THE ALLOWABLE STRESS FOR THE CONCRETE (f_c) WILL BE REACHED SIMULTANEOUSLY.
- DETERMINE THE NECESSARY AREA OF STEEL (A_s) TO ACHIEVE THIS, AND THE RESISTING MOMENT CAPACITY.

(II) USING THE "INTERNAL COUPLE" METHOD, DETERMINE THE ALLOWABLE RESISTING MOMENT FOR EACH SECTION. $A_s = 2.70$ IN.^{2} FOR EACH.

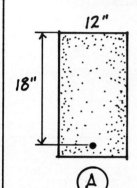

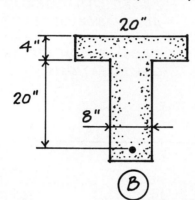

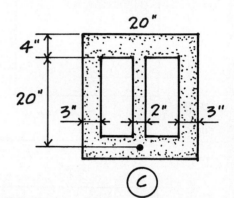

| A | B | C |

E (STEEL) = 30000 K.S.I.
E (CONC.) = 3333 K.S.I.

f_s ALLOW. = 20 K.S.I.
f_c ALLOW. = 1.5 K.S.I.

(III)

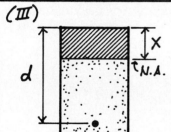

GIVEN: $A_s = 2.70$ IN.^{2}
(1) DETERMINE "X" AND "d" SO THAT ALLOW. f_s & f_c WILL BE REACHED SIMULTANEOUSLY.
(2) DETERMINE ALLOWABLE RESISTING MOMENT.

f_s ALLOW. = 24 K.S.I.
f_c ALLOW. = 1.5 K.S.I.

$n = \dfrac{E_s}{E_c} = 8.9$

(IV)

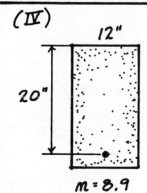

$n = 8.9$

DESIGN: DIMENSIONS GIVEN: FIND AREA OF STEEL.

BENDING MOMENT ON SECTION SHOWN = 50^{1K}
- DETERMINE A_s NEEDED AT ALLOW. $f_s = 20$ K.S.I.

NOTE: THE GIVEN INFORMATION CAUSES THIS MEMBER TO BE UNDERREINFORCED.

HINT: THINK ABOUT THE INTERNAL COUPLE AND SET UP THE APPROPRIATE EQUATIONS.

PROBLEM 13-3

REINFORCED CONCRETE - ALTERNATE METHOD

(I)

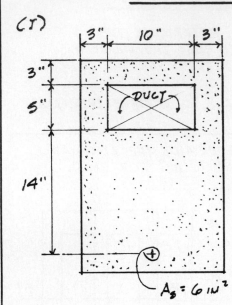

$E \text{ (CONC.)} = 3625 \text{ KSI}$
$E \text{ (STL.)} = 29000 \text{ KSI}$ $\Big\}$ $m = 8$

ALLOW. $f_c = 1.8 \text{ KSI}$
ALLOW. $f_s = 20 \text{ KSI}$

IS THE GIVEN SECTION BALANCED,
UNDERREINFORCED OR OVERREINFORCED
BASED ON MATERIAL REACHING
ALLOWABLE STRESS?

SHOW PROOF OF YOUR ANSWER BY
COMPUTATIONS.

(II)

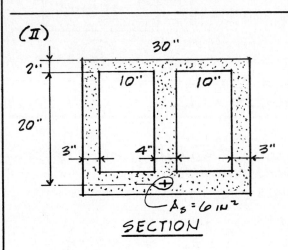

SECTION

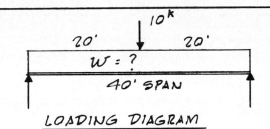

LOADING DIAGRAM

ALLOW. $f_c = 1.8 \text{ KSI}$
ALLOW. $f_s = 24 \text{ KSI}$.
$E \text{ (STEEL)} = 29 \times 10^3$
$E \text{ (CONC.)} = 3.65 \times 10^3$ $\Big\}$ $m = 8$

(1) DETERMINE THE LOCATION OF THE NEUTRAL AXIS FOR THE GIVEN SECTION.

(2) DETERMINE THE TRANSFORMED MOMENT OF INERTIA.

(3) DETERMINE THE MOMENT CARRYING CAPACITY, USING THE BENDING
STRESS EQUATION.

(4) DETERMINE THE UNIFORMLY DISTRIBUTED (W^k) THAT MAY BE ADDED
TO THE GIVEN CONCENTRATED LOAD.

(5) IS THE SECTION BALANCED, UNDERREINFORCED OR OVERREINFORCED?
EXPLAIN YOUR ANSWER.

PROBLEM 13–4

COMBINED MATERIALS - REINFORCED CONCRETE: W.S.D.

(I)

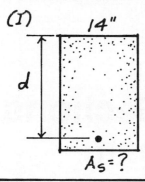

14"

d

$A_s = ?$

THE GIVEN SECTION MUST RESIST A MAXIMUM BENDING MOMENT OF 200^{lk}. DESIGN THE SECTION SO THAT IT IS EXACTLY BALANCED.

$E_s = 29000$ K.S.I.   ALLOW. $f_s = 24$ K.S.I.

$E_c = 3625$ K.S.I.   ALLOW. $f_c = 1.8$ K.S.I.

(II) ONE-WAY SLAB DESIGN

ONE-WAY SLAB

12'

PLAN VIEW

THE ONE-WAY SLAB SHOWN HAS AN EFFECTIVE DEPTH OF 4", AND AN ADDITIONAL 1" OF CONCRETE TO COVER THE STEEL (TOTAL DEPTH = 5").

L.L. = 150 P.S.F.
D.L. = (5" SLAB) = 62.5 P.S.F.

$E_s = 29000$ K.S.I.   ALLOW. $f_s = 20$ K.S.I.
$E_c = 3625$ K.S.I.   ALLOW. $f_c = 1350$ P.S.I.

DETERMINE THE REQUIRED AREA OF STEEL (A_s) PER FOOT OF WIDTH.

12"

4"
1"

SECTION

(III) PRE-CAST CONCRETE "T" SECTION

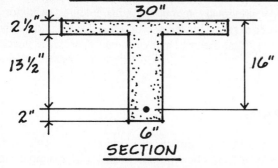

30"

2½"

13½"

2"

6"

16"

SECTION

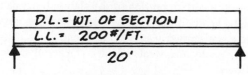

D.L. = WT. OF SECTION
L.L. = 200 #/FT.

20'

LOADING DIAGRAM

THE GIVEN PRE-CAST SECTION IS TO BE USED AS A SIMPLY SUPPORTED BEAM AS SHOWN IN THE LOADING DIAGRAM.

DETERMINE THE REQUIRED AREA OF STEEL (A_s).

IS THE COMPRESSIVE STRESS IN THE CONCRETE O.K.?

$$n = \frac{E_s}{E_c} = 9$$

ALLOW. $f_s = 24$ K.S.I.
ALLOW. $f_c = 1350$ P.S.I.

14

Composite Sections and Steel Beam Design

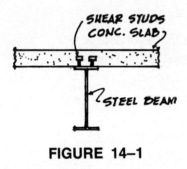

FIGURE 14–1

INTRODUCTION

Essentially, the purpose of this chapter is to present methods of analysis for a special kind of combined materials situation known as *composite construction*. The basic ideas of composite construction were presented in Chapter 5. In this chapter we will deal with numerical methods that will show the advantages of composite construction.

While all sections of combined materials may be properly thought of as composite sections, the term *composite,* in structures, is generally thought of as a combination of a concrete slab acting integrally with a steel beam upon which it rests, as shown in Fig. 14–1. In a case such as this, the concrete slab may be made to act integrally with the steel beam through the use of shear studs. As previously mentioned, this integral behavior of a concrete slab and a steel beam, with the use of shear studs, is described in Chapter 5. It is suggested that the reader review that chapter, if necessary, before proceeding.

BASIS FOR DATA

When dealing with composite construction, it is important to determine how much of the slab works effectively with the beam. For example, in the section of Fig. 14–2, only a certain portion of the slab will act as an integral flange on the steel beam. While it is difficult to determine the answer to this in absolute terms, there are certain useful recommendations, based on experience, that are provided by the steel industry. Generally, these recommendations are as follows:

When the slab extends on both sides of the steel beam:

1. The effective width of the concrete flange is to be taken as not more than one fourth of the span of the beam.

SECTION · SLAB & STEEL BEAM

FIGURE 14–2

2. The effective projection beyond the edge of the beam is to be not more than one half the clear distance to the next beam.
3. The effective projection beyond the edge of the beam shall be not more than eight times the slab thickness.

When these recommendations have been evaluated for the conditions at hand, the least value shall be used as the effective width of the slab.

When the slab exists on only one side of the beam (as in a spandrel beam):

1. The effective width of the concrete flange (the projection beyond the beam flange) is to be taken as not more than one twelfth of the beam span.
2. The projection of the slab beyond the beam flange is to be not more than six times the slab thickness.
3. The projection of the slab beyond the beam flange shall be not more than one half of the clear distance to the adjacent beam.

The least of the values obtained from the above is to be used in the design of a composite section.

For the purposes of this chapter, it will be necessary to make reference to certain data pertaining to steel sections. Data Sheets D–27 through D–36, in the Appendix, will provide all the data we'll need for structural steel beams.

DESIGN OF STEEL BEAMS

One of the purposes of this chapter is to present some comparisons between composite and non-composite sections. In a non-composite section the steel beam must carry the load without any benefit of the concrete slab (meaning that there is no positive attachment between the two to resist horizontal shear). Therefore, it is necessary, as a first step, that the student learn to design a steel beam. The wide flange steel beam is the most commonly used shape for architectural structures, so we will limit our selections to this shape.

Generally, steel beams are designed to resist stresses due to bending produced by the loads. Except for unusual conditions of loading, as pointed out in Chapter 10, shearing stresses will normally not govern in the design of a steel beam. Consequently, we will concern ourselves with the application of the bending stress equation:

$$f = \frac{Mc}{I} \quad \text{or} \quad M = \frac{f}{c} I$$

There is, however, a bit of a problem in the use of the bending stress equation in this form. It is the same problem that was encountered in the design process for wood and laminated timber beams, presented in Chapter 11. Specifically, in the design process the factors I and c are not known. Therefore, we will use another form of the bending stress equation that uses the section modulus (S), which is simply the ratio of I/c. Consequently, the bending stress equation may be restated as follows:

$$M = f \frac{I}{c} = fS$$

Using this form of the equation, we need only to solve for S based on the maximum moment, which is a function of the load and span, and the allowable stress of the material. While there are several grades of steel available, the most commonly used grade has an allowable bending stress of 24 k.s.i. We will limit ourselves to this grade of steel for the example problems.

Example 14–1 (Fig. 14–3)

In this example it is required that a wide flange steel beam be designed to resist the maximum stresses produced due to bending. These maximum stresses occur at the point of maximum bending moment. Therefore, the first step is to determine the maximum bending moment. For a simply supported, uniformly loaded beam, the maximum moment is determined by

$$M = \frac{WL}{8} = \frac{(20^K)(20')}{8} = 50'^K$$

Considering that the grade of steel used for this beam has an allowable bending stress of 24 k.s.i., then we need only to determine the section modulus required, using the bending stress equation:

$$S = \frac{M}{f} = \frac{(50'^K \times 12''/')}{24 \text{ k.s.i.}} = 25 \text{ in.}^3$$

We must now select a wide flange beam that has a section modulus of at least 25 in.[3] There are many sections made that will satisfy this requirement. We will be concerned with the selection of the most economical section, which is the section that has the least weight (see Data Sheet D–27 for the designations used for steel sections). This means that we want to select a member that satisfies the section modulus requirement and has the least weight in satisfying this requirement. Data Sheets D–35 and D–36 provide a list of wide flange beams and their section moduli. In scanning this data, it is found that the most economical section that will satisfy the requirements of this example is the W12 × 22, which has a section modulus slightly higher than that required.

Example 14–2 (Fig. 14–4)

In this example we have the same load and span as in Example 14–1, except that the load is concentrated at the midspan. Again, it is required to design a wide flange steel beam based on an allowable stress, in bending, of 24 k.s.i. The maximum moment produced by this loading is

$$M = \frac{PL}{4} = \frac{(20^K)(20')}{4} = 100'^K$$

The required section modulus is

$$S = \frac{M}{f} = \frac{(100'^K \times 12''/')}{24 \text{ k.s.i.}} = 50 \text{ in.}^3$$

In scanning the data given on Data Sheets D–35 and D–36, it will be found that the section that furnishes the closest section modulus to this (but not under this) is a W12 × 40, which has a section modulus of 51.9 in.[3] However, this is *not* the most economical section that will satisfy the requirements. If the data on Data Sheets D–35 and D–36 are carefully reviewed, it will be seen that a variety of sections are placed in distinct groups. At the top of each group there is a member designated in bold type. This member is the most economical of all members in the group. In other words, when we find a member with a section modulus that will just satisfy the requirements, this may not be the most economical section, although it will do the job. If we go to the top of the group, to the member designated in bold type, then we will find a beam that has more section modulus than required, but less weight. Therefore, the most economical section that will

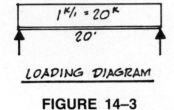

FIGURE 14–3

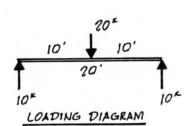

FIGURE 14–4

satisfy the problem of this example is not the W12 × 40, but the W18 × 35, which weighs 5#/′ less.

It must be mentioned here that there may be factors other than the weight of steel that may affect economy. In the previous example, it was shown that a W 12 × 40 would do the job. This member is 6 in. shallower in depth than the W 18 × 35, which was finally selected as the most economical because it weighed less. Considering that the depth of structural beams greatly influences the ceiling-to-floor thickness in a building, then there may be other issues involved in the economy of the building. For example, in a 20-story building, a 6-in. greater depth per floor for the structural thickness, based on the selection of the member with the least weight, will result in a 10-ft greater building height (considering fixed floor-to-ceiling requirements). This will result in greater costs for the "skin" of the building, increased height of stairs, elevator shafts, etc. The major point here is that questions of economy may often involve some very complex issues and may easily include considerations beyond the initial costs of material.

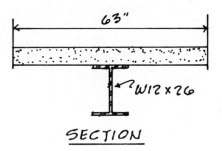

SECTION

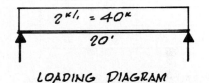

LOADING DIAGRAM

FIGURE 14–5

ANALYSIS OF COMPOSITE SECTIONS

Composite sections, where steel beams act integrally with concrete slabs, offer some advantages in terms of material savings. In order to demonstrate the advantages involved, we will immediately proceed to some example problems.

Example 14–3 (Figs. 14–5 and 14–6)

Let's consider the section loaded as shown in Fig. 14–5. The steel beam shown has the following properties:

Data:

Steel beam: W12 × 26
Steel f (all.): 24 k.s.i.
Conc. f (all.): 1.35 k.s.i.
Modular ratio $= \dfrac{E_c}{E_s} = \dfrac{1}{9}$

Note: The modular ratio given is based on the concrete being transformed to an equivalent area of steel.

To begin, we will first determine the most economical steel beam (W shape) that is necessary to carry the load without composite action. In order to do this, we must first determine the maximum bending moment that must be carried. For a simple beam with a uniformly distributed load:

$$M = \frac{WL}{8} = \frac{(40^K)(20')}{8} = 100'^K$$

and

$$S = \frac{M}{f} = \frac{(100'^K \times 12''/')}{24 \text{ k.s.i.}} = 50 \text{ in.}^3$$

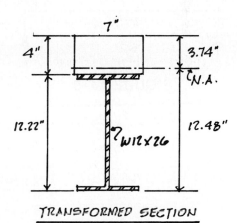

TRANSFORMED SECTION

FIGURE 14–6

In scanning Data Sheets D–35 and D–36, it will be found that the most economical section is a W18 × 35. However, Fig. 14–5 calls for a W12 × 26, which weighs 9#/′ less than the W18 × 35.

Let's now analyze the stresses that will be produced in the composite section that is made up of the W12 × 26 acting integrally with a concrete slab, as shown in the section of Fig. 14–5.

The first step necessary to do this is to establish the transformed section and determine its properties. The conventional method for developing the transformed section for a composite section such as the one under consideration is to transform the concrete to an equivalent area of steel by reducing only the width. Using the given data, the modular ratio is 1/9, and the transformed section is shown in Fig. 14–6. In order to determine the stresses in each material, we must first determine the location of the neutral axis and the transformed moment of inertia.

Location of Neutral Axis

Using the transformed section, as shown in Fig. 14–6 (Refer to Data Sheets D-27 through D-34):

$\underline{\Sigma Ax}$ (using the top as a reference)

$$
\begin{aligned}
\square \quad (28 \text{ in.}^2)(2 \text{ in.}) &= \quad 56 \text{ in.}^3 \\
\text{I} \quad (7.65 \text{ in.}^2)(10.11 \text{ in.}) &= \underline{\quad 77.34 \text{ in.}^3} \\
\Sigma Ax &= 133.34 \text{ in.}^3
\end{aligned}
$$

$$
\bar{x} = \frac{133.34 \text{ in.}^3}{35.65 \text{ in.}^2} = 3.74''
$$

Determine the transformed moment of inertia:

To begin with, it should be noted that the small amount of concrete below the neutral axis (which is the tension side) will be ignored in terms of its contribution to the moment of inertia of the section.

$\underline{I_{tr}}$

$$
\begin{aligned}
\square \quad \frac{(7)(3.74)^3}{3} &= 122.1 \text{ in.}^4 \\
\text{I} \quad 204 + (7.65)(6.37)^2 &= \underline{514.4 \text{ in.}^4} \\
I_{tr} &= 636.5 \text{ in.}^4
\end{aligned}
$$

Determine maximum stresses:
Steel:

$$
f = \frac{(100 \times 12)(12.48)}{636.5} = 23.5 \text{ k.s.i.}
$$

Concrete:

$$
f = \frac{(100 \times 12)(3.74)(1/9)}{636.5} = .78 \text{ k.s.i.}
$$

Both of these stresses are below the allowables specified for each material. The W12 × 26, acting compositely with the concrete slab, is suitable, thereby producing a savings in the weight of steel required, compared to the W18 × 35 required without composite action.

Example 14–4 (Figs. 14–7 and 14–8)

Referring to Fig. 14–7, and using the following data:

> Steel beams: W18 × 35 ($S = 57.6 \text{ in.}^3$)
> f (all.) = 24 k.s.i.
> 6'' thick conc. slab:
> f (all.) = 1.8 k.s.i.

$$
n = \frac{E_c}{E_s} = \frac{1}{8}
$$

Considering that the section shown in Fig. 14–7 is used as a simple span, as

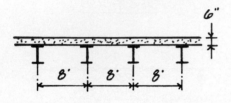

6"

PARTIAL SECTION THRU FLOOR

FIGURE 14–7

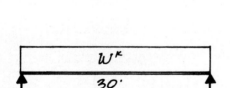

W^K

30'

FIGURE 14–8

shown in Fig. 14–8, let's determine the uniformly distributed load that may be safely carried by the given steel beam (W18 × 35) without the benefit of composite action. The maximum moment produced by this loading is

$$M = \frac{WL}{8} = \frac{W(30')}{8} = fS$$

$$\therefore W = \frac{(24 \text{ k.s.i.})(57.6 \text{ in.}^3)(8)}{(30')(12''/')} = 30.7^K$$

Now let's consider that this same beam is made to act compositely with the concrete slab it supports. In order to evaluate the uniformly distributed load-carrying capacity of this section, it is first necessary to determine the width of the concrete slab that will act effectively as an integral part of the section. To do this, we must refer to previously stated criteria. Therefore, the following must be evaluated:

1. The effective width of the concrete flange shall be not more than one fourth the span of the beam:

$$\therefore \text{Flange width} = 7.5' = 90''$$

2. The effective projection of the concrete flange beyond the edge of the beam shall be not more than one half the clear distance to the next beam:

$$\therefore \text{Flange width} > 7.5'$$

3. The effective projection of the concrete flange beyond the edge of the beam shall be not more than eight times the slab thickness:

$$\therefore \text{Flange width} > 7.5'$$

and criterion 1 governs:

$$\therefore \text{Effective flange width} = 7.5'$$

Based on the above criteria, the transformed section is as shown in Fig. 14–9. We will now proceed to determine the location of the neutral axis and the moment of inertia of the transformed section.

Determine location of neutral axis:

ΣAx (using the top as a reference)

$$(6 \times 11.25)(3) = 202.5 \text{ in.}^3$$

$$(10.3)(14.85) = \underline{153.0 \text{ in.}^3}$$

$$\Sigma Ax = 355.5 \text{ in.}^3$$

$$\bar{x} = \frac{355.5 \text{ in.}^3}{77.8 \text{ in.}^2} = 4.57''$$

Determine I_{tr} (ignoring concrete below the neutral axis):

$$\frac{(11.25)(4.57)^3}{3} = 357.9 \text{ in.}^4$$

$$(510) + (10.3)(10.28)^2 = \underline{1598.5 \text{ in.}^4}$$

$$I_{tr} = 1956.4 \text{ in.}^4$$

Determine the safe uniform load:
Steel:

$$\frac{WL}{8} = \frac{(24)(1956.4)}{(19.13)(12)}, \quad W = 54.5^K$$

Concrete:

$$\frac{WL}{8} = \frac{(1.8)(1956.4)}{(4.57)(1/8)(12)}, \quad W = 137^K$$

$$\therefore \text{Steel governs and } W = 54.5^K$$

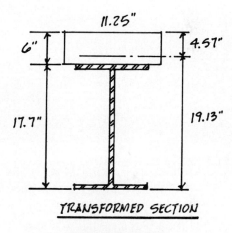

TRANSFORMED SECTION

FIGURE 14–9

Compared to $W = 30.7^K$ for the same beam acting alone, the advantages of composite construction, in terms of material savings, should be quite apparent.

It might well be emphasized at this point that several issues were not mentioned during the previous discussions. These have to do with the design of the shear studs that are necessary for true composite action to take place between the beam and slab, as well as certain building code requirements affecting composite design procedures. The examples shown were presented to demonstrate the fundamentals involved in the analysis of a composite section and the inherent advantages, in terms of structural efficiency. For detailed recommendations and procedures for the design of composite sections, it is suggested that the specifications of the American Institute of Steel Construction be used as a reference.

PROBLEM 14–1

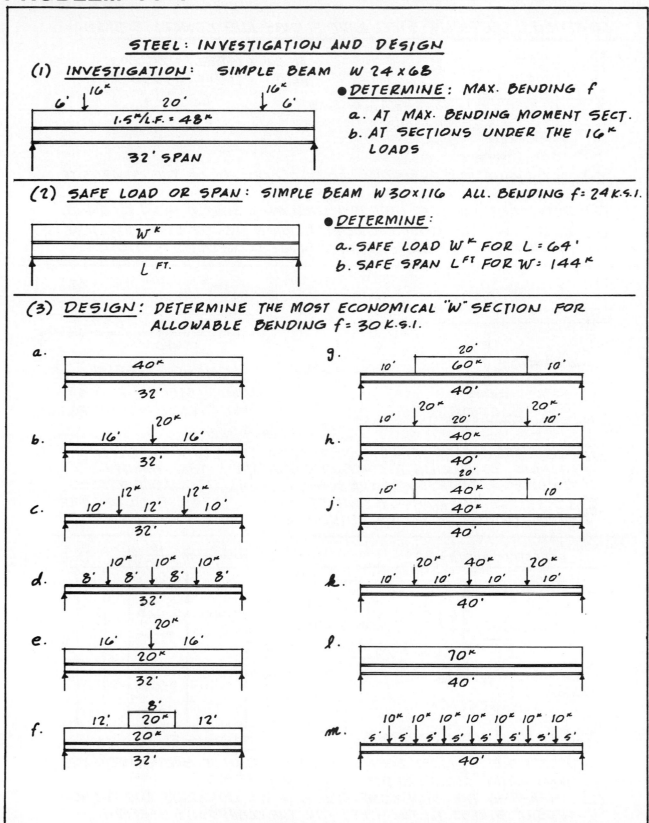

STEEL: INVESTIGATION AND DESIGN

(1) INVESTIGATION: SIMPLE BEAM W 24 x 68

- DETERMINE: MAX. BENDING f

  a. AT MAX. BENDING MOMENT SECT.
  b. AT SECTIONS UNDER THE 16ᴷ LOADS

6' 16ᴷ   20'   16ᴷ 6'

1.5"/L.F. = 48ᴷ

32' SPAN

(2) SAFE LOAD OR SPAN: SIMPLE BEAM W 30 x 116 ALL. BENDING f = 24 K.S.I.

- DETERMINE:

  a. SAFE LOAD Wᴷ FOR L = 64'
  b. SAFE SPAN Lꜰᵀ FOR W = 144ᴷ

Wᴷ

L ꜰᵀ.

(3) DESIGN: DETERMINE THE MOST ECONOMICAL "W" SECTION FOR ALLOWABLE BENDING f = 30 K.S.I.

a.
40ᴷ
32'

b.
16'   20ᴷ   16'
32'

c.
10' 12ᴷ 12' 12ᴷ 10'
32'

d.
8' 10ᴷ 8' 10ᴷ 8' 10ᴷ 8'
32'

e.
16'   20ᴷ   16'
20ᴷ
32'

f.
8'
12' 20ᴷ 12'
20ᴷ
32'

g.
20'
10' 60ᴷ 10'
40'

h.
10' 20ᴷ 20' 20ᴷ 10'
40ᴷ
40'

j.
20'
10' 40ᴷ 10'
40ᴷ
40'

k.
10' 20ᴷ 10' 40ᴷ 10' 20ᴷ 10'
40'

l.
70ᴷ
40'

m.
10ᴷ 10ᴷ 10ᴷ 10ᴷ 10ᴷ 10ᴷ 10ᴷ
5' 5' 5' 5' 5' 5' 5' 5'
40'

PROBLEM 14–2

COMPOSITE SECTIONS - STEEL WIDE FLANGE AND CONCRETE SLAB

(I)

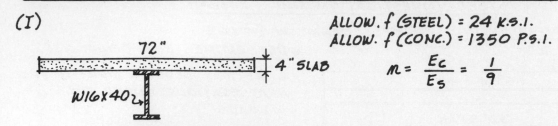

ALLOW. f (STEEL) = 24 K.S.I.
ALLOW. f (CONC.) = 1350 P.S.I.

$$m = \frac{E_c}{E_s} = \frac{1}{9}$$

72"
4" SLAB
W16×40

(1) DETERMINE THE RESISTING MOMENT CAPACITY OF THE COMPOSITE SECTION SHOWN.
(2) DETERMINE THE SAFE UNIFORM LOAD ON A SIMPLE SPAN OF 24 FT.
(3) DETERMINE THE MOST ECONOMICAL WIDE FLANGE STEEL SECTION NEEDED FOR THE SAME LOAD, WITHOUT COMPOSITE ACTION.

(II)

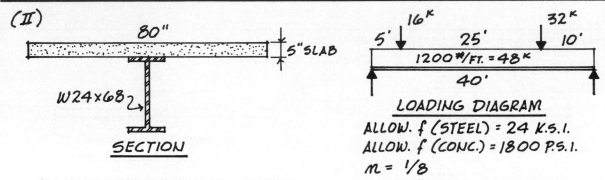

80"
5" SLAB
W24×68

SECTION

16^K 32^K
5'   25'   10'
1200 #/FT. = 48^K
40'

LOADING DIAGRAM
ALLOW. f (STEEL) = 24 K.S.I.
ALLOW. f (CONC.) = 1800 P.S.I.
$m = 1/8$

(1) CHECK THE SAFETY OF THE COMPOSITE SECTION, WITH THE GIVEN LOADING, BY MAKING A COMPARISON OF THE ACTUAL MAXIMUM TENSILE UNIT STRESS IN THE STEEL, WITH THE ALLOWABLE STRESS.
(2) DETERMINE THE MAXIMUM UNIT STRESS IN A W24×68, UNDER THE SAME LOADING, IF ACTING WITHOUT COMPOSITE ACTION.

(III)

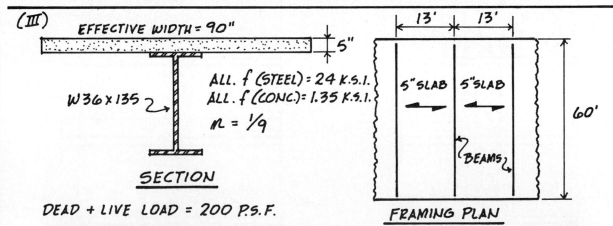

EFFECTIVE WIDTH = 90"
5"
W36×135

SECTION

ALL. f (STEEL) = 24 K.S.I.
ALL. f (CONC.) = 1.35 K.S.I.
$m = 1/9$

13'   13'
5" SLAB   5" SLAB
BEAMS
60'

FRAMING PLAN

DEAD + LIVE LOAD = 200 P.S.F.

(1) DETERMINE THE MOST ECONOMICAL WIDE FLANGE SECTION FOR THE BEAM ACTING ALONE, UNDER THE GIVEN LOADING.
(2) DETERMINE THE MAX. COMP. STRESS IN THE CONCRETE AND THE MAX. TENSILE STRESS IN THE STEEL FOR THE COMPOSITE SECTION.

PROBLEM 14–3

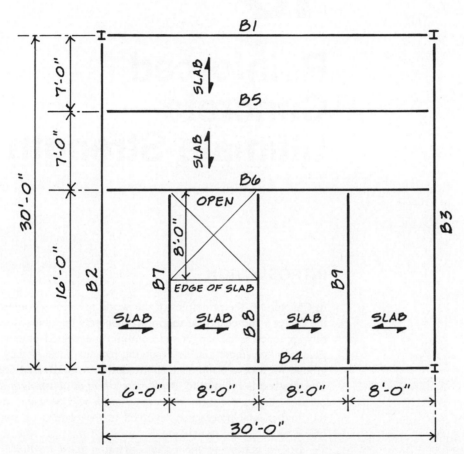

STEEL FRAMING

FLOOR FRAMING PLAN FOR A COMMERCIAL BUILDING

- TOTAL FLOOR LOAD = 200 P.S.F.
- WALL LOAD = 800#/FT. ON B1, B2, B3, B4
- MINIMUM ALLOWABLE BEAM DEPTH = 8"
- ONE-WAY CONCRETE SLABS DELIVER UNIFORM LOADING TO SUPPORTING STEEL BEAMS. SLABS ARE 4" THICK.

(1) DETERMINE THE MOST ECONOMICAL STEEL BEAMS (B1-B9) REQUIRED FOR BENDING AT ALLOWABLE f = 24 K.S.I.

(2) FOR THE BEAMS SELECTED FOR B1 AND B5, DETERMINE THE UNIFORM LOAD CARRYING CAPACITY BASED ON COMPOSITE ACTION.
STEEL: ALLOW. f = 24 K.S.I., CONCRETE: ALLOW. f = 1.35 K.S.I., n = 1/9

15

Reinforced Concrete— Ultimate Strength

INTRODUCTION

In Chapter 13, the principles of the working stress (or alternate) method for the analysis and design of reinforced concrete members were studied. While this approach is no longer the commonly accepted practical approach, the principles are important for academic reasons because they provide a background for a newer approach to the design of reinforced concrete members. This method is referred to as *ultimate strength* or *strength design*. Throughout this chapter the term *ultimate strength* will be used, and we will study the fundamental procedures involved in the context of analysis and design of beams subjected to bending.

Before getting to the basics, mention must be made of the fact that ultimate strength procedures are, for practical purposes, very closely controlled by minimum requirements set forth in the ACI (American Concrete Institute) Building Code. Many of these requirements are empirical in nature. It is the purpose of this chapter to provide the student with a *rudimentary* knowledge of the ideas that underlie the ultimate strength method and the fundamental procedures used in flexural analysis and design. Consequently, very little reference will be made to the details of building code requirements, although some widely accepted recommendations will be used. Students who wish to engage themselves in the structural design of a reinforced concrete structure should be prepared to carry their studies well beyond the scope of this chapter.

THE BASIS FOR ULTIMATE STRENGTH

In the ultimate strength method, the flexural analysis or design of a reinforced concrete beam is based on the ultimate compressive stress of the concrete and the yield stress of the reinforcing steel. The use of the ultimate strength of the materials suggests that beams designed by this procedure will actually fail. While this is the basis for the ultimate strength design method, beams, in fact, will not fail because there are a number of factors used in the design procedures that guard against this. One factor that guards against literal failure of a beam is the use of loads in the design process that are significantly higher than the anticipated loads. The anticipated loads are simply the dead loads and live loads, and will be referred to as *service loads*. In ultimate strength design, the service loads are inflated to *ultimate loads* by the use of a *load factor*. The load factor is a multiplier by which service loads are inflated to hypothetical design (or ultimate) loads. It is under these inflated loads that the beam is designed to "fail." In ultimate strength design in concrete, we use different load factors for dead load and live load. The recommended load factors are 1.4 for the service dead load and 1.7 for the service live load. The application of the load factors is quite a simple matter. For example, after we determine the service dead load (S.D.L.) and the service live load (S.L.L.) on a beam, we simply multiply these loads by the appropriate load factors. This yields ultimate loads, as shown in Fig. 15–1, which are used for design.

Another factor used in ultimate strength is called the "Φ (phi) factor." The Φ factor is used to further inflate the ultimate loads (or the ultimate design moment) in order to account for inaccuracies that are inevitable in the design and construction of a reinforced concrete structure. This includes any slight misalignment of forms, reinforcing bars not placed *exactly* as shown on the drawings, etc. The recommended Φ factor varies with the type of stress being considered. In this chapter we will only be concerned with bending; consequently, we will use a Φ factor of 0.90.

There is another important factor that guards against brittle compression failure of concrete when designing by the ultimate strength method. This additional protection comes from the idea that all sections must be designed as underreinforced sections. In ultimate strength language this means that sections will be designed so that the reinforcing steel will yield before the concrete fails in compression. Since the steel is ductile, it will stretch a great deal, forcing cracks in the concrete on the tension side of the beam, thereby giving visible warning that the member is being overloaded. The basis for the design of underreinforced sections will be discussed in subsequent pages of this chapter.

GENERAL EXPRESSIONS FOR ULTIMATE STRENGTH

In Chapter 13, which dealt with working stress methods for reinforced concrete, we made use of the fact that the early portion of the stress/strain curve for a concrete specimen is so close to being a straight line, as shown in Fig. 15–2, that we considered it, in fact, to be a straight line. Therefore, up to a certain level of stress it was safe to say that for all practical purposes, the material obeyed Hooke's Law, which indicated an elastic material. These assumptions allowed us to use standard elastic flexure theory and the bending stress equation (for the transformed section)

$$f = \frac{Mc}{I_{tr}}$$

It should be recalled from Chapter 7 that the concept of moment of inertia, and the determination of expressions that yield numerical values for I, was based strictly on the idea that stresses *vary proportionately* with the distance from the neutral axis, as shown in Fig. 15–3. This is the way we

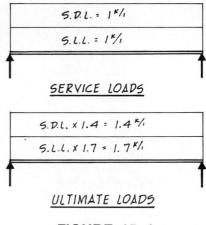

S.D.L. = 1 $^{k}/_{1}$

S.L.L. = 1 $^{k}/_{1}$

SERVICE LOADS

S.D.L. × 1.4 = 1.4 $^{k}/_{1}$

S.L.L. × 1.7 = 1.7 $^{k}/_{1}$

ULTIMATE LOADS

FIGURE 15–1

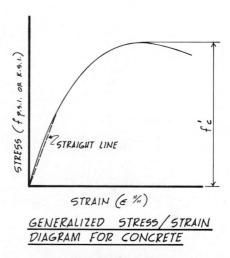

GENERALIZED STRESS/STRAIN DIAGRAM FOR CONCRETE

FIGURE 15–2

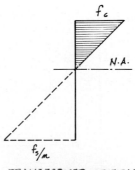

TRANSFORMED SECTION - STRESS DIAGRAM

FIGURE 15–3

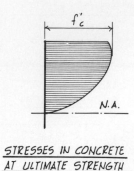

STRESSES IN CONCRETE
AT ULTIMATE STRENGTH

FIGURE 15–4

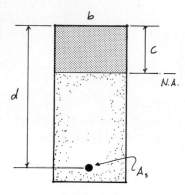

FIGURE 15–5

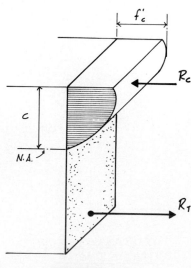

FIGURE 15–6

treated sections that were analyzed by working stress methods; consequently, we were able to compute values for the transformed moment of inertia.

When using ultimate strength procedures we deal with levels of stress up to the ultimate compressive stress of the concrete (f'_c). These stresses are well beyond the level where the stress/strain relationship is even close to being a straight line. Now, if we consider that *strains* will vary proportionately with the distance from the neutral axis, up to the ultimate strength of the concrete, then a corresponding stress diagram will look like that shown in Fig. 15–4.* It should be noted that the ultimate stress (f'_c) at failure occurs below the level of the outermost compression fiber. This is so because there tends to be a slight relaxation in the resistance (or stiffness) of concrete at high strain levels, which occurs at the outermost fibers. The moment of inertia concept, which is based on stresses below the elastic limit of a material, is invalid when dealing with stresses *beyond* the elastic limit. Therefore, when analyzing or designing beams by ultimate strength methods, we must use the statical relationship of the internal couple. The internal couple method for the design of reinforced concrete members was first presented in Chapter 13, where we dealt with the working stress method for design. We will now revisit these same ideas, but in the context of a much different level of stresses—i.e., the ultimate compressive stress of the concrete and the yield stress of the reinforcing steel.

In order to develop the general expressions used for analysis and design by the ultimate strength method, let's consider a simple rectangular cross section, as shown in Fig. 15–5, where:

b = width of the section
d = effective depth of the section (the dimension from the outermost compression fiber to the centerline of the reinforcing steel)
A_s = area of reinforcing steel
c = the distance from the outermost fiber to the neutral axis (as it has always been)

It should be recalled from Chapter 13 that concrete below the neutral axis (the tension side) is not considered part of the effective structural section. Therefore, the effective section consists of the compression zone (shaded area) and the reinforcing steel (A_s).

When a section like this is loaded and stressed to the ultimate stresses for both the concrete and steel, the stress distribution and the consequent resultant forces (the internal couple) will be as shown in Fig. 15–6, where:

f'_c = specified ultimate compressive stress†
R_C = resultant of the compressive stress (which is the volume of all of the compressive stress shown in Fig. 15–6)

---

*Experimentation reveals that strains will vary proportionately with the distance from the neutral axis all the way to failure. This means that the idea first presented in Chapter 7, which tells us that initially plane sections will remain plane during bending, is, for all practical purposes, true whether the stresses are within or beyond the elastic limit of the material.

†The ultimate compressive stress of concrete varies depending on the quality of the mix (i.e., amount of cement, water, etc.). The ultimate strength is verified by a compression test of a sample of the concrete that has aged for 28 days. For those interested in more information on this matter, see the current edition of *Design and Control of Concrete Mixtures*, prepared by the Portland Cement Association.

R_T = resultant of tension which, by ultimate strength methods, is simply the product of the area of steel and the yield stress of the steel

A two-dimensional side view of the stresses is shown in Fig. 15–7 for the purpose of clarity for the following discussion. Based on the laws of static equilibrium:

$$R_C = R_T$$

When dealing with the actual stress distribution for the concrete, which is in compression, the value of R_C (the resultant of compression) is difficult to determine because of the curvilinear variation of the stress diagram. Because of the nature of the stress variation, it is also difficult to determine the location of the center of gravity of the volume of stress, which is the point of application of R_C. This means that the dimension Z, as shown in Fig. 15–7, is not easily determined. It is this dimension that is a key factor in the resisting moment capacity of a section. Based on the internal couple method, the resisting moment capacity of a section is simply the product of the resultant (either R_C or R_T) and the distance Z between the two forces. Stated mathematically,

$$M = R_C Z = R_T Z$$

It should be clear, in light of the preceding discussion, that to deal with the actual stress distribution for analysis or design would be a nuisance at best.

Research and experimentation have shown that the actual stress distribution may be replaced by an equivalent rectangle, as shown in Fig. 15–8. The important factors in developing a rectangular stress block equivalent are that the volume of the stress block (which is the value of R_C) is the same as the volume of the actual stress distribution, and that the location of R_C not be changed. Research has shown that these values will be maintained in the rectangular stress block if the intensity of stress is taken as .85 f'_c, as shown in Fig. 15–8 and in the two-dimensional side view shown in Fig. 15–9. Also, the depth of the stress block, shown as a in the figures, must be taken as some fraction of the true distance to the neutral axis *(c)* in order to maintain the resultant of compression in its true location. Expressed mathematically, the value of a is

$$a = k_1 c$$

where a = the depth of the rectangular stress block equivalent
k_1 = a factor that varies with the ultimate compressive strength (f'_c) of the concrete as follows:

—for $f'_c \lesseqgtr$ 4000 p.s.i., k_1 = .85
—beyond f'_c = 4000 p.s.i., k_1 reduces linearly at the rate of .05 for every 1000 p.s.i. increase in f'_c
c = the distance from the outermost compression fiber to the neutral axis

Using the rectangular stress block equivalent, we will now develop general expressions based on the internal couple, which we will use for purposes of analysis and design. It should be noted that in Fig. 15–7, the notation used for the lever arm was Z. In Fig. 15–9 this notation was changed to jd. We will continue to use the latter notation throughout this chapter; the reason for this seemingly strange notation will be made clear in the subsequent design problems.

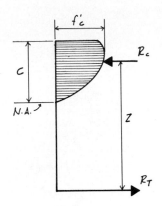

FIGURE 15–7

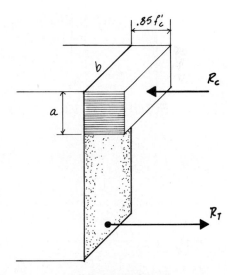

FIGURE 15–8

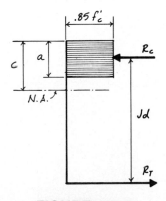

FIGURE 15–9

For a rectangular section as shown in Fig. 15–10, subjected to stresses shown in Fig. 15–8 (or Fig. 15–9):

$$R_C = R_T \qquad (\Sigma F_x = 0)$$

Furthermore,

R_C = volume of the rectangular stress block
R_T = the product of the area of steel (A_s) and the yield stress of the steel (f_y)

Therefore,

$$R_C = .85f'_c \, ab = R_T = A_s f_y \qquad (15\text{–}1)$$

and

$$a = \frac{A_s f_y}{.85f_c b} \qquad (15\text{–}2)$$

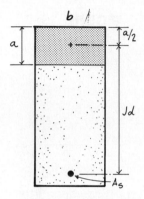

FIGURE 15–10

For a cross section where the compression zone is something other than rectangular, such as the shaded area of the T-beam shown in Fig. 15–11, the only difference from the rectangular section is that the area of the compression zone must be generalized, since b varies. We will, therefore, refer to the area of a non-rectantular compression zone as A_c. Therefore, when dealing with an irregular compression zone:

$$R_C = .85f'_c \, A_c = R_T = A_s f_y \qquad (15\text{–}3)$$

and

$$A_c = \frac{A_s f_y}{.85f'_c} \qquad (15\text{–}4)$$

It might be useful to point out, based on the lessons of Chapter 4, that given a uniform unit stress applied to an area, the resultant of that uniform unit stress will act at the centroid of the area.

We'll now develop a general expression for the resisting moment, which is a reasonably simple matter. The resisting moment capacity of a section, based on ultimate strength methods, will be expressed in two ways:

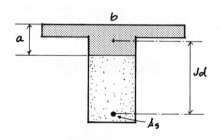

FIGURE 15–11

1. $\overline{M}$ ("bar M"), which will be referred to as the "ideal" ultimate moment capacity of a section. This would be the ultimate moment capacity if everything were *absolutely* accurate (i.e., alignment of formwork, placement of reinforcing, etc.).

2. M_u, which is referred to as the "practical" ultimate moment capacity of a section. It is a reduced value of $\overline{M}$ which accounts for slight inaccuracies in construction and design. The factor by which $\overline{M}$ is reduced to obtain M_u is, as discussed earlier in this chapter, called the Φ factor, and is equal to 0.90 for flexural computations.

$$\therefore M_u = 0.90 \, \overline{M}$$

Referring to Fig. 15–9, the "ideal" resisting moment $(\overline{M})$ is simply equal to the product of R_C or R_T (since they are numerically equal), and the lever arm jd.

$$\therefore \overline{M} = R_C jd = R_T jd$$

and, substituting values from Eq. (15–1) and Eq. (15–3):

$$\overline{M} = .85f'_c abjd = .85f'_c A_c jd = A_s f_y jd \qquad (15\text{–}5)$$

These are the expressions necessary for dealing with both rectangular and irregular compression zones. In conclusion of this discussion, it should be noted that in the development of the preceding expressions, the actual location of the neutral axis is not involved, since we have replaced the true stress distribution with an equivalent rectangular stress block.

Earlier in this chapter it was mentioned that by ultimate strength methods, all beams must be designed as underreinforced sections. This means that sections must be designed so that the reinforcing steel will yield before the concrete fails in compression. In order to assure this situation, we must first develop an understanding of the balanced condition at the level of ultimate strength. In the ultimate strength design method, a balanced section is defined as one where concrete is at its crushing *strain* and the reinforcing steel is simultaneously at its initial yield stress. Research has shown that generally, the crushing *strain* of concrete (ϵ_c) may be taken as $.003''/''$. The strain in the steel at initial yield (see Chapter 7 for Hooke's Law) will be:

$$\epsilon_s = \frac{f_y}{E_s}$$

where ϵ_s = strain in the steel
  f_y = yield stress of the steel
  E_s = modulus of elasticity of steel = 29,000 k.s.i.

Considering that strains will vary proportionately with the distance from the neutral axis, the balanced condition may be defined based on the similar triangles of the strain diagram shown in Fig. 15–12. Based on similar triangles,

$$\frac{c}{.003} = \frac{d}{.003 + f_y/29,000}$$

and

$$\frac{c}{d} = \frac{.003}{.003 + f_y/29,000} = \frac{87}{87 + f_y} \tag{15–6}$$

and, as established earlier in this chapter,

$$a = k_1 c \text{ or } c = a/k_1$$

Therefore, rewriting Eq. (15–6)

$$\frac{a}{d} = k_1 \left[\frac{87}{87 + f_y} \right] \tag{15–7}$$

Note: The units of f_y in Eq. (15–7) are in terms of k.s.i.

Equation (15–7) represents the *balanced condition,* as a ratio of the depth of the rectangular stress block to the effective depth of the section. It should be noted that the expression, which defines the balanced condition based on strain relationships, may be expressed (in other reference sources) in altered form. Specifically, the expression is based on a percentage of reinforcing steel necessary to produce the balanced condition, where the percentage of steel is taken as

$$P = A_s/bd$$

where P = percentage of reinforcing steel
  A_s = area of reinforcing steel
  b = width of the cross section
  d = effective depth of the section

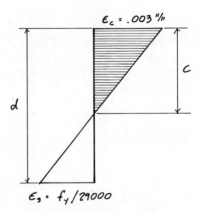

STRAIN DIAGRAM FOR THE BALANCED CONDITION

FIGURE 15–12

It is generally recommended that for ultimate strength procedures, the percentage of steel in a cross section be limited to a maximum of 75% of that which will produce a balanced condition. This also means that the area of the compression zone, in a beam subjected to bending, is limited to 75% of the compression area for a balanced section. In essence, the 75% limitation guards against any possibility of a brittle compression failure. In a rectangular cross section, the 75% limitation may be applied directly to the depth a of the rectangular stress block. Therefore, Eq. (15–7) may be restated, for practical purposes,

$$\text{Limit } a = k_1 \left[\frac{87}{87 + f_y} \right] d(.75) \qquad (15\text{–}8)$$

To use the limiting value of a as a guard against the balanced condition (which signifies compression failure in the concrete) in sections where the compression zone is irregular would not, in the strictest sense, be completely accurate based on the 75% limit for the percentage of steel. We will explore this idea in slightly more detail in the context of the example problems.

ANALYSIS BY ULTIMATE STRENGTH METHODS

Now that the necessary general expressions have been developed, we can look at a few simple example problems. We will first deal with analysis (or investigation) problems, where all information about the cross section is known and the moment-carrying capacity is to be determined.

Example 15–1 (Fig. 15–13)

In this problem we want to determine the ultimate moment-carryng capacity for the rectangular section shown in Fig. 15–13. This section has an area of steel $(A_s) =$ 3 in.^{2}, $f_c' = 4$ k.s.i., $f_y = 50$ k.s.i.

We must first determine the maximum allowable depth for the stress block based on Eq. (15–8):

$$\text{Limit } a = k_1 \left[\frac{87}{87 + f_y} \right] d(.75)$$

where $k_1 = .85$ for $f_c' = 4000$ p.s.i. or less, and decreases by 0.05 for every 1000 p.s.i. increase in f_c'. Values for k_1 for several strengths of concrete are shown on Data Sheet D–37.

$$\text{Limit } a = .85 \left[\frac{87}{87 + 50} \right] (20)(.75) = 8.1''$$

This is the limiting value for the depth of the compression zone in order to guard against a compression failure. Assuming that the reinforcing steel is at yield, we must now determine the actual value of a. Using the internal couple relationship that

$$R_C = R_T$$

And, using Eq. (15–2),

$$a = \frac{A_s f_y}{.85 f_c' b} = \frac{(3 \text{ in.}^2)(50 \text{ k.s.i.})}{.85(4 \text{ k.s.i.})(12'')} = 3.68''$$

This is less than the limit of $a = 8.1''$ and, therefore, the entire amount of steel is considered to be yielding, without any danger of producing a compression failure in the concrete. In order to determine the moment capacity, we have the following relationships:

$$\overline{M} = A_s f_y jd = .85 f_c' abjd$$

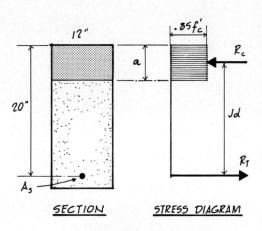

SECTION   STRESS DIAGRAM

FIGURE 15–13

In this situation we can use either of these expressions. For a rectangular section, $jd = d - a/2$. Therefore,

$$\overline{M} = A_s f_y jd = \frac{(3 \text{ in.}^2)(50 \text{ k.s.i.})(18.16'')}{12''/'}$$

$$= 227'^K$$

$$M_u = \Phi\overline{M} = 0.90(227'^K) = 204.3'^K$$

Let's reconsider this rectangular section, but this time with $A_s = 8$ in.^{2}. Using the same specifications for steel and concrete, the limiting value a remains the same.

$$\therefore \text{ Limit } a = 8.1''$$

To determine the actual value of a, based on the assumption that the steel will be at yield,

$$a = \frac{A_s f_y}{.85 f_c b} = \frac{(8 \text{ in.}^2)(50 \text{ k.s.i.})}{.85(4 \text{ k.s.i.})(12'')} = 9.8''$$

This is greater than the limiting value of a, which tells us, in fact, that we cannot use the steel at full yield because the demand being made for the depth of the compression zone (so that $R_C = R_T$) is such that the concrete will be close to failure. In this case the limiting value of $a = 8.1''$ is the greatest depth we can use. Since at this point we don't know what is the stress in the steel, we will use:

$$\overline{M} = .85 f_c' abjd$$

where $a = 8.1''$ and $jd = d - a/2$:

$$\overline{M} = \frac{.85(4 \text{ k.s.i.})(8.1'')(12'')(15.95'')}{12''/'} = 439.3'^K$$

$$M_u = \Phi\overline{M} = 0.90(439.3'^K) = 395.4'^K$$

This example problem is now complete.

In a cast-in-place concrete structure the slabs are cast integrally with the beams; consequently, a beam cross section will include a portion of the slab, thereby making it a T-beam, as shown in Fig. 15–14. The portion of slab (width b) that will act effectively, for an interior beam, is taken as the least of the following values:

1. Eight times the slab thickness on each side of the web.
2. One half of the clear distance to the next beam, on each side of the web.
3. A total width equal to one fourth of the beam span.

The flange width for a T-beam will normally be large; consequently, the depth of the compression zone will be relatively shallow in order to carry a given moment. If the depth of the compression zone is less than the thickness of the slab (t), then the T-beam is, in effect, a rectangular section of width b, as shown in Fig. 15–14. If the compression zone is deeper than the slab thickness, then the compression area will be T-shaped, as shown by the shaded area in Fig. 15–15. Because of the varying width of the compression zone, the analysis of this sort of situation is slightly different than for a rectangular section. We will discuss this further in the example problems.

Example 15–2 (Figs. 15–16 and 15–17)

Let's determine the moment-carrying capacity (M_u) for the section of Fig. 15–16, based on the following data:

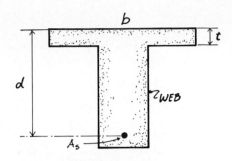

FIGURE 15–14

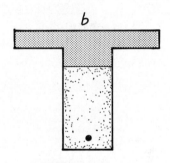

FIGURE 15–15

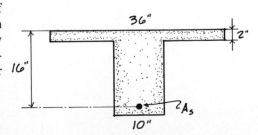

FIGURE 15–16

$f'_c = 3000$ p.s.i.

$f_y = 60$ k.s.i.

$A_s = 2$ #8 bars $= 1.58$ in.^{2} (See Data Sheet D–23 for reinforcing steel data.)

The first thing to do is to determine the area of the compression zone (A_c), assuming that the steel is at yield. Therefore, using the internal couple relationship:

$$A_c = \frac{A_s f_y}{.85 f'_c} = \frac{(1.58 \text{ in.}^2)(60 \text{ k.s.i.})}{.85(3 \text{ k.s.i.})}$$

$$= 37.2 \text{ in.}^2$$

Since the total flange area is 72 in.^{2}, the rectangular stress block falls entirely within the flange and, as far as we're concerned, this is a rectangular section, 36″ wide. Referring to Fig. 15–17,

$$a = \frac{37.2 \text{ in.}^2}{36 \text{ in.}} = 1.03''$$

$$\text{Limit } a = .85\left[\frac{87}{87 + 60}\right](.75)(16'') = 6.04'' \text{ O.K.}$$

$$\therefore jd = d - a/2 = 16 - \frac{1.03}{2} = 15.49''$$

$$\overline{M} = A_s f_y jd = \frac{(1.58 \text{ in.}^2)(60 \text{ k.s.i.})(15.49'')}{12''/'}$$

$$= 122.4'^K$$

$$M_u = 0.90(122.4'^K) = 110.1'^K$$

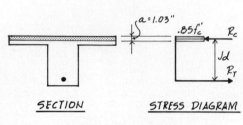

SECTION   STRESS DIAGRAM

FIGURE 15–17

Example 15–3 (Figs. 15–16 and 15–18 through 15–20)

In order to illustrate what happens when the compression zone is not rectangular, let's use the T-beam of Example 15–2 and increase the area of steel (A_s) to 4 in.^{2}. Assuming that $f_y = 60$ k.s.i., then

$$A_c = \frac{A_s f_y}{.85 f'_c} = \frac{(4 \text{ in.}^2)(60 \text{ k.s.i.})}{.85(3 \text{ k.s.i.})} = 94.1 \text{ in.}^2$$

The area of the flange is only 72 in.^{2}. Therefore, the compression zone must come down into the web; it will be T-shaped, as shown by the shaded area in Fig. 15–18. The area needed in the web is 94.1 in.^{2} – 72 in.^{2} = 22.1 in.^{2}. Since the web has a 10″ width, a depth of 2.21″ into the web is required.

It was previously mentioned that the percentage of steel $(P = A_s/bd)$ in a section should be limited to 75% of that which will produce a balanced condition. In essence, this means that the area of the compression zone must be limited to 75% of the balanced condition. When dealing with *rectangular* sections, this requirement is most easily evaluated by using 75% of the depth (a) of the rectangular stress block. When dealing with an irregular compression area, such as the case being considered here, we cannot use 75% of a as a limit because of the varying width. Since dealing with the percentage of steel is generally awkward, at best, for an irregular section, we will make the necessary checks based on 75% of the balanced compression area. To do this for the case under consideration, we will determine the depth of the stress block (a) for the balanced condition, as follows:

$$\text{Balanced } a = k_1\left[\frac{87}{87 + f_y}\right]d = .85\left[\frac{87}{147}\right]16''$$

$$= 8.05''$$

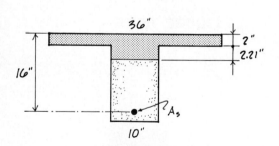

FIGURE 15–18

Reminder: k_1 is based on the strength of concrete and is equal to .85 for $f'_c \lesseqgtr 4000$ p.s.i. Values of k_1 for other grades of concrete are given on Data Sheet D–37.

Based on the value of a for a balanced section, the area of the compression zone is

$$A_{\text{Balanced}} = (2'')(36'') + (6.05'')(10'')$$

$$= 132.5 \text{ in.}^2$$

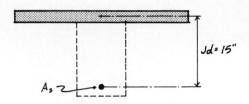

FIGURE 15–19

The usable compression area is based on 75% of this value. Therefore,

$$A = .75(132.5 \text{ in.}^2) = 99.4 \text{ in.}^2$$

Going back to the beginning of this problem, an area of 94.1 in.^{2} was required based on the steel at yield. This is less than the limiting compression area of 99.4 in.^{2}; consequently, all values determined to this point are usable. In order to evaluate the resisting moment capacity for this irregular section, the easiest approach is to break the irregular compression zone (shown as the shaded area in Fig. 15–18) into two rectangular pieces, as shown by the shaded areas in Figs. 15–19 and 15–20. The results of these may then be added for the final answer. Referring to Fig. 15–19 for the moment capacity provided by the flange (let's call it $\overline{M}_1$):

$$\overline{M}_1 = .85f'_c A_c jd = \frac{(.85)(3 \text{ k.s.i.})(72 \text{ in.}^2)(15'')}{12''/'}$$

$$= 229.5'^K$$

For the part shown in Fig. 15–20 (call it $\overline{M}_2$):

$$\overline{M}_2 = .85f'_c A_c jd$$

$$= \frac{(.85)(3 \text{ k.s.i.})(22.1 \text{ in.}^2)(12.9'')}{12''/'}$$

$$= 60.6'^K$$

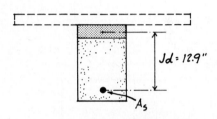

FIGURE 15–20

Therefore, the total "ideal" moment capacity equals

$$\overline{M}_1 + \overline{M}_2 = 229.5'^K + 60.6'^K = 290.1'^K$$

and

$$M_u = 0.90(290.1'^K) = 261.1'^K$$

This completes the investigation problems, and now we are ready to deal with the flexural *design* of beams by ultimate strength methods.

FLEXURAL DESIGN BY ULTIMATE STRENGTH

In an investigation problem, all of the information about the cross section is known, which allows us to determine the ultimate moment capacity. In a design problem, however, the approach is quite different, since we know nothing about the cross-sectional properties at the outset. The only information we would have at the time we're ready to design individual members would be the framing plan, which is related to the design of the building. Based on the framing plan and, consequently, the manner in which floor loads will be distributed, we can determine the load and the moment to be carried. We would then have to determine the cross-sectional dimensions and the area of steel required. This means that there are three unknowns that have to be determined, as indicated in the rectangular cross section of Fig. 15–21. Since each of these unknowns has an effect on the magnitude of the

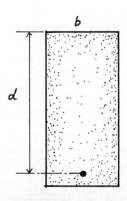

FIGURE 15–21

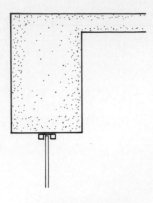

FIGURE 15-22

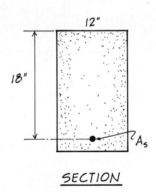

SECTION

FIGURE 15-23

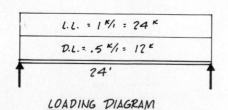

LOADING DIAGRAM

FIGURE 15-24

others, we must eliminate at least one, and this can be done as a matter of judgment. For example, the width of the section (b) may be set based on formwork dimensions. If stock lumber is used for forming the concrete, the width may be set at $11\frac{1}{4}$ in., which is the actual dimension of 12-in. stock. Other actual dimensions of stock lumber may also be used, or dimensions may be based on the use of plywood, which can be cut to satisfy a variety of dimensions. In any case, the width of the section will normally be set based on formwork efficiency or architectural requirements. In addition, the depth of a reinforced concrete beam may often be set by architectural design requirements. For example, if a concrete beam spans over a glass wall, as shown in Fig. 15-22, the total depth may be set, most sensibly, by the architectural requirements, unless such requirements require a depth that is too shallow to carry the load efficiently. Very often, the width and depth of a reinforced concrete beam are set by a variety of factors, and the computational part of the design problem is the determination of the area of steel required. Let's look at a few example problems to see how we deal with questions of design. We will consider several issues in the flexural design of beams; then we will discuss issues other than bending, which eventually must be considered in the design of a reinforced concrete building.

Example 15–4 (Figs. 15–23 and 15–24)

The rectangular section shown in Fig. 15–23 is simply supported and must carry the service dead and live loads shown in Fig. 15–24. Since the cross-sectional dimensions are set, the design problem becomes one of determining the required area of reinforcing steel. The specifications are: $f'_c = 4000$ p.s.i., $k_1 = .85$, $f_y = 60$ k.s.i. The first thing we must do is inflate the given service loads to ultimate loads. We'll use load factors of 1.4 for dead load and 1.7 for live load. Therefore:

$$\text{Ultimate dead load} = (1.4)(12^K) = 16.8^K$$
$$\text{Ultimate live load} = (1.7)(24^K) = \underline{40.8^K}$$
$$\text{Ultimate total load } (W_u) = 57.6^K$$

The next step is to determine the ultimate moment to be carried, based on the ultimate loading. This loading will yield M_u, which must be increased by the Φ factor (0.90) to get the design moment, $\overline{M}$. Designing for $\overline{M}$ will account for inaccuracies in construction and design and will assure that an M_u capacity will be provided.

$$M_u = \frac{W_u L}{8} = \frac{(57.6^K)(24')}{8} = 172.8'^K$$

The design moment $\overline{M} = \dfrac{M_u}{\Phi} = \dfrac{172.8'^K}{.90} = 192'^K$.

For a rectangular section,

$$\overline{M} = .85\, f'_c abjd = A_s f_y jd$$

Using the expression for steel will not get us anywhere since we don't know A_s or jd. In a rectangular section $jd = d - a/2$. Therefore, if we rearrange the expression for concrete, we can solve this problem.

$$\overline{M} = .85\, f'_c abjd = .85 f'_c ab(d - a/2)$$

$$(192'^K)(12''/') = .85(4)(a)(12)(18 - a/2)$$

$$2304 = 734.4a - 20.4a^2$$

Solving the quadratic:

$$a = 3.47''$$

We must check this against the limiting value of a:

$$\text{Limit } a = .85\left[\frac{87}{87 + 60}\right]18(.75) = 6.79'' \text{ O.K.}$$

Now that we have a value for *a*, we can determine a value for *jd*, and then we can solve for A_s.

$$a = 3.47'', \therefore jd = 18 - \frac{3.47}{2} = 16.27''$$

$$A_s = \frac{\overline{M}}{f_y jd} = \frac{(192'^K)(12''/')}{(60 \text{ k.s.i.})(16.27'')} = 2.36 \text{ in.}^2$$

Referring to Data Sheet D–23, three #8 reinforcing bars will do the job.

Earlier in this chapter, it was suggested that the reason for the notation *jd* would be clarified in the study of design procedures. The reason for the use of *jd*, which is the distance between R_C and R_T (the internal couple), will now be explained.

In the design of the rectangular section that we've just concluded, it was shown that we had to solve a quadratic equation in order to determine *a*, which would then allow us to determine *jd*. It would be time-consuming and, at best, a nuisance to have to set up and solve a quadratic equation every time we designed a concrete beam. Fortunately, there are design aids that can make the process much more palatable. One such aid is given on Data Sheet D–37; it is a nomograph for rectangular sections.* It should be recognized by now that the values of *jd* must always be less than the depth of a cross section, as shown in the generalized stress diagram of Fig. 15–25. In fact, the value of *j* is a fraction which, when multiplied by *d*, will give the value of the lever arm between R_C and R_T. With this in mind, a range of values for *j* is given on the nomograph along with certain other related values that are needed for the design process. You must refer to Data Sheet D–37 for the notations used and the basis for the arrangement of the nomograph. In essence, the nomograph solves the quadratic equation for us. Let's do a design problem with the aid of the nomograph.

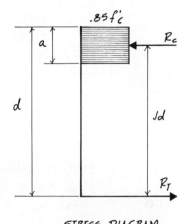

STRESS DIAGRAM

FIGURE 15–25

Example 15–5 (Fig. 15–26)

In this problem we'll design a one-way slab with $f'_c = 4500$ p.s.i., $k_1 = .825$, $f_y = 50$ k.s.i. As discussed in Chapter 13, one-way slabs are designed by using a typical one-foot-wide strip of slab, as shown in Fig. 15–26, which makes it a rectangular section. The slab is to be designed for an ultimate moment, $\overline{M} = 6'^K$.

Referring to Data Sheet D–37:

$$K = \frac{\overline{M}}{f'_c bd^2} = \frac{(6'^K)(12''/')}{(4.5 \text{ k.s.i.})(12'')(4'')^2} = .083$$

Note: Because of the magnitude of the numbers involved on the nomograph, it is recommended that *no fewer* than two decimal places be used.

Going to the nomograph, the corresponding value of *j* = .948 and the ratio of *a*/*d* = .103, which is less than the limiting ratio of *a*/*d* of .393. Limiting values of the ratio *a*/*d* based on 75% of the balanced condition are given on Data Sheet D–37. In this case, there is no danger of a compression failure. Using *j* = .948, we can now determine the necessary A_s.

$$A_s = \frac{\overline{M}}{f_y jd} = \frac{(6'^K)(12''/')}{(50 \text{ k.s.i.})(.948)(4'')} = .38 \text{ in.}^2$$

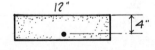

SECTION OF SLAB

FIGURE 15–26

This is the area of steel that must be provided as an *average* for every foot of width for the slab, since the slab was designed as a typical one-foot-wide strip. To select bars, we'll refer to Data Sheet D–23 in the Appendix. While there is a variety of choices of bar sizes that will provide the required *average* of steel for every foot of width, it is generally recommended that closely spaced smaller bars are a better

---

*A nomograph is a chart that gives values for two or more related variables.

solution than large bars spaced far apart. Generally, the spacing should be kept within 12 in. For this example we'll use #5 bars, with each bar having an area of .31 in.^{2}. Since we need $A_s = .38$ in.^{2} per foot of width, we can determine the spacing of #5 bars based on a direct proportion, as follows:

$$\frac{.31 \text{ in.}^2}{X} = \frac{.38 \text{ in.}^2}{12''}, \quad X = 9.8''$$

This is the maximum spacing for #5 bars in order to provide the minimum area of 0.38 in.^{2} per foot of width of the slab. Should one choose to use #4 bars or #3 bars, the spacing would be determined by direct proportioning, as demonstrated. In any case, it should be noted that for practical purposes, the indicated spacing would be rounded off to, at most, the nearest 1/2 in. (on the safe side). If we used #5 bars for the design of the slab, the spacing indicated on structural drawings would probably be, at most, 9.5 in. This example problem is now complete.

It was mentioned earlier that the depth of a section is often dictated by architectural requirements. On the other hand, as perhaps in the case of a slab thickness, the only requirement for the depth of the section would be that of reasonable structural efficiency. The nomograph on Data Sheet D-37 can be a good guide to determine the depth of a section where no other criteria are available. In general, it is suggested that depths be chosen that will yield values of K, as defined on the nomograph, in roughly the upper half of the scale (but not too close to the top). Specifically, values of K based on depths that fall in the lower part of the top half of the scale will provide for reasonably efficient cross sections, without excessive depths. Staying within this recommended range will generally help to avoid deflection problems which, in the case of a slab, could cause a "bouncy" situation . . . always to be avoided.

Although the nomograph we used for the preceding example problem is referred to as a "rectangular section nomograph," its use is not really restricted to rectangular sections. This will be demonstrated in the following example problem.

Example 15–6 (Figs. 15–27 through 15–30)

In this problem we'll deal with the design of a cross section where the compression zone is irregular. For this purpose we'll determine the area of reinforcing steel (A_s) required for the simply supported, pre-cast concrete beam shown in Fig. 15–27. The beam is to be designed to carry service loads, as shown in Fig. 15–28. The indicated service dead load includes the weight of the beam. For the purpose of this problem we will use the following specifications:

$$f'_c = 4000 \text{ p.s.i.,} \quad k_1 = .85, \quad f_y = 60 \text{ k.s.i.}$$

In order to determine the ultimate moment to be carried, we must first increase the loads to the ultimate loading, using load factors of 1.4 for dead load and 1.7 for live load.

$$
\begin{aligned}
\text{U.L.L.} &= 1.7 \times 800\#/' = 1360\#/' \\
\text{U.D.L.} &= 1.4 \times 600\#/' = \underline{840\#/'} \\
\text{Ultimate Total Load} &= 2200\#/' = 2.2^{\text{K}}/'
\end{aligned}
$$

and

$$W_u = (2.2^{\text{K}}/')(24') = 52.8^{\text{K}}$$

The maximum moment is

$$M_u = \frac{W_u L}{8} = \frac{(52.8^{\text{K}})(24')}{8} = 158.4'^{\text{K}}$$

$$\overline{M} = \frac{158.4'^{\text{K}}}{.90} = 176'^{\text{K}}$$

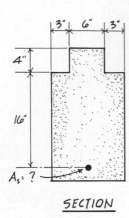

SECTION

FIGURE 15–27

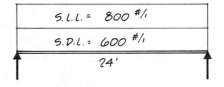

S.L.L. = 800 $^\#/'$

S.D.L. = 600 $^\#/'$

24'

LOADING DIAGRAM

FIGURE 15–28

In order to determine the required area of steel for this cross section, we must first determine the moment-carrying capacity of the part shown shaded in Fig. 15–29 and the area of steel (A_{s_1}) that will be required for this. If the piece shown shaded does not carry the total moment, then more area is required to carry the excess moment, along with additional steel (A_{s_2}) as shown in Fig. 15–30. For the part shown in Fig. 15–29, the moment capacity is

$$\overline{M} = .85f'_c A_c jd = \frac{.85(4 \text{ k.s.i.})(24 \text{ in.}^2)(18'')}{12''/'}$$

$$= 122.4'^K$$

This means that more compression zone is needed to carry the additional moment. This part is shown in Fig. 15–30. The moment to be carried by this *rectangular* section is

$$\overline{M}_2 = 176'^K - 122.4'^K = 53.6'^K$$

We can now determine the total area of steel required by simply adding the results of the parts.

$$A_{s_1} = \frac{\overline{M}}{f_y jd} = \frac{(122.4'^K)(12''/')}{(60 \text{ k.s.i.})(18'')} = 1.36 \text{ in.}^2$$

Since the portion of the cross section shown in Fig. 15–30 is a rectangular section, we can use the nomograph (Data Sheet D–37) to determine A_{s_2} required to carry the excess moment of $53.6'^K$. Based on a rectangular section,

$$K = \frac{\overline{M}_2}{f'_c bd^2} = \frac{(53.6'^K)(12''/')}{(4 \text{ k.s.i.})(12'')(16'')^2} = .052$$

Referring to the nomograph, the corresponding $J = .968''$. Consequently,

$$A_{s_2} = \frac{\overline{M}_2}{f_y jd} = \frac{(53.6'^K)(12''/')}{(60 \text{ k.s.i.})(.968)(16'')} = .69 \text{ in.}^2$$

The corresponding compression area is

$$A_c = \frac{A_{s_2} f_y}{.85 f'_c} = \frac{(.69 \text{ in.}^2)(60 \text{ k.s.i.})}{.85(4 \text{ k.s.i.})}$$

$$= 12.2 \text{ in.}^2$$

Therefore, the total compression zone area = 24 in.^{2} + 12.2 in.^{2} = 36.2 in.^{2}.
The total area of steel required is

$$A_s = A_{s_1} + A_{s_2} = 1.36 \text{ in.}^2 + .69 \text{ in.}^2$$

$$= 2.05 \text{ in.}^2$$

We must now check to see that this area of steel does not exceed the recommended limit of 75% of the balanced condition. This is most easily done by checking the compression area, which also must be limited to 75% of the balanced condition. Using the balanced condition expression:

$$a_{\text{Balanced}} = .85\left[\frac{87}{87 + 60}\right]20'' = 10.1''$$

The balanced compression area is:

$$A_{c(\text{Balanced})} = (4'' \times 6'') + (6.1'' \times 12'') = 97.2 \text{ in.}^2$$

The usable compression area is:

$$A_c = .75(97.2 \text{ in.}^2) = 72.9 \text{ in.}^2 > 36.2 \text{ in.}^2$$

The actual compression area is considerably less than the area for 75% of the balanced condition; consequently, there is no danger of a compression failure in this case.

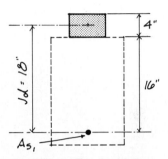

FIGURE 15–29

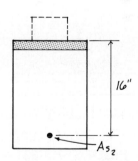

FIGURE 15–30

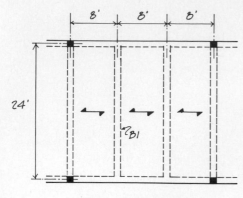

FRAMING PLAN

FIGURE 15–31

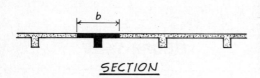

SECTION

FIGURE 15–32

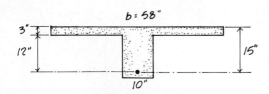

FIGURE 15–33

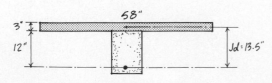

FIGURE 15–34

Example 15–7 (Figs. 15–31 through 15–33)

In this case we'll design the beam B1, which supports a one-way slab, as shown in the framing plan of Fig. 15–31. The ultimate design moment ($\overline{M}$) to be carried by the beam has been determined to be 200'^{K}. The material specifications are

$$f'_c = 3000 \text{ p.s.i.}, \quad k_1 = .85, \quad f_y = 60 \text{ k.s.i.}$$

Since the slab is poured monolithically with the beam, we must determine how much of the slab will act as a flange with the beam. This will provide us with a T-shaped cross section, as shown in Fig. 15–32. The width of the web, depth of the section, and thickness of the slab have been set as shown in Fig. 15–33. In order to determine the width of the flange, we will evaluate the three criteria for T-beams stated earlier in this chapter. They are:

1. Eight times the slab thickness on each side of the web:
$$b = 48'' + 10'' = 58''$$

2. One fourth of the beam span:
$$b = 6' = 72''$$

3. One half of the clear distance to the next beam on each side of the web:
$$b = 8' = 96''$$

The least of these values is the one to use; therefore, $b = 58''$.

We must now determine if the required compression zone will be greater or less than can be furnished by the thickness of the slab. If the stress block comes down into the web, then we will have a T-shaped compression zone. If the stress block falls entirely within the thickness of the slab, then as far we we're concerned, we have a rectangular section with the width of the section being the flange width b. The simplest way to determine this is to calculate the moment-carrying capacity of the flange and compare it with the required moment. Therefore, assuming the compression zone to be the shaded area of Fig. 15–34,

$$\overline{M} = .85 f'_c A_c jd = \frac{.85(3 \text{ k.s.i.})(58'')(3'')(13.5'')}{12''/'}$$

$$= 499'^K$$

This is far greater than the required $\overline{M}$ of 200 '^{K}; therefore, the entire compression stress block will be within the flange. This section will be, for our purposes, designed as a rectangular section, 58'' wide. Using the design equations of Data Sheet D–37,

$$K = \frac{\overline{M}}{f'_c bd^2} = \frac{(200'^K)(12''/')}{(3 \text{ k.s.i.})(58'')(15)^2} = .061$$

Going to the nomograph on Data Sheet D–37, the corresponding value of $J = .962$.

$$A_s = \frac{\overline{M}}{f_y jd} = \frac{(200'^K)(12''/')}{(60 \text{ k.s.i.})(.962)(15'')} = 2.77 \text{ in.}^2$$

Referring to Data Sheet D–23, three #9 bars would probably be the best choice for the reinforcing bars to meet the requirement.

In this case, as would be true in most T-sections, the compression zone is so shallow, because of the wide flange, that there is no danger of even being close to a brittle compression failure. In T-beams where the compression zone comes down into the web of the beam, which is a rare occurrence except where loads are very heavy, deflection would probably be a serious problem even if the percentage of steel limitation of 75% of the balanced condition were satisfied.

OTHER ISSUES IN REINFORCED CONCRETE

In the introductory portion of this chapter it was suggested that the primary purpose of this chapter was to provide the student with a rudimentary knowledge of the procedures (and the basis for these procedures) involved in the analysis and design of reinforced concrete beams by the ultimate strength method. It must be emphasized that these procedures cover only the flexural behavior of beams. The design procedures should provide the designer of a building with the knowledge to make preliminary and sensible estimates (unlike many "rule of thumb" procedures) of the sizes of members. It must, however, be understood that there are issues other than bending which must be considered in the final analysis or design of beams made of reinforced concrete. One of the issues that has not been presented in terms of computational procedures is that of shear in concrete. Shear in concrete involves considerations quite different from those presented in Chapter 10. In that chapter we dealt with shear in members made of elastic materials. Concrete, of course, is a brittle material; for all practical purposes, the true behavior of concrete members subjected to shear is not well known. Consequently, the design procedures for dealing with shear are largely empirical and require following recommendations of the ACI Code. This involves procedures well beyond the scope or intent of this chapter on fundamentals and is most appropriate for advanced studies in concrete. It must be said here, however, that concrete is weak in its resistance to shear; consequently, all beams should contain shear reinforcing in addition to the longitudinal reinforcing that is provided for flexure. Shear reinforcing is referred to as *stirrups,* and they are placed perpendicular to the longitudinal reinforcing, as shown in Fig. 15–35. The spacing of the stirrups throughout the length of a beam will vary with the intensity of the shearing force. Unlike beams, which normally have substantial depth, slabs cannot have stirrups because they are relatively thin; consequently, slabs should be designed so the concrete alone will resist the shear. The maximum shearing stress to which concrete may be subjected, as well as a variety of other issues about shear, are specified in the ACI Code.

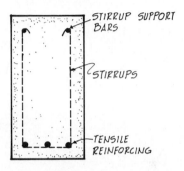

SECTION

FIGURE 15–35

Another issue of great importance in the design of reinforced concrete beams, as well as beams of any material, is that of deflection. We will consider deflection for elastic materials in some detail in the following chapter. Since concrete is a brittle material, the procedures for estimating deflection are somewhat different and, as with shear, empiricism guides us. Let it suffice for now to say that if the effective depth of a reinforced concrete beam or slab is such that it produces a value of K on the nomograph of Data Sheet D–37 somewhere above a line through the mid-height of the scale, then deflections will probably be within tolerable limits.

In the design examples presented earlier in this chapter, the recommended limitation of 75% of the area of reinforcing steel (and, consequently, 75% of the compression area) of that which would produce a balanced condition was in no case exceeded. What does it mean (and what do we do) if, in a design problem, the 75% limit of the balanced condition is exceeded? Basically, it means that the depth of the member is too shallow to carry the design moment efficiently. The simplest remedy is, of course, to increase the depth of the section. If for some reason the depth of a member cannot be increased, all is not lost. It is possible to develop a "doubly reinforced beam," as shown in Fig. 15–36. In a doubly reinforced beam, compression reinforcing is provided to reduce the demand being made on the compression concrete, thus avoiding any possibility of brittle compression failure. The computational aspects of the design of a doubly reinforced concrete beam are best left to a more advanced treatment of the subject.

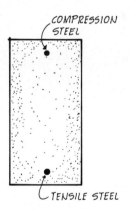

SECTION

FIGURE 15–36

There are an extraordinary number of other issues to be dealt with in the design of a reinforced concrete structure, such as minimum amounts of reinforcing, maximum amounts of reinforcing, placement of reinforcing bars, etc. It was said in the introduction to this chapter, and it bears repeating here, that "Students who wish to engage themselves in the structural design of a reinforced concrete structure should be prepared to carry their studies well beyond the scope of this chapter."

PROBLEM 15–1

REINFORCED CONCRETE - ULTIMATE STRENGTH-ANALYSIS

DETERMINE THE ULTIMATE MOMENT CAPACITY (M_u) FOR EACH OF THE FOLLOWING SECTIONS. LIMIT TO NO MORE THAN 75% OF THE BALANCED CONDITION. f'_c = 3000 P.S.I., k_1 = .85, f_y = 60 K.S.I.

Ⓐ
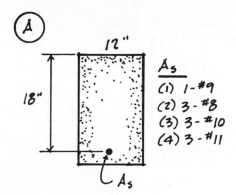

A_s
(1) 1-#9
(2) 3-#8
(3) 3-#10
(4) 3-#11

Ⓓ
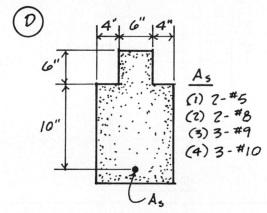

A_s
(1) 2-#5
(2) 2-#8
(3) 3-#9
(4) 3-#10

Ⓑ
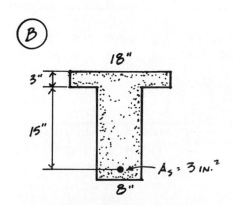

A_s = 3 IN.²

Ⓔ
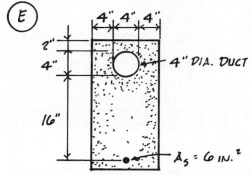

4" DIA. DUCT

A_s = 6 IN.²

Ⓒ
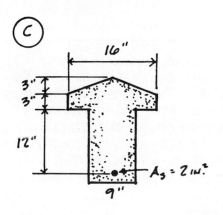

A_s = 2 IN.²

Ⓕ
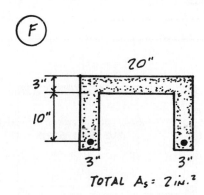

TOTAL A_s = 2 IN.²

PROBLEM 15–2

REINFORCED CONCRETE - ULTIMATE STRENGTH

DETERMINE THE ULTIMATE MOMENT CAPACITY (Mu) FOR EACH OF THE FOLLOWING SECTIONS. $f'_c = 3000$ P.S.I, $k_1 = .85$, $f_y = 60$ K.S.I.

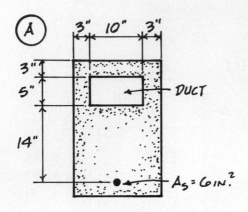

A SECTION

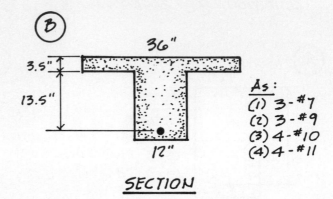

B SECTION

As:
(1) 3 - #7
(2) 3 - #9
(3) 4 - #10
(4) 4 - #11

C DETERMINE THE ULTIMATE MOMENT CAPACITY (Mu) FOR THE PRE-CAST CHANNEL SECTION SHOWN. $f'_c = 5000$ P.S.I., $k_1 = .80$, $f_y = 60$ K.S.I.

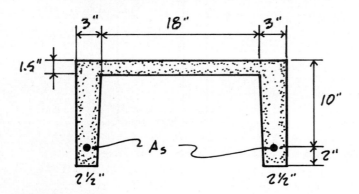

A_s IN EACH LEG:
(1) 1 - #7
(2) 1 - #9
(3) 1 - #10

D

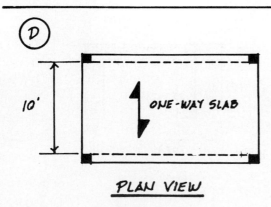

PLAN VIEW

THE ONE-WAY SLAB SHOWN IS SIMPLY SUPPORTED AND IS REINFORCED WITH #3 BARS AT 8" O.C.. THE EFFECTIVE DEPTH OF THE SLAB IS 3" WITH AN ADDITIONAL 1" OF CONCRETE TO COVER THE STEEL. DETERMINE THE SERVICE LIVE LOAD THAT MAY BE ADDED TO THE SLAB DEAD LOAD. $f'_c = 4000$ P.S.I., $k_1 = .85$, $f_y = 60$ K.S.I.

LOAD FACTORS:
L.L. = 1.7
D.L. = 1.4 $\phi = .90$

PROBLEM 15-3

REINFORCED CONCRETE - ULTIMATE STRENGTH DESIGN

DETERMINE THE UNKNOWN FOR EACH OF THE FOLLOWING BASED ON ULTIMATE STRENGTH PROCEDURES.

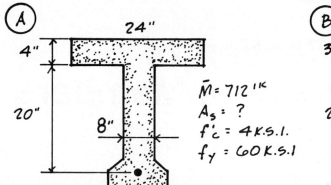

(A) 24" 4" 20" 8"

$\bar{M} = 712^{1K}$
$A_s = ?$
$f'_c = 4 \text{ K.S.I.}$
$f_y = 60 \text{ K.S.I}$

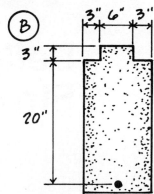

(B) 3" 6" 3" 3" 20"

$\bar{M} = 470^{1K}$
$A_s = ?$
$f'_c = 6 \text{ K.S.I.}$
$f_y = 40 \text{ K.S.I.}$

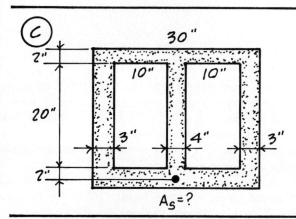

(C) 30" 2" 10" 10" 20" 3" 4" 3" 2"
$A_s = ?$

THE REINFORCED CONCRETE "BOX GIRDER" IS USED ON A 40' SIMPLE SPAN.
SERVICE LIVE LOAD = 1250 #/ₗ
SERVICE DEAD LOAD = 500 #/ₗ (BEAM WT.)
$f'_c = 4.5 \text{ K.S.I.}, \; k_1 = .825, \; f_y = 50 \text{ K.S.I.}$

LOAD FACTORS:
1.7 X SERVICE LIVE LOAD
1.4 X SERVICE DEAD LOAD
$\phi = .90$

(D) THE PRE-CAST SECTION SHOWN IS USED AS A DOUBLE CANTILEVER WITH THE GIVEN LOADING. DETERMINE A_s FOR THE POSITIVE MOMENT AND THE MAXIMUM NEGATIVE MOMENT. $f'_c = 5 \text{ K.S.I.}, \; k_1 = .80, \; f_y = 60 \text{ K.S.I.}$

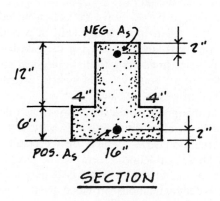

NEG. A_s
12"
4" 4"
6"
POS. A_s 16"
2"
2"

SECTION

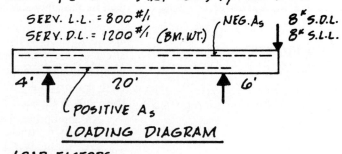

SERV. L.L. = 800 #/ₗ
SERV. D.L. = 1200 #/ₗ (BM. WT.)
NEG. A_s
8^K S.D.L.
8^K S.L.L.
4' 20' 6'
POSITIVE A_s

LOADING DIAGRAM

LOAD FACTORS
1.7 X SERVICE LIVE LOAD
1.4 X SERVICE DEAD LOAD
$\phi = .90$

16

Deflection of Structural Members

INTRODUCTION

In preceding chapters we learned to design members based on allowable stresses for bending and for shear. In the design of a structural member there is, however, a further consideration of great importance. When a beam is subjected to loads, a bending moment is produced, causing compression in the top fibers (as in the midspan of a simply supported beam) and tension in the bottom fibers. The consequences of these stresses are shortening of the fibers above the neutral axis and elongation of the fibers below the neutral axis, resulting in a curved shape, as shown in Fig. 16–1. The distance the beam bends from its original horizontal position is called *deflection*. The amount of deflection that can be tolerated in a beam is largely dictated by the function of the member, and this consideration often can be the governing criterion in its design. In many cases, deflection limits may be small enough so that the design of a beam will be dictated by deflection rather than by the allowable stress of the material.

The problems that can be caused by excessive deflection include damage to finishes, water penetration when the exterior skin of a building is affected, and water puddling on roofs, thereby causing increased load and compounding the deflection problem. In addition to these physical problems, there is a psychological problem of a "bouncy" floor system, which can be the result of the lack of proper stiffness in a beam.

An appropriate question at this point would be, "How much deflection is too much?" When dealing with allowable stresses we have guidelines provided by the various materials industries, which are the results of testing and experience. While these guidelines seem to be absolute, the amount of deflection that can be tolerated in a beam is largely a matter of judgment. There are certain recommendations provided by the steel, timber, and concrete industries that help guide the designer's judgment. The most notable of these recommendations, and the one most frequently used as a guide, is:

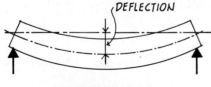

FIGURE 16–1

Deflection should not exceed 1/360 of the span (in inches), due to the live load.

This guideline was developed in the days when plaster ceilings were common, and it was determined that deflections in excess of this would cause the plaster to crack. While plaster ceilings are no longer common, this value for limiting deflection is still in general use for the design of a relatively stiff floor system.

There is no reason, however, why the designer of a structure could not use guidelines that give more or less limiting deflections. In general, limiting deflections are taken from as large as 1/180 of the span in inches to as small as, perhaps, 1/2000 of the span in inches. The larger value, 1/180 of the span, may be used for roof systems of utilitarian structures, so long as proper drainage is provided. The smaller value, 1/2000 of the span, may be used where the floor system supports sensitive machinery or equipment. A range of limiting deflection values between these may be used depending on the function of the building and the judgment of the structural designer.*

SLOPE AND DEFLECTION DIAGRAMS

We'll now study the procedures necessary to determine the maximum deflection in structural members under loads. Let's begin with simply supported beams with loads symmetrically placed; then we'll move on to more complex considerations.

The first thing we must do is to establish the various terms we will be using and their meanings. When a member is subjected to a load, as shown in Fig. 16–2, it bends, with the result being a curved shape (shown greatly exaggerated by the dotted line). The curved shape is called the *elastic curve* of the beam. The distance that the member deflects from the horizontal will be called Δ *(delta)*, which will be measured in inches. The slope of a tangent to the elastic curve, which is an influencing factor on Δ, will be referred to as θ*(theta)*, which is simply a ratio, i.e., a pure number. As we go through several examples, the relationship between Δ and θ should become obvious. While, for practical purposes, we will be interested in determining deflections (Δ) in beams, we cannot get this directly without dealing with the slope of the tangent (θ) to the elastic curve. With the terminology now established, we can go on to the procedures.

Symmetrically Loaded Beams

In Chapter 8 the development of shear and bending moment diagrams was presented. These techniques have been used extensively since that point, and it is assumed that the reader has sufficiently learned the lessons involved. In Fig. 16–3 a simply supported beam is shown along with the shear and bending moment diagrams produced by the loading. It will be recalled that the shear diagram is used to instruct us in the drawing of the bending moment diagram. In precisely the same manner, the bending moment diagram will tell us how to draw the next diagram in the sequence, which is the *slope diagram*. This yields values of the slope of a tangent to the elastic curve at

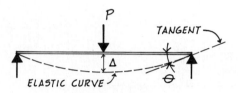

FIGURE 16–2

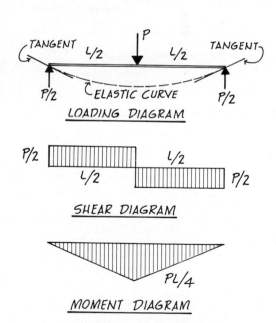

LOADING DIAGRAM

SHEAR DIAGRAM

MOMENT DIAGRAM

FIGURE 16–3

*For more detailed recommendations, see: 1. AISC Specifications, *Manual of Steel Construction,* American Institute of Steel Construction. Current edition. 2. AITC Specifications, *Timber Construction Manual,* prepared by The American Institute of Timber Construction, 2nd ed. John Wiley & Sons, New York, 1974. 3. *Building Code Requirements for Reinforced Concrete,* American Concrete Institute. Current edition.

any point along the length of the beam. The slope diagram, in turn, leads to the final diagram in the sequence, which is the *deflection diagram*. These diagrams are shown in Fig. 16–4, with values for maximum slope and deflection. Let's analyze the diagrams to see how the maximum values were determined.

Referring to Figs. 16–3 and 16–4, it can be seen that the bending moment diagram varies linearly (first degree), which means that the slope diagram will have a second-degree variation. In order to draw the slope diagram properly, certain obvious physical conditions must be used. If the elastic curve is studied (the dotted line in the load diagram), it should be obvious that the maximum slope of a tangent to the elastic curve occurs at the ends of the beam. These are opposite in direction and, therefore, opposite in sign. In *a simply supported and symmetrically loaded beam* it may, in fact, be said that:

> The maximum slopes of tangents to the elastic curve will occur at the ends, and they will be equal and opposite to each other.

Another condition that should be obvious from studying the elastic curve is that the slope of the tangent to the elastic curve is zero at the midspan. In *a simply supported and symmetrically loaded beam:*

> The point of zero slope of a tangent to the elastic curve occurs at the midspan.

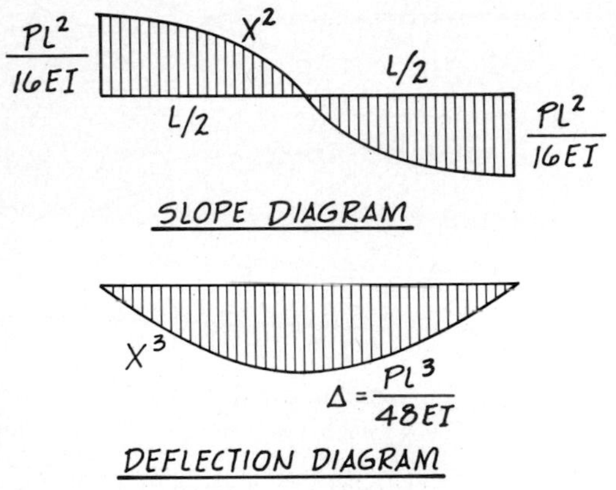

With these rules established, we can now proceed with the derivation of the values of slope shown in Fig. 16–4. Using the shape of the structure (horizontal) as a base line on which to construct the diagram, we can plot three points based on obvious conditions. We can plot points at each end, which will represent the maximums. While these values are equal to each other, they are opposite in sign and, therefore, must be plotted on opposite sides of the base line. We can now plot a third point on the base line at the midspan, which represents the fact that the slope of the tangent to the elastic curve at this point is zero. We know, based on the bending moment diagram, which varies linearly, that the three points established will be connected with a second-degree variation. The bending moment diagram will dictate the exact character of this second-degree variation in the same way that the shear diagram dictates the exact character of the curve on the bending moment diagram. Also, it will be recalled that the difference in values between any two points on the bending moment diagram can be determined by computing the area under the shear diagram between those same two points. In the same manner, the *difference* in value between any two points on the slope diagram can be determined by computing the area under the bending moment diagram between the same two points. This is the basis for the maximum value of slope shown on the slope diagram.

A few comments are now necessary about the product of E and I appearing in the denominator of the slope value. It should be recognized that the actual numerical value of the slope of a tangent to the elastic curve will be influenced by the strength of a member. The factors that indicate strength are the modulus of elasticity *(E)*, which is a measure of the material quality, and the moment of inertia *(I)*, which is a measure of strength based on cross-sectional geometry. In order to get a proper numerical value, the value of slope determined by the diagram must be divided by *EI*. Therefore, in the example being considered,

$$\theta = \frac{PL^2}{16EI} \quad \text{or} \quad EI\theta = \frac{PL^2}{16}$$

$$\frac{PL^2}{16EI} \qquad x^2 \qquad L/2$$

$$L/2 \qquad \qquad \frac{PL^2}{16EI}$$

SLOPE DIAGRAM

$$x^3 \qquad \Delta = \frac{PL^3}{48EI}$$

DEFLECTION DIAGRAM

FIGURE 16–4

Now that we have information on the slope diagram, we can proceed with the *deflection diagram*. In essence, the deflection diagram is a mathematical picture of the elastic curve. This idea provides us with at least a hint regarding the general shape of the deflection diagram. Specifically, the curve will be a third-degree variation, since the slope diagram is a second-degree variation. Just as we used obvious conditions as a starting point for the slope diagram, we will use obvious conditions to help us develop the deflection diagram. In the beam with which we are dealing, it should be obvious that the maximum deflection will occur at the midspan, where the slope of the tangent to the elastic curve is zero. In a *simply supported and symmetrically loaded beam*:

The maximum deflection occurs at the midspan.

In addition to this, it should also be obvious that the deflection at the supports is zero. We will assume, in every case, that we are dealing with nonsettling supports. Now that we have three points established, we can connect these with a third-degree curve, as shown in Fig. 16–4. In order to determine the maximum value of deflection, we simply evaluate the area under the slope diagram between the end and the midspan. The reader is referred to Data Sheet D–24, which gives general expressions for areas under various curves. Just as the factors E and I were included in the slope value, their product must be applied to the deflection expression in order to account for the material and the cross-sectional geometry of the member. Therefore,

$$EI\Delta = \frac{PL^3}{48}$$

The expressions we have developed for the beam of Fig. 16–3 are given on Data Sheet D–25, along with general expressions for other loadings that are frequently encountered.

It would be useful at this point to summarize, in short form, the principles used for the development of the diagrams and the determination of values.

Summary of Principles for Drawing Slope and Deflection Diagrams

Slope diagram ordinates in terms of $EI\theta$. Deflection diagram ordinates in terms of $EI\Delta$.

Use obvious physical conditions, such as the positions of zero or maximum amounts, symmetry, etc.

The following relationships exist throughout the entire sequence of diagrams:

1. *Ordinate* (amount at a position) *of any diagram controls pitch* (rate of change in ordinates, at the same position) *of the next diagram.* This enables the drawing of correctly shaped diagrams.

Note: Zero amount at any position compels a maximum or minimum in the next diagram at the same position. *Location is vital.*

2. Between any two positions: The area of any diagram equals the *difference* of ordinates in the next diagram between the two positions. (Opposite quantities must be treated algebraically.) This enables computation of amounts in one diagram from areas in the previous diagram.

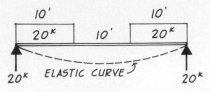

LOADING DIAGRAM

FIGURE 16–5

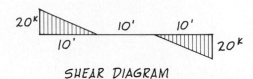

SHEAR DIAGRAM

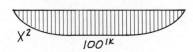

MOMENT DIAGRAM

FIGURE 16–6

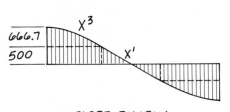

SLOPE DIAGRAM

FIGURE 16–7

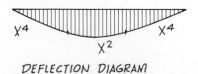

DEFLECTION DIAGRAM

FIGURE 16–8

Let's look at a numerical example dealing with a simply supported and symmetrically loaded beam.

Example 16–1 (Figs. 16–5 through 16–8)

It is desired to determine the maximum deflection for the steel beam (W16 × 36) under the loading as shown in Fig. 16–5. The shear and bending moment diagrams are shown in Fig. 16–6, with the maximum values.

Now, in going from the moment diagram to the slope diagram shown in Fig. 16–7, we can use the obvious conditions that the maximum slopes occur at the ends of the beam, and zero slope of a tangent to the elastic curve occurs at the midspan. Having established the three points to be connected, we can now draw the slope diagram based on information from the moment diagram. Referring to Figs. 16–6 and 16–7, and working from the left to the right, we see that there is a second-degree variation in the moment diagram for the first 10′. Therefore, there will be a third-degree variation in the slope diagram. For the next 10′ there is a constant value on the moment diagram and, therefore, a linear variation on the slope diagram going through the base line at the midspan, which represents the point of zero slope. The diagram is then completed with another third-degree variation. We can now evaluate the maximum value on the slope diagram by determining the difference in ordinates between the point of zero slope and the end of the member. This is done by computing the area under the moment diagram between those two points. Therefore:

$$A = \frac{2}{3}(10)(100) = 666.7^{\text{K} \cdot \text{ft}^2}$$

$$A = (5)(100) = \underline{500.0^{\text{K} \cdot \text{ft}^2}}$$
$$EI\theta = 1166.7^{\text{K} \cdot \text{ft}^2}$$

A few words are necessary at this point regarding the units indicated for $EI\theta$. It was previously mentioned that θ is a pure number. The student can verify that this is so if the units "ft.²" in the value for $EI\theta$ are converted to "in.²" by multiplying by 144 in.²/ft² and then dividing by the units of $E(\text{k/in.}^2)$ and $I(\text{in.}^4)$.

The deflection diagram can now be drawn by using the obvious conditions that the deflection at the ends is zero and the maximum deflection occurs at the midspan. These three points can now be connected with the curves, as shown in Fig. 16–8, which are dictated by the previous diagram, i.e., the slope diagram. The maximum value of deflection is computed by determining the difference in ordinates between the end, where the deflection is zero, and the midspan, where the deflection is maximum. This is done by evaluating the area under the slope diagram between the end and the midspan. Since the area under the slope diagram is bounded by different degrees of curves, this is best done by breaking the area into regular pieces, each of which is simple to evaluate, and then adding them.

Evaluating the maximum deflection:

$$A = (500^{\text{K} \cdot \text{ft}^2})(5')\left(\frac{1}{2}\right) = 1250^{\text{K} \cdot \text{ft}^3}$$

$$A = (500^{\text{K} \cdot \text{ft}^2})(10') = 5000^{\text{K} \cdot \text{ft}^3}$$

$$A = (666.7^{\text{K} \cdot \text{ft}^2})(10')\left(\frac{5}{8}\right) = \underline{4167^{\text{K} \cdot \text{ft}^3}}$$

$$EI\Delta = 10417^{\text{K} \cdot \text{ft}^3}$$

In order to find the maximum deflection (Δ) in inches, we must change the units of ft³ to in.³ by multiplying by 1728 in.³/ft³ and dividing by EI. The moment of inertia for the W 16 × 36 steel beam is 448 in.⁴, as shown in Fig. 16–5, and the modulus of elasticity of steel will be taken as 30 × 10³ k.s.i. Therefore,

$$\Delta = \frac{(10417^{\text{K} \cdot \text{ft3}})(1728 \text{ in.}^3/\text{ft}^3)}{(30 \times 10^3 \text{ k/in.}^2)(448 \text{ in.}^4)} = 1.34 \text{ in.}$$

Unsymmetrically Loaded Beams

The unsymmetrically loaded, simply supported beam presents a slightly more difficult problem than the symmetrically loaded members in the preceding discussion. Nevertheless, the principles involved in the relationship between the diagrams are still applicable. We will now develop the ideas necessary to evaluate deflections in unsymmetrically loaded members.

Consider that a simply supported beam is to be used to carry the loads as shown in Fig. 16–9. It should be obvious that under the unsymmetrical loading pattern, the elastic curve will be unsymmetrical. The maximum slope of a tangent to the elastic curve will occur at the left-hand support, and the slope of a tangent will be somewhat less at the right-hand support. Somewhere within the span, the slope of a tangent to the curve is zero, but it is not at all obvious exactly where this occurs. This presents a problem that doesn't exist in dealing with symmetrically loaded beams where the maximum slope and zero slope locations were obvious, which formed the basis for drawing the slope diagram.

The unsymmetrical loading pattern presents us with the problem of not knowing where the point of zero slope occurs. Without this knowledge, we cannot know where the maximum deflection occurs; consequently, we cannot evaluate the maximum deflection. In order to develop a solution for this sort of problem, let's look at the generalized slope diagram of Fig. 16–10. Since we do not know the location of zero slope (where the slope diagram crosses the base line), we will start out by *assuming* a location of zero slope and then making certain adjustments. The location of this assumed point of zero slope is arbitrary, although for the sake of convenience, we normally choose a point that has been dimensioned on the loading diagram, such as under a concentrated load. When this is done, it will be seen that the areas below and above the base line are not equal to each other. For example, in Fig. 16–10, suppose the area marked A is greater than the area marked B. Now, referring to the generalized deflection diagram of Fig. 16–11, it can be seen that the deflection at the ends of the member is zero. This means that the *difference in ordinates* on the deflection diagram measured at the ends is zero. The principle that will be used here is the one that tells us that the difference in ordinates between any two points on a diagram can be determined by computing the area under the preceding diagram between the same two points. Now, since the difference in ordinates on the deflection diagram from one end to the other is zero, the area under the slope diagram between the two ends must have a net value of zero. Considering that areas on opposite sides of the base line must be treated as algebraic opposites, this means that the area above the base line *must be equal* to the area below the base line. Using our assumed point of zero, with area A greater than area B, we have not satisfied this requirement. In essence, we have not shown the base line in its proper position in relation to the curve. This means that an adjustment must be made in the position of the base line. Since we are considering that area A is larger than area B, it should be clear that the base line must shift upward in relation to the curve. Now we must determine the amount by which the base line must be shifted in order to bring the areas above and below the base line into balance. Referring to Fig. 16–12, the area A, which was created by the assumed base line location, must be reduced somewhat. We'll take an area out of this, which we'll call a, shown as the unshaded portion under the dotted base line. Likewise, the original area called B must have some area added, which we'll call b. The reduction (area a) and the increase (area b) are achieved by moving the base line upward by a distance D. It is the value D that we want to determine in

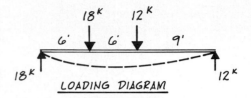

LOADING DIAGRAM

FIGURE 16–9

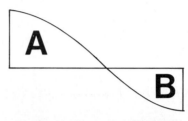

GENERALIZED SLOPE DIAGRAM

FIGURE 16–10

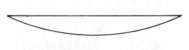

GENERALIZED DEFL. DIAGRAM

FIGURE 16–11

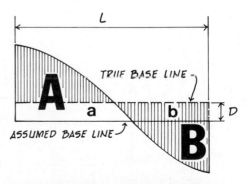

GENERALIZED SLOPE DIAGRAM

FIGURE 16–12

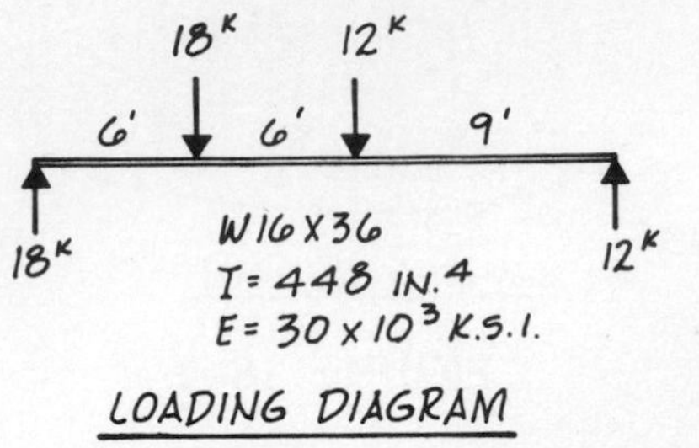

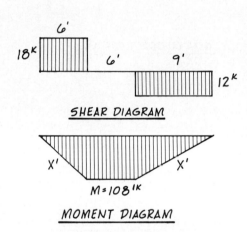

18^K 12^K

6'   6'   9'

18^K W16×36 12^K
$I = 448 \; IN.^4$
$E = 30 \times 10^3 \; K.S.I.$

LOADING DIAGRAM

FIGURE 16–13

order to locate the true base line. Let's now state, in mathematical terms, what we have established thus far:

$$A - a = B + b$$

and, rearranging this expression,

$$A - B = a + b$$

Further, it should be recognized that a and b are simply the areas of the rectangle DL. Therefore,

$$a + b = DL$$

and substituting this in the previous expression,

$$A - B = DL$$

and

$$D = \frac{A - B}{L}$$

which makes the evaluation of the true base line location a simple matter. What this expression tells us is to take the original area A, created by the assumed base line, subtract from it the original area B, and divide the result by the length of the member (in feet). This will yield the value of $EI\theta$ by which the base line must be shifted.

In studying Fig. 16–12, it can also be seen that, with the vertical shift of the base line, there is a horizontal shift of the assumed point of zero slope to its true location. This locates the actual point of maximum deflection. We will need to know the location of this point in order to compute the deflection, but this procedure can best be shown in the specific example problem.

6'

18^K 6'   9'

                12^K

SHEAR DIAGRAM

X'   X'

$M = 108^{1K}$

MOMENT DIAGRAM

FIGURE 16–14

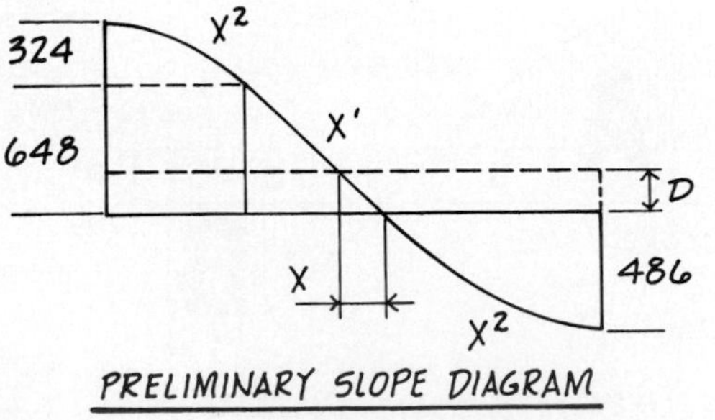

Example 16–2 (Figs. 16–13 through 16–17)

In this problem we want to determine the deflection for the steel beam shown in Fig. 16–13. The shear and bending moment diagrams produced by the given loading are shown in Fig. 16–14. Referring to Fig. 16–15, the slope diagram is drawn based on information from the moment diagram. The assumed point of zero slope has been taken under the 12^K load. In order to determine the true base line location, we will use the expression

$$D = \frac{A - B}{L}$$

Therefore, we must first evaluate the areas to either side of the assumed point of zero slope. The first step in this process is to determine the values of $EI\theta$ based on the assumed base line location. This is done by computing the areas under the moment diagram between the points where the values of ordinate are desired.

324   X^2

648

         X'

                D

X 486

       X^2

PRELIMINARY SLOPE DIAGRAM

FIGURE 16–15

Left side:

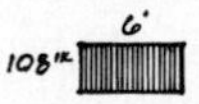

108^{1K} 6' $(6')(108'^K) = 648^{K \cdot ft^2}$

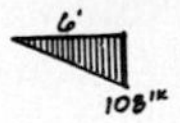

     6' $(6')(108'^K)\left(\dfrac{1}{2}\right) = \underline{324}^{K \cdot ft^2}$
108^{1K}

                $EI\theta = 972^{K \cdot ft^2}$

Right side:

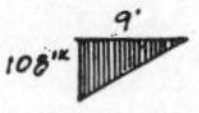

108^{1K} 9' $(9')(108'^K)\left(\dfrac{1}{2}\right) = 486^{K \cdot ft^2}$

Now that we have values on the slope diagram, we can evaluate areas A and B.

Area A

$(6')(648^{K \cdot ft^2})\left(\dfrac{1}{2}\right) = 1944^{K \cdot ft^3}$

$(6')(648^{K \cdot ft^2}) = 3888^{K \cdot ft^3}$

$(6')(324^{K \cdot ft^2})\left(\dfrac{2}{3}\right) = \underline{1296^{K \cdot ft^3}}$

$\text{Area } A = 7128^{K \cdot ft^3}$

Area B

$(9')(486^{K \cdot ft^2})\left(\dfrac{2}{3}\right) = 2916^{K \cdot ft^3}$

and

$$D = \frac{A - B}{L} = \frac{7128 - 2916}{21} = 201^{K \cdot ft^2}$$

This is the value by which the base line must be shifted in order to produce true values of slope on the left and right ends. The corrected slope diagram is shown in Fig. 16–16.

As previously mentioned, when the base line shifts vertically, there is a horizontal shift in the location of the point of zero slope. Referring back to Fig. 16–15, this horizontal distance will be called x, and this can easily be determined using the relationships between the slope diagram and the moment diagram. Specifically, the relationship involved is that where the difference in ordinate between any two points on the slope diagram can be determined by evaluating the area under the moment diagram between the same two points. The difference in ordinates over the horizontal distance x has already been established as D. The area under the moment diagram over the horizontal distance x is simply $(M)(x)$. Therefore, stated mathematically,

$$D = Mx \quad \text{or} \quad x = \frac{D}{M}$$

In our problem,

$$D = 201^{K \; ft^2}$$

$$M = 108'^{K}$$

Therefore,

$$x = \frac{201}{108} = 1.86$$

Since the assumed base line was shifted upward, it should be obvious that the actual point of zero slope is to the left of the assumed point. Now that we know the actual point of zero slope and, consequently, the point of maximum deflection, the deflection can be computed by evaluating the areas under the slope diagram between the end of the member and the zero point.

Evaluating max. $Ei\Delta$

$\left(\dfrac{2}{3}\right)(6')(324^{K \cdot ft^2}) = 1296^{K \cdot ft^3}$

$(6')(447^{K \cdot ft^2}) = 2682^{K \cdot ft^3}$

$(4.14')(447^{K \cdot ft^2})\left(\dfrac{1}{2}\right) = \underline{925^{K \cdot ft^3}}$

$EI\Delta = 4903^{K \cdot ft^3}$

Using the data given on the loading diagram of Fig. 16–13:

$$\Delta = \frac{(4903^{K \cdot ft^3})(1728 \text{ in.}^3/ft^3)}{(30 \times 10^3 \text{k.s.i.})(448 \text{ in.}^4)} = .63''$$

The deflection diagram is shown in Fig. 16–17, and the problem is now complete.

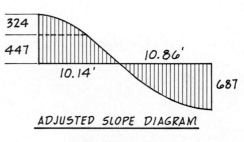

324
447
10.86'
10.14'
687

ADJUSTED SLOPE DIAGRAM

FIGURE 16–16

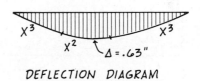

x^3 x^3
x^2 $\Delta = .63''$

DEFLECTION DIAGRAM

FIGURE 16–17

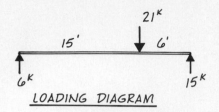

LOADING DIAGRAM

FIGURE 16–18

In Example 16–2 the determination of the horizontal shift from the point of assumed zero slope was very direct because the area under the portion of the moment diagram with which we were concerned was a pure rectangle. Often this is not so, and an approximation must be made based on trial-and-error procedures. To illustrate, let's consider another example where some judgment must be introduced in determining the amount of horizontal shift.

Example 16–3 (Figs. 16–18 through 16–21)

It is desired to determine the maximum $EI\Delta$ for the beam shown in Fig. 16–18. The corresponding shear and moment diagrams are shown in Fig. 16–19. In this case, the assumed point of zero slope is taken at the concentrated load and the preliminary slope diagram is shown in Fig. 16–20. Determining the values of slope by using the areas under the moment diagram:

Left side:

$$(15')(90'^K)\left(\frac{1}{2}\right) = 675^{K \cdot ft^2}$$

Right side:

$$(6')(90'^K)\left(\frac{1}{2}\right) = 270^{K \cdot ft^2}$$

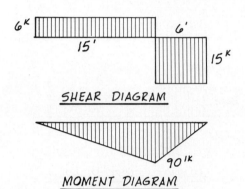

SHEAR DIAGRAM

MOMENT DIAGRAM

FIGURE 16–19

Now, with values on the preliminary slope diagram, we can compute the area to either side of the assumed point of zero slope:

Area A:

$$(15')(675^{K \cdot ft^2})\left(\frac{2}{3}\right) = 6750^{K \cdot ft^3}$$

Area B:

$$(6')(270^{K \cdot ft^2})\left(\frac{2}{3}\right) = 1080^{K \cdot ft^3}$$

and

$$D = \frac{A - B}{L} = \frac{6750 - 1080}{21} = 270^{K \cdot ft^2}$$

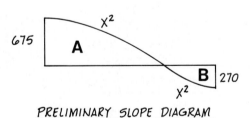

PRELIMINARY SLOPE DIAGRAM

FIGURE 16–20

The adjusted slope diagram is shown with the correct values in Fig. 16–21. In order to determine the horizontal shift, x, from the assumed point of zero slope, we must make some judgment regarding the area under the moment diagram over this distance. The moment value is decreasing and, since we do not know the value of x, we do not know the value of the moment at this distance from the point of maximum moment. The easiest thing to do in such a case is to try a value that will be considered as an average of the maximum moment and the moment at a distance x. Once we have determined a trial value for x, we must then check to see if the areas to either side of the zero slope point are in balance or close enough so any error will be negligible. Let's use this procedure and see what happens:

First trial—Average $M = 80'^K$

$$x = \frac{D}{M} = \frac{270}{80} = 3.38'$$

We must now check the areas using $x = 3.38'$:

Left side:

$$(11.62)(405)\left(\frac{2}{3}\right) = 3137^{K \cdot ft^3}$$

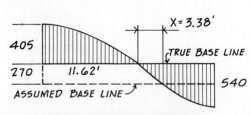

ADJUSTED SLOPE DIAGRAM

FIGURE 16–21

Right side:

Referring to Fig. 16–21, the area on the right side of zero slope must be broken into segments. For the first 3.38' the curvature of the diagram is changing direction, and we have no readily available piece of data to evaluate the area bounded by this segment. However, the curvature over this short distance is so slight that the area can be taken as a triangle without any significant error. Therefore:

$$(3.38)(270)\left(\frac{1}{2}\right) = 456^{K \cdot ft^3}$$

$$(6)(270) = 1620^{K \cdot ft^3}$$

$$(6)(270)\left(\frac{2}{3}\right) = \underline{1080^{K \cdot ft^3}}$$

$$3156^{K \cdot ft^3}$$

Comparing area A with area B:

$$A = 3137^{K \cdot ft^3} \qquad B = 3156^{K \cdot ft^3}$$

Considering the magnitude of the numbers involved, these values are very close; for all practical purposes, they are well within the range of acceptable error. The maximum $EI\Delta$ may be taken as the average of these values. Therefore,

$$EI\Delta = 3147^{K \cdot ft^3}$$

Actually, when using this technique, it is the author's judgment that differences in the areas may be as much as 5%, depending on the size of the number being dealt with, without discernible error. This is so because deflections (Δ) are generally small values, we never really know what the loads will be to any great degree of accuracy, and, in the end, the tolerable deflection is largely a matter of individual judgment. Therefore, any attempts at absolute precision would, it seems, be futile.

RESTRAINED ENDS

While we have dealt with simply supported beams thus far, there are other conditions with which we must deal—namely, where end conditions do not allow rotation. Such end conditions are referred to as *restrained ends*. When dealing with restrained ends, there are two categories that must be considered. They are the *fixed end condition,* where 100% restraint is provided at the support, and the *continuous end,* which occurs when using a member over two or more spans. The fixed end condition, as shown in Fig. 16–22, is one where the ends are completely restrained from rotation; therefore, the slope ($EI\theta$) at the ends is zero. In the continuous beam, as shown in Fig. 16–23, negative moment is developed at the interior support, thereby providing some restraint at this point. Unlike the completely fixed end condition, the continuous condition is thought of as a beam going over a "knife-edge" support. In this condition, the beam can rotate at the support, but it is not *free* to rotate as is the end of a simply supported beam, which is completely unrestrained. We will deal with continuity in some detail in the following chapter. Before doing so, however, we must develop a set of principles and corresponding data for restrained end conditions. While the ideas we are about to look at seem abstract, they must be developed before it is possible to analyze and design continuous members. Essentially, continuous members will be analyzed and designed by the process of superposition, whereby a complex structure can be visualized as broken down into simple parts, each of which we have data for, and the final results are obtained by the simple algebraic addition of the parts. We will now develop data necessary for the superposition process.

The "Weightless" Stick

Consider a member with pinned ends which carries no load and itself is weightless, as shown in Fig. 16–24. We will refer to this kind of condition

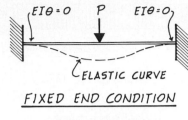

FIXED END CONDITION

FIGURE 16–22

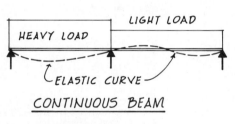

CONTINUOUS BEAM

FIGURE 16–23

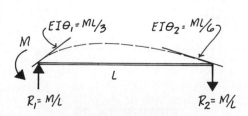

FIGURE 16–24

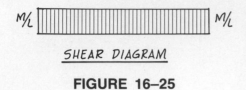

SHEAR DIAGRAM

FIGURE 16–25

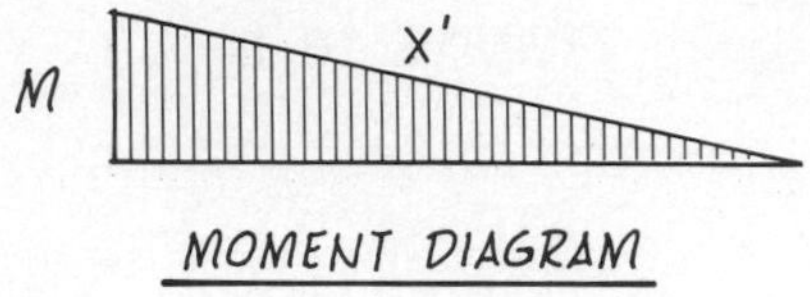

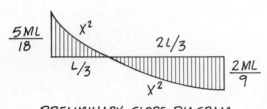

MOMENT DIAGRAM

FIGURE 16–26

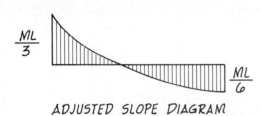

PRELIMINARY SLOPE DIAGRAM

FIGURE 16–27

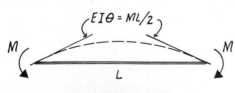

ADJUSTED SLOPE DIAGRAM

FIGURE 16–28

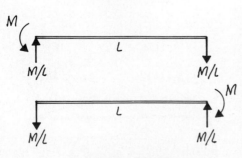

FIGURE 16–29

as a "weightless" stick. Now apply a pure moment at one end and the member will deform unsymmetrically, as shown by the elastic curve. The student can verify this by taking a flexible stick (such as a length of balsa wood), holding it at the ends, and applying a twist at one end. If this experiment is performed, it can also be seen that the end opposite to where the moment is applied must be held down in order to prevent the entire stick from rotating. Therefore, referring to Fig. 16–24, there are reactions produced in response to the applied moment. The reaction on the right-hand side is in the downward direction, to resist rotation of the stick, and the reaction at the left must be equal and opposite to this ($\Sigma F_y = 0$). To solve for the magnitude of these reactions, take moments about the left-hand end:

$$\Sigma M_{R_1} = 0 = M - R_2 L$$

and

$$R_2 = \frac{M}{L}, \quad \therefore \quad R_1 = \frac{M}{L}$$

Since these reactions are equal and opposite, the shear diagram, shown in Fig. 16–25, will have a constant value. Proceeding to the moment diagram of Fig. 16–26, we will have a value of M, which is the applied moment, and this will diminish linearly to zero at the unrestrained end. Based on the moment diagram, the slope diagram will be a second degree variation, as shown in Fig. 16–27. Since we are dealing with an unsymmetrical elastic curve, we must go through a base line shift process in order to determine the true value of the slopes at the ends. To do this, an assumed point of zero slope was taken, as shown in the preliminary slope diagram of Fig. 16–27. From this point a series of geometrical manipulations was performed in order to determine the true base line location and the values of the slopes, as shown in Fig. 16–28. Let it be a challenge to the student to verify these values.

We have developed two important pieces of data for our future work. That is, for a "weightless" stick the values of slope at the ends are:

At the moment end: $\quad EI\theta = \dfrac{ML}{3}$

At the opposite end: $\quad EI\theta = \dfrac{ML}{6}$

The actual location of zero slope and the maximum deflection for this condition will serve no useful purpose for us; consequently, we will not proceed to this step.

Another condition we will want to develop data for is that of a "weightless" stick with equal values of moment applied at each end, as shown in Fig. 16–29. In order to determine the end reactions for this, we will use the principle of superposition along with data we have already established. With this in mind, consider the "weightless" stick with equal end moments broken into two "weightless" sticks, each with a value M applied at opposite ends, as shown in Fig. 16–30. The addition of these two conditions, in fact, is the same as the original condition of Fig. 16–29. Again referring to Fig. 16–30, it was previously established that in a "weightless" stick with a moment applied at one end, reactions are produced that are equal and opposite to each other. When the two conditions of Fig. 16–30 are added, the net reactions at the ends are zero. Therefore, in a "weightless" stick with equal end moments there are no reactions and, consequently, a constant zero shear, as shown in Fig. 16–31. The moment diagram, shown in Fig. 16–

FIGURE 16–30

32, will be a constant value *M,* which is the value of the applied end moments. In order to draw the slope diagram, we can use obvious conditions, since this is a symmetrical case. The slopes at the ends of the member will be equal and opposite to each other; the slope at the center of the member will be zero. With the three points established, we will connect them with a linear variation, as shown in Fig. 16–33, based on the fact that the moment diagram is a constant. The value of slope at the ends is determined by computing the area in the moment diagram between the center and the end. Therefore,

$$EI\theta = \frac{ML}{2}$$

Now that we have developed general expressions for the slope at the ends of a "weightless" stick with a moment applied at the ends, we can deal with members that are not simply supported and unrestrained. In this chapter we will look at fixed end conditions; in the following chapter we will deal with continuous members. To do this we will study several examples of fixed end conditions using the principles of superposition and established data.

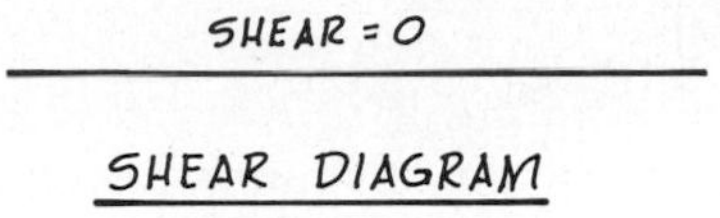

Example 16–4 (Figs. 16–34 through 16–37)

The uniformly loaded beam shown in Fig. 16–34 is supported at fixed ends. By definition, a fixed end is one that completely restrains the member from rotation at the support. In other words, $EI\theta = 0$ at the supports. Because of this restraint, there is a *negative moment* at the support, which is called the *fixed end moment*. For the case being studied, it is desired to determine the general expression for the fixed end moment. In order to do this, consider first that the restraints are removed, allowing the member to rotate at the ends, which is shown in Fig. 16–35. For the simple span condition, we know what the value of slope is at the ends by referring to the uniform load condition on Data Sheet D–25. This is given as

$$EI\theta = \frac{WL^2}{24}$$

But in reality, we don't have a simple span. Therefore, we must add to this condition the effects of the restraint at the end due to the fixed end moment (*M*). This condition is shown as the "weightless" stick of Fig. 16–35. Now when we add the two conditions shown in Fig. 16–35, we will have the original picture shown in Fig. 16–34. We know that at the fixed ends the slope ($EI\theta$) must be zero. Referring to Fig. 16–35, this means that the slope that would occur at the ends of the simple span condition must be completely negated by the slope produced by the fixed end moment (*M*). Therefore,

$$\frac{ML}{2} = \frac{WL^2}{24} \quad \text{and} \quad M = \frac{WL}{12}$$

which is the value of the fixed end moment.

Let's continue with this problem and draw the shear and bending moment diagrams. In order to draw the shear diagram, we can use the two conditions shown in Fig. 16–35 and add the results. It should be recalled that there is zero shear in the "weightless" stick with equal end moments. Therefore, the net shear diagram is the same as it would be for a simple span, and this is shown in Fig. 16–36. In the bending moment diagram, shown in Fig. 16–37, we must start at the ends with the value of the negative fixed end moment. The moment diagram will be a second degree curve, and the value of the positive moment is determined by computing the area under the shear diagram between the end and the midspan. It must be remembered at this point that the area under the shear diagram between any two points is equal to the *difference* in ordinate on the moment diagram. Therefore, in computing the value of the positive moment,

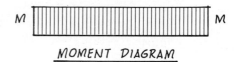

SHEAR = 0

SHEAR DIAGRAM

FIGURE 16–31

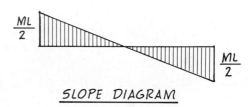

MOMENT DIAGRAM

FIGURE 16–32

$ML/2$ ... $ML/2$

SLOPE DIAGRAM

FIGURE 16–33

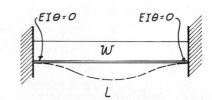

FIXED END BEAM

FIGURE 16–34

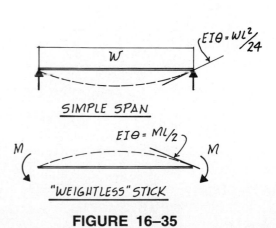

SIMPLE SPAN

"WEIGHTLESS" STICK

FIGURE 16–35

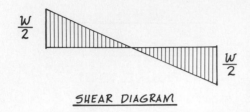

SHEAR DIAGRAM

FIGURE 16–36

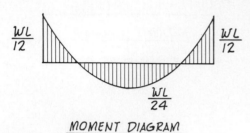

MOMENT DIAGRAM

FIGURE 16–37

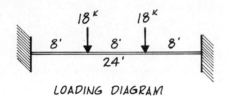

LOADING DIAGRAM

FIGURE 16–38

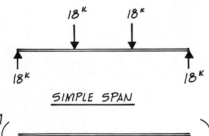

SIMPLE SPAN

"WEIGHTLESS" STICK

FIGURE 16–39

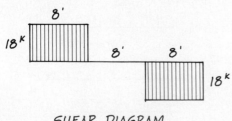

SHEAR DIAGRAM

FIGURE 16–40

$$\frac{WL}{8} - \frac{WL}{12} = \frac{WL}{24}$$

which is the value of the positive moment.

Values for fixed end moments for a variety of loading conditions are given on Data Sheet D–26. It is recommended that the student attempt to derive as many of these as possible.

Example 16–5 (Figs. 16–38 through 16–41)

In this numerical example, we will determine the maximum negative moment, positive moment, and the maximum $EI\Delta$ for the member of Fig. 16–38. In order to do this, we will superimpose the two conditions shown in Fig. 16–39. For the simple span, Data Sheet D–25 gives the slope at the ends for equal concentrated loads at the third points. The value is

$$EI\theta = \frac{PL^2}{9}$$

For the "weightless" stick with equal end moments,

$$EI\theta = \frac{ML}{2}$$

And for the fixed end condition shown we know that the net $EI\theta = 0$. Therefore,

$$\frac{ML}{2} = \frac{PL^2}{9} \quad \text{and} \quad M = \frac{2PL}{9}$$

The maximum negative moment is

$$-M = \frac{2PL}{9} = \frac{(2)(18)(24)}{9} = 96'^{K}$$

The shear diagram is shown in Fig. 16–40 with appropriate values. In order to determine the value of the maximum positive moment, compute the area under the shear diagram and subtract the negative moment:

$$+M = (18^{K})(8') - 96'^{K} = 48'^{K}$$

The moment diagram is shown in Fig. 16–41.

Computing the net maximum deflection for this case is a relatively simple matter. This will be done by the superposition of the conditions shown in Fig. 16–39. Data Sheet D–25 gives the maximum deflection for the simple span case as

$$EI\Delta = \frac{23PL^3}{648}$$

This causes a downward deflection, and we must subtract the upward deflection due to the effect of negative moments. In Fig. 16–33 the slope diagram for the "weightless" stick was shown. From this we can develop, quite easily, a general expression for $EI\Delta$. Since this is a symmetrical case, it is obvious that the maximum $EI\Delta$ occurs at the midspan. Therefore, we need only to compute the area under the slope diagram:

$$EI\Delta = \left(\frac{ML}{2}\right)\left(\frac{L}{2}\right)\left(\frac{1}{2}\right) = \frac{ML^2}{8}$$

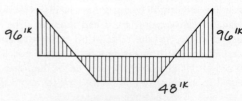

FIGURE 16–41   MOMENT DIAGRAM

Therefore, the net $EI\Delta$ for our case is

$$\frac{23PL^3}{648} - \frac{ML^2}{8} = \frac{23(18)(24)^3}{648} - \frac{(96)(24)^2}{8} = 1920^{K \cdot ft^3}$$

which is the net deflection, and it is downward (below the horizontal). The example is now complete.

CONCLUSION

The principles of slope and deflection studied thus far have a wide application in the design and analysis of structures. This will become even clearer in the following chapter where we will deal with continuous members. As will be seen, a continuous member is one which we will not be able to analyze or design using the basic equations of static equilibrium ($\Sigma F_x = 0$, $\Sigma F_y = 0$, $\Sigma M = 0$). Such structures are called *statically indeterminate*. In order to analyze structures such as these, which are more complex than those we have dealt with in the past, we will need to use the idea of superposition and the relationships between slopes and deflections. In this chapter, we have formed the basis for dealing with statically indeterminate structures.

PROBLEM 16–1

SLOPE AND DEFLECTION

(1) DRAW THE COMPLETE SERIES OF DIAGRAMS – EVALUATE MAXIMUM
    VALUES OF BENDING MOMENT, $EI\theta$ (K.FT.^{2}) & $EI\Delta$ (K.FT.^{3}).
(2) DETERMINE THE MAXIMUM DEFLECTION (Δ") FOR THE GIVEN STEEL
    "W" SECTION ($E = 29,000$ K.S.I.).

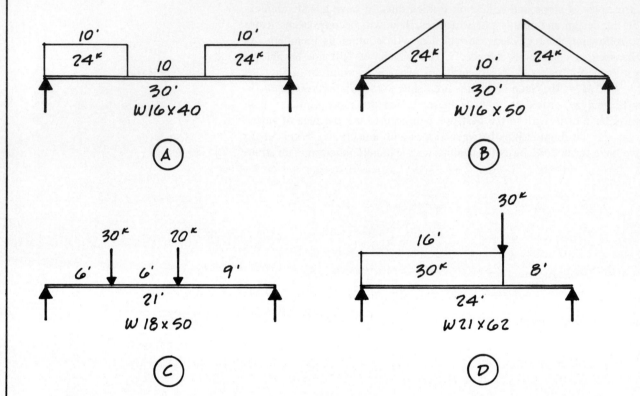

DRAW THE COMPLETE SERIES OF DIAGRAMS AND DETERMINE THE GENERAL
EXPRESSIONS (IN TERMS OF W AND/OR P) FOR THE FOLLOWING:

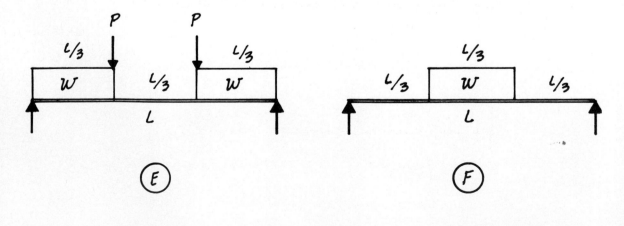

PROBLEM 16–2

SLOPE & DEFLECTION

(I) FOR EACH OF THE FOLLOWING, DRAW THE SHEAR, BENDING MOMENT, EIθ AND EIΔ DIAGRAMS AND DETERMINE THE VALUE OF EIΔ AT POINT A. SHOW ALL FEATURES CLEARLY (SUCH AS DEGREE OF CURVE, INTERSECTIONS, ETC.

(A)

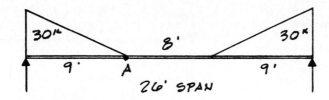

(B)

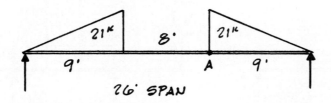

(C)

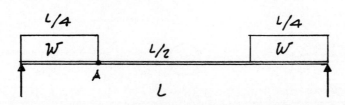

(II) FOR EACH OF THE FOLLOWING, DRAW THE SHEAR, BENDING MOMENT, EIθ AND EIΔ DIAGRAMS AND DETERMINE THE MAXIMUM EIΔ.

(A)

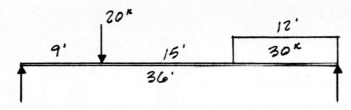

(B)

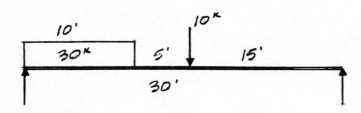

PROBLEM 16–3

STATICALLY INDETERMINATE BEAMS

FOR EACH OF THE FOLLOWING, DETERMINE THE FIXED END MOMENT, AND DRAW THE SHEAR AND BENDING MOMENT DIAGRAMS. SHOW VALUES AT ALL POINTS OF CHANGE AND SHOW ALL FEATURES CLEARLY.

Ⓐ
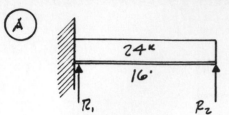

Ⓕ
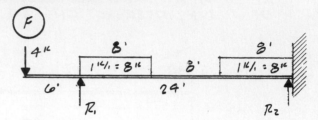

Ⓑ
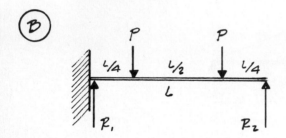

Ⓖ
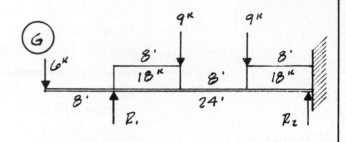

Ⓒ
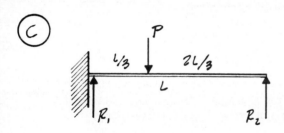

Ⓗ
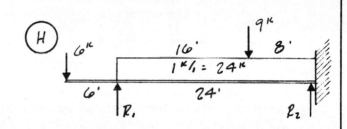

Ⓓ
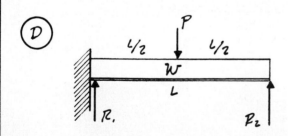

Ⓙ
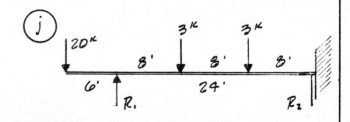

Ⓔ
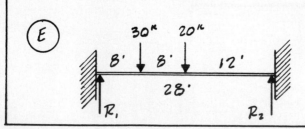

Ⓚ
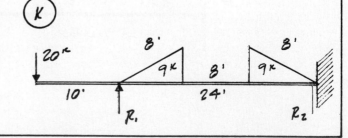

17

Structural Continuity

INTRODUCTION

The idea of the continuous member was first mentioned in the preceding chapter. It was said in the conclusion of that chapter that continuous members are statically indeterminate; consequently, their analysis and design is somewhat more complex than that of simply supported beams. The point in getting involved in the more complex problems of continuity is that there are important advantages to be derived through the design of a framing system of continuous beams. The primary advantage is that beam sizes will normally be smaller (meaning less material) for continuous beams than they would be for simply supported beams to do a given job.

To elaborate on this, let's consider Figs. 17–1 and 17–2. In Fig. 17–1 two spans are achieved by two independent simply supported beams. These members share a common interior support, but each is unrestrained at both ends. The only character of bending moment in these beams is *positive*. The maximum value of this moment would be determined by methods already established and the appropriate size would be selected. In Fig. 17–2 the same two spans are achieved with one continuous beam. In studying the generalized deflected shape of the beam, it should be obvious that there will be a *negative* moment at the interior support. This negative moment will reduce the positive bending moment within the span, as compared to the magnitude of moment that would exist if these were two simply supported beams. Generally, the *negative* and *positive* moments in a continuous beam will each be less in magnitude than the maximum moment produced in simply supported beams under the same load and span conditions. This leads to smaller size requirements for the beam, with generally substantial savings in material.

It was mentioned previously that the analysis of continuous beams is somewhat more complex than that of simply supported beams because they are statically indeterminate. To elaborate this, consider the two-span continuous beam shown in Fig. 17–3. In order to design this member, we must

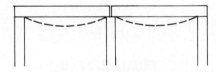

TWO SPANS - SIMPLY SUPPORTED

FIGURE 17–1

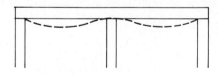

TWO SPANS - CONTINUOUS

FIGURE 17–2

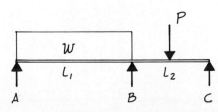

2 - SPAN CONTINUOUS BEAM

FIGURE 17–3

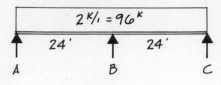

FIGURE 17–4

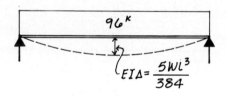

FIGURE 17–5

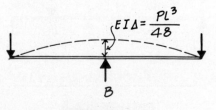

FIGURE 17–6

first determine the maximum bending moment, which may turn out to be the negative moment at the interior support or the positive moment in spans L_1 or L_2. In order to determine these values, we must be able to draw the bending moment diagram, which means that we must first draw the shear diagram. In order to do this, we must determine reactions A, B, C; this is where we encounter the problem of statical indeterminacy. There are three unknown reactions and only two equations of statics that can be written ($\Sigma M = 0$ and $\Sigma F_y = 0$). In order to solve the three unknowns, we must have three equations. Such a problem is said to be statically indeterminate to the first degree, since there is one more unknown than there are equations. These kinds of problems are solvable, but involve something more than the equations of statics. We will now look at the various techniques to be used in the solution of continuous beams. It must be emphasized at this point that all of the conditions we will deal with are based on the assumption that the supports are nonsettling, meaning that the tops of the supports will all remain at the same level.

METHODS OF ANALYSIS

We will look at three approaches to the analysis of continuous members. In doing so, constant reference will be made to already established data on Data Sheet D–25 in the Appendix, and the information developed in Chapter 16 under the heading of "The 'Weightless' Stick."

Deflection Solution Method

This approach is limited to two-span conditions with symmetrical loads and equal spans. This is not a very severe limitation, since these conditions are not uncommon in buildings.

Example 17–1 (Figs. 17–4 through 17–9)

In order to determine the maximum bending moments for the case shown in Fig. 17–4, we must first sketch the shear diagram, which means that reactions A, B, and C must be evaluated. In order to do this by the process called the *deflection solution method*, first imagine that the center support (B) is removed, as shown in Fig. 17–5. This creates a simply supported beam with a uniform load. Previous experience tells us that the maximum deflection for this case occurs at the center of the span; the value of this is given on Data Sheet D–25 as

$$EI\Delta = \frac{5WL^3}{384}$$

But the fact remains that we do not have a simply supported beam, since there is actually a support at the center. Therefore, now visualize the center support placed back in position, as shown in Fig. 17–6. The center support is an upward point load, which causes an upward deflection. Referring again to Data Sheet D-25, the deflection for this case is given as

$$EI\Delta = \frac{PL^3}{48}$$

We know that the deflection at support B is, in fact, zero. Therefore, the deflection produced with this support removed (Fig. 17–5) must be negated by the deflection produced by the condition of Fig. 17–6. Therefore

$$\frac{5WL^3}{384} = \frac{PL^3}{48}$$

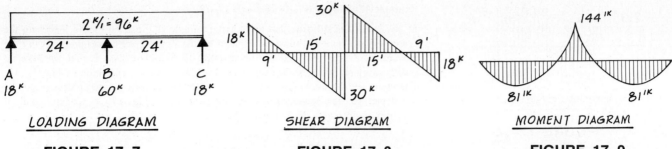

LOADING DIAGRAM

FIGURE 17-7

SHEAR DIAGRAM

FIGURE 17-8

MOMENT DIAGRAM

FIGURE 17-9

and, solving this for P (which is the reaction called B in this case)

$$P = B = 60^K$$

Now that we have solved for one of the unknowns, we can use the fact that symmetry exists and that $\Sigma F_y = 0$. Therefore

$$A + C = 96^K - 60^K = 36^K$$

and

$$A = C = 18^K$$

In order to complete the analysis, we must now sketch the shear and bending moment diagrams. The loading diagram with the appropriate reactions is shown in Fig. 17–7. The shear diagram, shown in Fig. 17–8, is constructed in precisely the same manner as that for simply supported beams. That is, the load diagram is followed by going up the amount of the end reaction and then down by the rate of loading. The uniform load yields a linear variation on the shear diagram. The moment diagram shown in Fig. 17–9 will be a second degree variation, with the maximum values of positive and negative moment computed from the areas under the shear diagram. If we were now to size a member of constant and symmetrical cross section it would, of course, be sized for the largest value, which in this case is the negative moment of $144'^K$.

Example 17–2 (Figs. 17–10 through 17–15)

In order to determine the maximum moment for the case of Fig. 17–10 by the deflection solution method, imagine the center support (B) removed, as shown in Fig. 17–11. This is a case of quarter-point loading; the deflection for this condition is given on Data Sheet D–25 as

$$EI\Delta = \frac{19PL^3}{384}$$

Now visualize the center support back in place, as shown in Fig. 17–12. This must cause a deflection equal and opposite to that produced by the quarter-point loading. Therefore

$$\frac{19PL^3}{384} = \frac{BL^3}{48}$$

and

$$B = 19^K$$

By symmetry

$$A = C = 2.5^K$$

The completed loading diagram, shear diagram, and bending moment diagram with maximum values are shown in Figs. 17–13 through 17–15. The maximum value of moment for which this member would be designed (symmetrical and constant cross section) is the negative moment of $27'^K$.

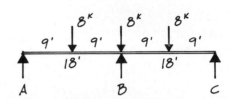

FIGURE 17-10

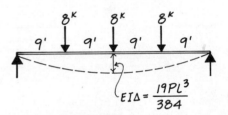

FIGURE 17-11

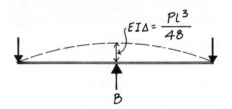

FIGURE 17-12

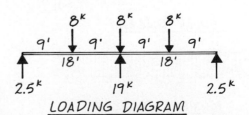

LOADING DIAGRAM

FIGURE 17-13

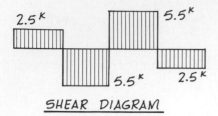

SHEAR DIAGRAM

FIGURE 17–14

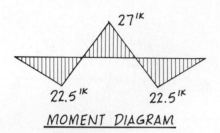

MOMENT DIAGRAM

FIGURE 17–15

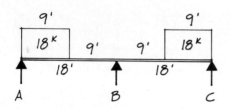

FIGURE 17–16

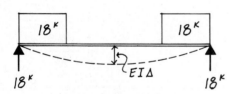

FIGURE 17–17

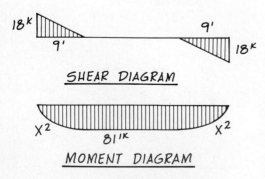

SHEAR DIAGRAM

MOMENT DIAGRAM

FIGURE 17–18

Example 17–3 (Figs. 17–16 through 17–19)

In this case, it is still true that the maximum deflection caused by the load with the center support removed must be negated by the upward deflection produced by the upward point load provided at this support. The problem here is that we have no readily available piece of data on Data Sheet D–25 for the condition of Fig. 17–17. Therefore, we must determine the value of $EI\Delta$ for this case by the process presented in Chapter 16. The sequence of diagrams is shown in Fig. 17–18, and the value of $EI\Delta$ at the midspan is determined to be

$$EI\Delta = 12575^{K \cdot ft^3}$$

In order to solve for the reaction B

$$\frac{BL^3}{48} = 12575, \quad B = 13^K$$

and, by symmetry,

$$A = C = 11.5^K$$

The completed load, shear, and bending moment diagrams are shown in Fig. 17–19, with the maximum moment being the negative moment of $36'^K$.

It should be reasonably clear now that the deflection solution method is most handy when dealing with two-span conditions for which there are readily available data. In the previous example we encountered a condition that did not involve a "standard" sort of loading, such as those given on Data Sheet D–25. In this case we had to determine a value for $EI\Delta$, which was not difficult, but did involve a bit of extra computation time. Hopefully, it is also clear by now that the deflection solution method will only work conveniently with symmetrically placed loads and equal spans. This is so because the maximum deflection produced with the center support removed must occur at the same location as the deflection produced by the center support. Obviously, an infinite variety of loading and span possibilities exist for two-span continuous members. Therefore, we will now develop the procedures necessary for dealing with *any span and load* relationship for two-span continuous beams, whether symmetrical or not.

Slope Solution Method

In order to develop the general expression required to deal with any span and load relationship for a two-span continuous beam, consider the case of Fig. 17–20. Load 1 and Load 2 may be of any configuration, not necessarily the same, and spans L_1 and L_2 may be of any dimensions, not necessarily equal. Now we'll break this into two free-body diagrams of span L_1 and span L_2. In doing so, the negative moment at support B is released and must be shown acting externally on the free bodies, as shown in Fig. 17–21. The next step is to take each of these free bodies and consider them as individual simple spans, without the negative moment (M) acting, as shown in Fig. 17–22, and with Load 1 and Load 2 acting. Each of these simple spans will deflect under the loads. At the center support (support B), the slope of the tangent to the elastic curve is recorded as $EI\theta_1$ (for span L_1) and $EI\theta_2$ (for span L_2). These are the slopes that would occur if each of these spans were truly simply supported and independent of each other. The precise values of $EI\theta_1$ and $EI\theta_2$ can be computed for whatever the loads and spans may be, based on the lessons of Chapter 16. *However,* the fact is that we do not have two independent simply supported spans. Therefore, to correct the pictures, we must superimpose the effect of the negative moment (M) which exists at support B, due to continuity. Since the effect of the loads has been accounted for in Fig. 17–22, we will show the negative moment acting on two ''weightless'' sticks of span L_1 and span L_2, as shown in Fig. 17–23. We learned in Chapter 16 that when a moment is applied at the end of a ''weightless'' stick, the slope of the tangent to the elastic curve at the end where the moment is applied is $ML/3$. Since we are dealing with two different spans (L_1 and L_2), the values of slopes are

$$\frac{ML_1}{3} \text{ for span } L_1$$

$$\frac{ML_2}{3} \text{ for span } L_2$$

as shown in Fig. 17–23.

What must be recognized at this point is that we are dealing with a continuous beam, which has a common interior support. This means that there can be only one value of the slope of the tangent to the elastic curve at support B. What this tells us is that the algebraic summation of the slopes for span L_1 at support B must be the same as the algebraic summation of the slopes for span L_2 at support B. Adding the slope values for L_1, from Figs. 17–22 and 17–23, and equating them to the slope values for span L_2, and treating slopes in opposite directions with opposite signs, we get

$$EI\theta_1 - \frac{ML_1}{3} = \frac{ML_2}{3} - EI\theta_2$$

and

$$\frac{M(L_1 + L_2)}{3} = EI\theta_1 + EI\theta_2$$

Solving for M:

$$M = \frac{3}{L + L_2}[EI\theta_1 + EI\theta_2]$$

which is the general expression for analyzing two-span continuous members with any load and span relationships. Since this expression was developed based on the idea of a single value of slope at the common interior support, it will be referred to as the *slope solution method*. This expression yields the value of the negative moment at the interior support; we will see, in the example problems, how the knowledge of this value reduces the indeterminacy of the beam so that an analysis can be performed.

Example 17–4 (Figs. 17–24 through 17–28)

The two-span continuous beam of Fig. 17–24 is unsymmetrical and will, therefore, be analyzed using the slope solution method. In order to do this, we will use the expression

$$M = \frac{3}{L_1 + L_2}[EI\theta_1 + EI\theta_2]$$

In this case,

$$L_1 + L_2 = 32' + 27' = 59'$$

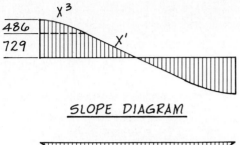

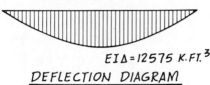

SLOPE DIAGRAM

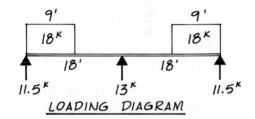

$EI\Delta = 12575 \ K.FT.^3$

DEFLECTION DIAGRAM

FIGURE 17–18 contd.

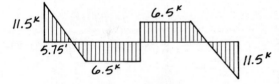

LOADING DIAGRAM

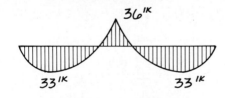

SHEAR DIAGRAM

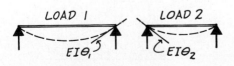

MOMENT DIAGRAM

FIGURE 17–19

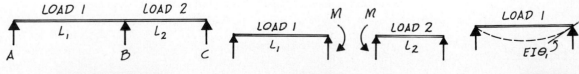

FIGURE 17–20

FIGURE 17–21

FIGURE 17–22

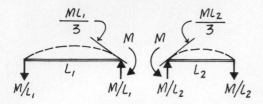

FIGURE 17–23

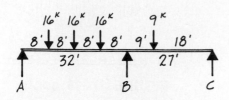

FIGURE 17–24

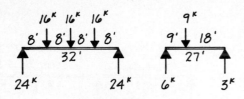

FIGURE 17–25

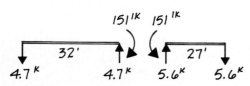

FIGURE 17–26

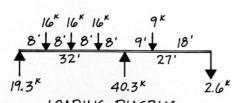

LOADING DIAGRAM

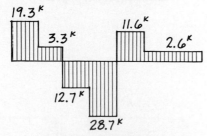

SHEAR DIAGRAM

FIGURE 17–27

Remembering now that the values of $EI\theta$ in the equation are the simple span slopes for each span, where they come together at the common support, and referring to Data Sheet D–25,

$$EI\theta_1 = \frac{5PL^2}{32} = \frac{(5)(16)(32)^2}{32} = 2560^{K \cdot ft^2}$$

$$EI\theta_2 = \frac{5PL^2}{81} = \frac{(5)(9)(27)^2}{81} = 405^{K \cdot ft^2}$$

$$\therefore EI\theta_1 + EI\theta_2 = 2965^{K \cdot ft^2}$$

and

$$M = \frac{3}{59}(2965) = 151'^K$$

Now that we know the value of the negative moment, we can determine the reactions by using free bodies of each span and the principle of superposition. To begin with, consider each span as being independent and simply supported, as shown in Fig. 17–25. The computation of the reactions for these cases is a simple matter and the values are shown in the figure. But we really don't have two simply supported spans. Therefore, we must add the effect of the negative moment at the interior support. We must superimpose two "weightless" sticks with a moment acting at the ends, as shown in Fig. 17–26. Since the moment at the unrestrained ends (*A* and *C*) is zero, there is nothing more to add to the series of diagrams. It was shown in Chapter 16 that the reactions at the ends of a "weightless" stick are equal to *M/L*. These values are shown in Fig. 17–26, in the proper directions. In order to determine the reactions *A*, *B*, and *C*, we must take the summation of all reactions of Figs. 17–25 and 17–26 at each of the supports, treating opposite directions as opposite in sign. Therefore,

$$A = 24^K - 4.7^K = 19.3^K$$

$$B = 24^K + 6^K + 4.7^K + 5.6^K = 40.3^K$$

$$C = 3^K - 5.6^K = -2.6^K$$

The minus sign for the net reaction *C* means that this force is in the *downward* direction.

The analysis may now be completed by sketching the shear and bending moment diagrams, which are shown in Fig. 17–27. In this case, the maximum moment is the positive moment of $180.8'^K$. It can also be seen, on the moment diagram, that positive bending is never developed in span L_2. This is due to the heavy loading and longer span of L_1 and the comparatively light loading of span L_2. This condition points out the fact that adjacent spans have an effect on each other, when dealing with continuous beams. The deflected shape for this condition would be, generally, that shown by the dotted line in Fig. 17–28.

Example 17–5 (Figs. 17–29 through 17–32)

In order to analyze the case of Fig. 17–29, the slope solution method will be used.

$$L_1 + L_2 = 24' + 20' = 44'$$

Referring to Data Sheet D–25, the slopes for each span as a simple beam are

$$EI\theta_1 = \frac{WL^2}{24} = \frac{(48)(24)^2}{24} = 1152^{K \cdot ft^2}$$

$$EI\theta_2 = \frac{PL^2}{16} = \frac{(32)(20)^2}{16} = 800^{K \cdot ft^2}$$

$$\therefore EI\theta_1 + EI\theta_2 = 1152 + 800 = 1952^{K \cdot ft^2}$$

and, using the slope solution equation,

$$M = \frac{3}{44}(1952) = 133'^K$$

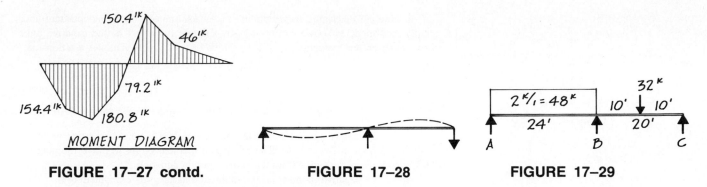

MOMENT DIAGRAM

FIGURE 17–27 contd.

FIGURE 17–28

FIGURE 17–29

To determine the reactions *A, B,* and *C,* consider each span as independent and simply supported, as shown in Fig. 17–30. Then we must superimpose the effect of the moment at the interior support, as shown in Fig. 17–31. The net reactions are the algebraic summation of the forces at each support point. Therefore,

$$A = 24^K - 5.5^K = 18.5^K$$

$$B = 24^K + 16^K + 5.5^K + 6.7^K = 52.2^K$$

$$C = 16^K - 6.7^K = 9.3^K$$

The shear and bending moment diagrams are shown in Fig. 17–32; the maximum moment is the negative moment of $133'^K$.*

A point might be made, at this time, regarding Example 17–4 and Example 17–5. In Example 17–4, it was found that the positive moment was the critical value for designing. In order to discover this, it was necessary to draw the shear and moment diagrams. In Example 17–5, the negative moment was the critical value, and this was determined from the slope solution equation. Had we known that this would turn out to be the critical moment value, we would not have needed to draw the shear and bending moment diagrams. For "normal" loading and span conditions, the negative moment will generally be the critical value. The problem here is that it is not possible to give any kind of absolute guidelines as to what "normal" loadings and span relationships are, and at what point these relationships cease to be "normal." The only safe way to analyze continuous beams and determine the critical moment value is to sketch the shear and bending moment diagrams. It is recommended that this be done for all problems.

Example 17–6 (Figs. 17–33 through 17–37)

Referring to Fig. 17–33, we have another case that must be analyzed using the slope solution method. Unlike the previous two example problems, however, we do not have readily available data for the simple span slope of span *B-C*. Therefore, the first step in this analysis will be to take span *B-C* as a simple span and determine the value of $EI\theta$ at support *B,* using the principles presented in Chapter 16. The necessary series of diagrams is shown in Fig. 17–34. Proceeding with the analysis, we have

$$L_1 + L_2 = 48'$$

$$EI\theta_1 = \frac{PL^2}{9} = 576^{K \cdot ft^2} \text{ (Data Sheet D–25)}$$

$$EI\theta_2 = 597^{K \cdot ft^2} \text{ (Fig. 17–34)}$$

$$M = \frac{3}{48}(597 + 576) = 73'^K$$

*The moment diagram doesn't quite close out due to rounding off of reactions.

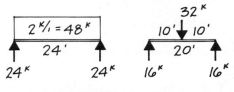

FIGURE 17–30

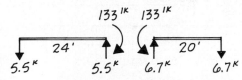

FIGURE 17–31

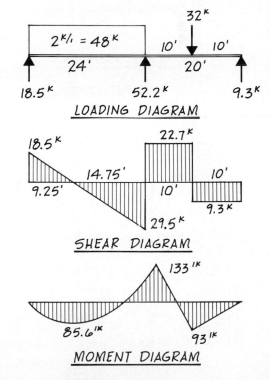

LOADING DIAGRAM

SHEAR DIAGRAM

MOMENT DIAGRAM

FIGURE 17–32

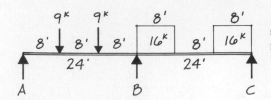

FIGURE 17-33

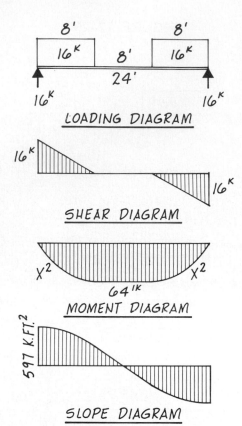

LOADING DIAGRAM

SHEAR DIAGRAM

MOMENT DIAGRAM

SLOPE DIAGRAM

FIGURE 17-34

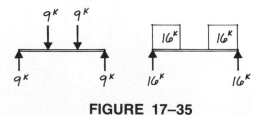

FIGURE 17-35

In order to determine the reactions, take each span as being independent and simply supported, as shown in Fig. 17-35, and add the effect of the negative moment acting on the "weightless" sticks, shown in Fig. 17-36. The net reactions are

$$A = 9^K - 3^K = 6^K$$

$$B = 9^K + 16^K + 3^K + 3^K = 31^K$$

$$C = 16^K - 3^K = 13^K$$

The analysis may now be completed by drawing the shear and bending moment diagrams, shown in Fig. 17-37. The maximum moment is the negative moment of $73'^K$, which was determined from the slope solution equation.

Note: The negative moment on the moment diagram was computed as $72'^K$ due to the rounding off of reactions.

Both the deflection solution method and the slope solution method deal with only *two-span* continuous beams. Oftentimes more than two spans are involved in continuous beams. We will now develop the general expressions for analyzing continuous beams with any number of spans and any loading pattern.

The Three-Moment Theorem

The general case, shown in Fig. 17-38, will be considered in the development of the approach required for dealing with the infinite variety of conditions that may be encountered when dealing with continuous beams. In order to develop the general expressions that will enable an analysis of this sort of problem, we will cut two spans out as a free-body diagram. It is arbitrary as to which two spans are used. Let's take the second and third spans from the left and generalize the condition by referring to the loads as Load 1 and Load 2 and the spans as L_1 and L_2. This is shown in the free-body diagram of Fig. 17-39. This picture looks very much like the one used to develop the slope solution (Fig. 17-20) except for one important factor. In this case, supports A and C are interior supports; consequently, there will be negative moments at these points which must be shown in the free-body diagram. We will now make free-body diagrams of spans L_1 and L_2, which will release the negative moment at B, as shown in Fig. 17-40. The next step is to consider each of these spans as a simply supported beam, shown in Fig. 17-41, with the simple span slopes at the common support recorded as $EI\theta_1$ and $EI\theta_2$. However, we do not have simple spans. Span L_1 is affected by the negative moment at A (M_A) and span L_2 is affected by the negative moment at C (M_C). Therefore, we will add two "weightless" sticks to the simply supported conditions, with M_A and M_C acting, as shown in Fig. 17-42. The slopes on a "weightless" stick at the end opposite to where the moment is applied was determined in Chapter 16 to be $ML/6$. These values are shown in the figure with the appropriate subscripts. In addition to the negative moments at A and C, there is also the effect of the negative moment at B (M_B). This is shown in Fig. 17-43. The slope of a "weight-

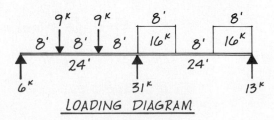

FIGURE 17-36   **FIGURE 17-37**   LOADING DIAGRAM

less'' stick at the end where the moment is applied is $ML/3$, and this value is shown in the figure with the appropriate subscripts.

Since this continuous beam has a common support at B, there can be only one value of slope at that point. This means that the net value of the algebraic summation of all the slopes in span L_1 at point B must be the same as that for span L_2. Therefore, summing the slopes of Figs. 17–41 through 17–43 and treating opposite directions with opposite signs:

$$EI\theta_1 - \frac{M_A L_1}{6} - \frac{M_B L_1}{3} = \frac{M_C L_2}{6} + \frac{M_B L_2}{3} - EI\theta_2$$

$$\frac{M_A L_1}{6} + \frac{M_B (L_1 + L_2)}{3} + \frac{M_C L_2}{6} = EI\theta_1 + EI\theta_2$$

Multiplying both sides by 6, we get

$$M_A L_1 + 2M_B(L_1 + L_2) + M_C L_2 = 6[(EI\theta_1 + EI\theta_2)]$$

which is known as the *three-moment equation* because the three negative moments are involved. Let's go through some example problems to see how this expression is used.

Example 17–7 (Figs. 17–44 through 17–49)

In order to analyze the continuous beam shown in Fig. 17–44, the three-moment equation must be used. The slope solution method will not work because this is something more than a two-span condition. While there are two main spans involved here, there is also a cantilever, which introduces a negative moment at C. Using the three-moment equation,

$$M_A L_1 + 2M_B (L_1 + L_2) + M_C L_2 = 6[EI\theta_1 + EI\theta_2]$$

where $M_A = 0$ (unrestrained end)

$\quad M_B = ?$

$\quad M_C = 90'^K$ (the cantilever moment)

$\quad L_1 = 30'$

$\quad L_2 = 20'$

$\quad L_1 + L_2 = 50'$

$\quad EI\theta_1 = \dfrac{PL^2}{9} = \dfrac{(18)(30)^2}{9} = 1800^{K \cdot ft^2}$

$\quad EI\theta_2 = \dfrac{WL^2}{24} = \dfrac{(48)(20)^2}{24} = 800^{K \cdot ft^2}$

$\hspace{4cm} \overline{\Sigma EI\theta = 2600^{K \cdot ft^2}}$

$$\therefore 2M_B (50) + (90)(20) = 6(2600)$$

and

$$M_B = 138'^K$$

Now that we know the values of the negative moments, we can solve for the reactions by using free bodies of each span and superimposing the effects of loads and moments. To begin with, consider each span as an independent simply supported beam, as shown in Fig. 17–45. To this we will add the load due to the cantilever which must go to C, shown in Fig. 17–46, since it is the only thing supporting the cantilever. Now we will add the effect of the negative moment at B, as shown in Fig. 17–47. The final step is to add the effect of the negative moment at C, which

FIGURE 17–41

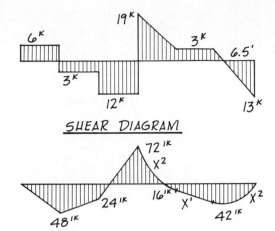

SHEAR DIAGRAM

MOMENT DIAGRAM

FIGURE 17–37 contd.

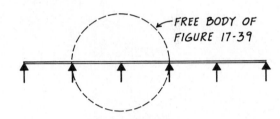

FIGURE 17–38

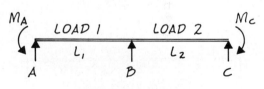

FIGURE 17–39

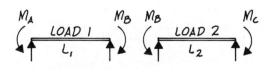

FIGURE 17–40

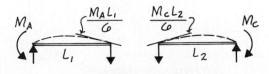

FIGURE 17–42

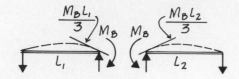

FIGURE 17–43

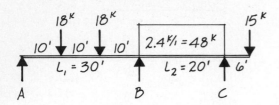

FIGURE 17–44

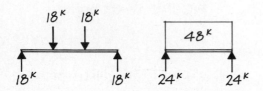

FIGURE 17–45

FIGURE 17–46

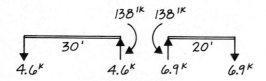

FIGURE 17–47

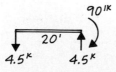

FIGURE 17–48

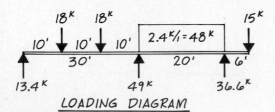

LOADING DIAGRAM

FIGURE 17–49

acts on span L_2. This is shown in Fig. 17–48. The algebraic summation of the reactions, shown in Figs. 17–45 through 17–48, gives the net reactions:

$$A = 18^K - 4.6^K = 13.4^K$$

$$B = 18^K + 24^K + 4.6^K + 6.9^K - 4.5^K = 49^K$$

$$C = 24^K + 15^K - 6.9^K + 4.5^K = 36.6^K$$

The shear and bending moment diagrams are shown in Fig. 17–49, and the analysis is complete.

At the beginning of the derivation of the three-moment equation, it was mentioned that this approach can be used to analyze conditions with any number of spans. We'll now see how this works.

Example 17–8 (Figs. 17–50 through 17–55)

In looking at the problem presented in Fig. 17–50, it should be obvious that there are four unknowns involved. These are the reactions A, B, C, and D. In order to analyze this beam, we must use the three-moment equation. However, what must be remembered is that the three-moment equation was derived based on two spans of a multiple-span condition. In order to analyze this beam, we will first write the equation for spans A–B and B–C (which we will call L_1 and L_2), using the appropriate values.

$$M_A L_1 + 2M_B(L_1 + L_2) + M_C L_2 = 6[EI\theta_1 + EI\theta_2]$$

where $M_A = 72^{\prime K}$ (the cantilever moment)

$M_B = ?$

$M_C = ?$

$L_1 = 24'$

$L_2 = 24'$

$$EI\theta_1 = \frac{WL_1^2}{24} = \frac{(24)(24)^2}{24} = 576^{K \cdot ft^2}$$

$$EI\theta_2 = \frac{WL_2^2}{24} = \frac{(24)(24)^2}{24} = 576^{K \cdot ft^2}$$

$$\therefore (72)(24) + 96(M_B) + (24)M_C = 6912$$

We have two unknowns in this equation, therefore we must write another equation with the same two unknowns. What we must do is write another three-moment equation using spans B–C and C–D (which we will call spans L_2 and L_3), and simply change the subscripts in the basic form of the three-moment equation:

$$M_B L_2 + 2M_C(L_2 + L_3) + M_D L_3 = 6[EI\theta_2 + EI\theta_3]$$

where $M_B = ?$

$M_C = ?$

$M_D = 0$ (unrestrained end)

$L_2 = 24'$

$L_3 = 24'$

$$EI\theta_2 = \frac{WL_2^2}{24} = \frac{(24)(24)^2}{24} = 576^{K \cdot ft^2}$$

$$EI\theta_3 = \frac{WL_3^2}{24} = \frac{(24)(24)^2}{24} = 576^{K \cdot ft^2}$$

$$\therefore 24M_B + 96M_C = 6912$$

We now have two independent equations with the unknowns M_B and M_C, which can be solved by simultaneous equations. Restating the equations, we have

1. $1728 + 96M_B + 24M_C = 6912$

2. $\qquad 24M_B + 96M_C = 6912$

The simultaneous solution yields

$$M_B = 38.4'^K$$

$$M_C = 62.4'^K$$

In order to solve for the reactions, we will use a series of free-body diagrams and the principle of superposition. For the sake of clarity, we will break with the usual format used in this book and extend the diagrams across the width of the entire page.

To solve for the reaction at *A*, take span *A–B* as a simply supported beam, as shown in Fig. 17-51, which also shows the cantilever delivering its total load to *A*. Then we must add the effect of the negative moments to span *A–B*. This is shown in the two "weightless" sticks of Figs. 17–52 and 17–53. There is enough information now to determine the reaction at *A*:

$$A = 12^K + 12^K + 3^K - 1.6^K = 25.4^K$$

The reaction at *B* is affected by span *B–C*, which must be added to the information already developed for span *A–B*. This is shown in Figs. 17–51, 17–52, and 17–53.

$$B = 12^K + 12^K - 3^K + 1.6^K - 2.6^K + 1.6^K$$

$$= 21.6^K$$

To determine the reaction at *C* and *D*, we must add the values shown in the series of figures (17–51 through 17–53):

$$C = 12^K + 12^K - 1.6^K + 2.6^K + 2.6^K = 27.6^K$$

$$D = 12^K - 2.6^K = 9.4^K$$

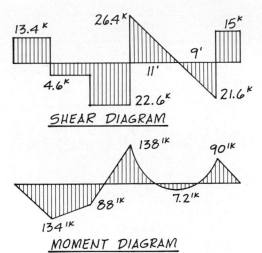

SHEAR DIAGRAM

MOMENT DIAGRAM

FIGURE 17–49 contd.

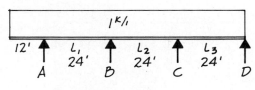

FIGURE 17–50

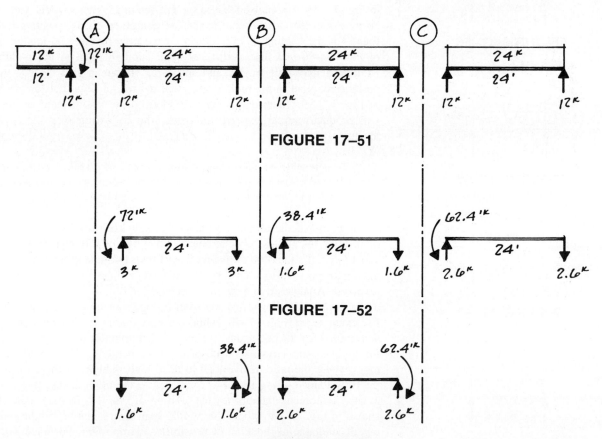

FIGURE 17–51

FIGURE 17–52

FIGURE 17–53

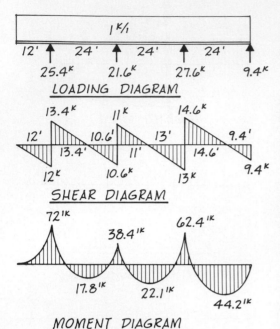

LOADING DIAGRAM

SHEAR DIAGRAM

MOMENT DIAGRAM

FIGURE 17–54

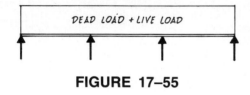

DEAD LOAD + LIVE LOAD

FIGURE 17–55

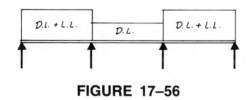

D.L. + L.L.   D.L.   D.L. + L.L.

FIGURE 17–56

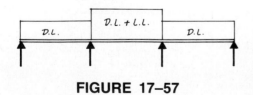

D.L.   D.L. + L.L.   D.L.

FIGURE 17–57

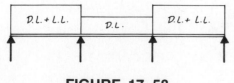

D.L. + L.L.   D.L.   D.L. + L.L.

FIGURE 17–58

To complete the analysis, the shear and bending moment diagrams are drawn. These are shown in Fig. 17–54. From this analysis it is determined that the cantilever moment at A is the maximum moment.

LOAD ARRANGEMENTS FOR MAXIMUM MOMENT

When analyzing continuous beams to determine the maximum moment for which the beam is to be designed, there is a further issue of which the student should be aware. This has to do with the fact that when there is continuity from one span to the next, the loads on each span will have an effect on the maximum moments produced in the adjacent spans. It may, in fact, be said that in a continuous beam, all spans are "on speaking terms with each other."

In the techniques of analysis presented to this point, the loads on continuous beams have been given as total loads. In an architectural structure the total loads, in fact, are made up of dead loads and live loads. In reality, only the dead loads are predictable in terms of permanence. Live loads, on the other hand, may or may not be present at any given time. In order to consider the effects of the magnitude of load that may be on a span at any time, we separate the dead and live loads and use "checkerboard" loading patterns to determine maximum moments. For example, in the three-span continuous beam shown in Fig. 17–55, an analysis based on full loading on all spans (dead load plus live load) would yield values for negative and positive moments. If, however, we considered that the live load might not be present on the middle span, then the loading pattern would be as shown in Fig. 17–56. An analysis based on this loading pattern would yield higher positive moments in the end spans, as compared to the positive moments found for the loading of Fig. 17–55. If we now reverse the loading pattern of Fig. 17–56 and place full dead load and live load on the middle span with dead load only on the end spans, as shown in Fig. 17–57, then the positive moment in the middle span will be greater than that for the loading of Fig. 17–55. Therefore, when considering the arrangement of loads that will produce critical positive moments, the following rule should be followed:

> To determine the maximum positive moment in a span, load that span and alternate spans on each side with the total load (live load plus dead load). The spans in between are loaded with only the dead load.

Because of the alternating pattern of full load and dead load only, we often refer to this as "checkerboard" loading. In referring to Figs. 17–56 and 17–57, it should be clear that it will take two analyses to determine the maximum positive moments in all spans, regardless of the number of spans involved. Analyses based on checkerboard loading may actually reveal other useful information regarding the span (or spans) supporting only dead load. For example, in Fig. 17–58, while the maximum positive moments will be determined for the end spans, we may find that under this loading condition, the middle span stays in a state of negative bending. This could happen in cases where the ratio of live load to dead load is high, or the dimensions of the fully loaded spans are large compared to the other spans. In a case such as this, a moment diagram for the loading of Fig. 17–58 may look like that shown in Fig. 17–59, where the middle span is in negative bending throughout. Knowledge of this kind of possibility is important. For example, if this was a reinforced concrete beam, appropriate reinforcing would have to be provided in the *top* of the beam throughout the entire span even though the loading condition is not permanent.

In order to determine the worst possible *negative* moment at a support, other loading patterns must be considered. The maximum negative moment at a support can be found by using the following rule:

> Use full load (live load plus dead load) on spans adjacent to the support and on alternate spans beyond. All other spans are to be loaded with dead load only.

To illustrate the above rule, consider the continuous beam shown in Fig. 17–60. The loading pattern shown will yield the maximum negative moment at support *C*. In order to determine the maximum negative moment at support *B*, the loading pattern should be as shown in Fig. 17–61 . . . and so on.

While the procedures indicated for determining the maximum negative and positive moments in continuous members may seem a bit tedious because of the number of different analyses necessary, the situation normally isn't quite as bad as it seems. For most situations, only two analyses would be necessary (one for maximum positive moment and one for negative moment). This is especially so in a case where a beam of constant cross section (such as a steel wide flange) is to be selected. In unusual situations where, for instance, span dimensions vary a great deal from span to span, or where loading patterns are unusual, it may be that more than two or three analyses are necessary.

In Chapter 20 we will look at the issue of critical load arrangements in a quantitative way. Let it suffice for now that the student be aware of the potential consequences of the fact that live loads may not exist at any given time.

TWO-HINGED FRAMES

In Chapter 9 the three-hinged frame was presented and discussed. In the analysis of this, it was only necessary to use the equations of static equilibrium in order to determine the reactions. Once the reactions were determined, we could determine the bending moments at any point within the structure. The three-hinged frame is a relatively simple structure to analyze, since it is a statically determinate structure.

A two-hinged frame (often called a rigid frame) offers some advantages as compared to a three-hinged frame, but the analysis of such a structure is a bit more complicated. To demonstrate the issues involved, consider the two-hinged frame shown in Fig. 17–62. For this discussion we will consider only frames that are symmetrical in configuration and symmetrically loaded, as shown in the figure. In a situation such as this, the determination of the vertical reactions is a simple matter. However, as in a three-hinged frame, there are also horizontal (thrust) reactions produced that cannot be determined using the equations of static equilibrium, because of the absence of the third hinge (a known point of zero bending moment).

In order to consider a way to analyze a two-hinged frame, which is a statically indeterminate structure, we must first recognize how this structure will deform under the load. The general shape of the deformed structure is shown in Fig. 17–63 (greatly exaggerated). The rigid connection at the knee of the frame means that the original angle between the vertical and horizontal members will be maintained when the structure deforms under the load. Let's now develop a general expression that will give us the necessary in-

BENDING MOMENT DIAGRAM

FIGURE 17–59

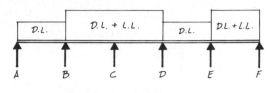

FIGURE 17–60

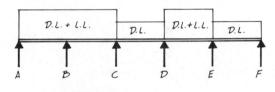

FIGURE 17–61

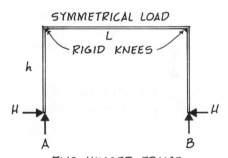

TWO-HINGED FRAME

FIGURE 17–62

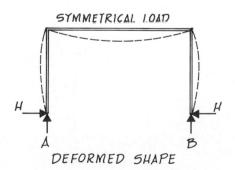

DEFORMED SHAPE

FIGURE 17–63

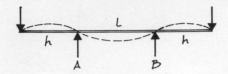

FIGURE 17–64

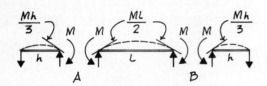

FIGURE 17–65

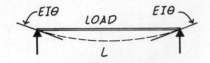

FIGURE 17–66

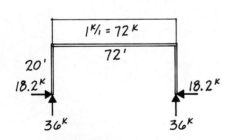

FIGURE 17–67

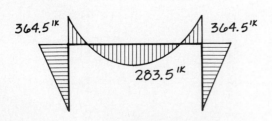

FIGURE 17–68

formation to analyze this sort of statically indeterminate structure.

To begin with, let's consider that the frame shown in Fig. 17–63 is bent out straight, as shown in Fig. 17–64. The deformed shape of this structure would be as shown in the figure. If Figs. 17–63 and 17–64 are studied carefully, it will be realized that no relationships between the vertical and horizontal members have really been changed. We will now use the fact that the slope of a tangent to the elastic curve (the dotted line in the figure) is the same as support A (or B) whether this value is related to span h or span L. With this in mind, using the principle of superposition, we will now develop a general expression that will give us the information necessary to complete the analysis of such a structure.

First, let's consider span L, with the applied load, as a simply supported beam. As such, the elastic curve is shown in Fig. 17–65. However, this span does not exist as a simply supported beam; consequently, the effects of the moments at the ends must be included. This is shown in Fig. 17–66. Since there is a common value for the slope at point A (and B) for span h and span L, then

$$\frac{Mh}{3} = EI\theta - \frac{ML}{2}$$

$$\frac{Mh}{3} + \frac{ML}{2} = EI\theta$$

and

$$M = \frac{6EI\theta}{2h + 3L}$$

This gives the moment at the knee of the frame, which must be equal to Hh, and now we can solve for H (the horizontal reaction).

Example 17–9 (Fig. 17–67)

Let's consider the two-hinged frame as shown in Fig. 17–67. Because of the symmetry involved, it is a simple matter to determine the vertical reactions; these are shown in the figure. In order to determine the horizontal reactions, we must first determine the moment at the knee of the frame using the general expression previously established:

$$M = \frac{6EI\theta}{2h + 3L}$$

The value of $EI\theta$ must be determined for the 72' span, as a simple beam. Referring to Data Sheet D–25,

$$EI\theta = \frac{WL^2}{24} = \frac{(72)(72)^2}{24} = 15{,}552^{K \cdot ft^2}$$

and

$$M = \frac{6(15{,}552)}{256} = 364.5'^K$$

$$H = \frac{364.5'^K}{20'} = 18.2^K$$

The bending moment variation for this frame is shown in Fig. 17–68, using the same convention established for drawing bending moment diagrams for three-hinged frames. Because of the absence of a third hinge in the structure, there is a positive bending moment at the midspan.

Let's now compare this with a three-hinged frame having the same dimensions and the same load. The third hinge will be located at the middle of the 72' span. The three-hinged frame under consideration is shown in Fig. 17–69. The vertical reactions and the horizontal reactions are shown in the figure. The horizontal reactions are determined using the lessons presented in Chapter 9. The bending moment diagram for this three-hinged frame is shown in Fig. 17–70. The maximum bending moment in the three-hinged frame is far greater than that produced in the two-hinged frame under the same conditions of span, load, and configuration. This comparison suggests that more material will be required for the three-hinged frame.

It must be emphasized that the above procedures have been developed for, specifically, symmetrical loadings on symmetrical two-hinged frames. This has been presented for the purpose of comparison and to show the advantages of the two-hinged frame. When dealing with two-hinged frames that are not symmetrically loaded or are unsymmetrical in configuration, there are other issues involved in the determination of the maximum moments which are beyond the scope of these studies or the intent of this section of the book.

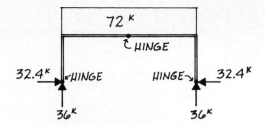

FIGURE 17–69

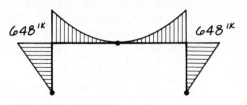

FIGURE 17–70

PROBLEM 17-1

CONTINUOUS BEAMS

(CONSTANT SECTION - HOMOGENEOUS MATERIAL)

I EQUAL SPANS AND LOADING SYMMETRY

(1) DETERMINE REACTIONS: <u>DEFLECTION SOLUTION</u> ($EI_\Delta = 0$ AT THE SUPPORTS)

(2) DRAW THE SHEAR AND BENDING MOMENT DIAGRAMS AND DETERMINE THE MAXIMUM POSITIVE AND NEGATIVE BENDING MOMENTS.

(3) DETERMINE THE MOST ECONOMICAL STEEL "W" BEAM NEEDED FOR THE CONTINUOUS BEAM. (ALLOWABLE BENDING $f = 24^{KSI}$)

(4) DETERMINE THE MOST ECONOMICAL STEEL "W" BEAMS REQUIRED IF THIS WERE TWO SIMPLE SPANS. (ALLOWABLE BENDING $f = 24^{KSI}$)

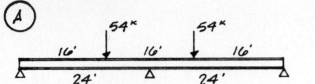

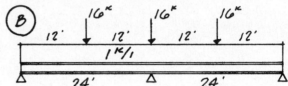

II ANY SPANS AND LOADING

(1) DETERMINE NEGATIVE BENDING MOMENT AT INTERIOR SUPPORT: SLOPE SOLUTION ●

(2) DETERMINE REACTIONS

(3) DRAW SHEAR AND BENDING MOMENT DIAGRAMS AND DETERMINE THE MAXIMUM POSITIVE AND NEGATIVE BENDING MOMENTS

(4) DETERMINE THE MOST ECONOMICAL STEEL "W" BEAM NEEDED FOR THE CONTINUOUS BEAM. (ALLOWABLE BENDING $f = 24^{KSI}$)

(5) DETERMINE THE MOST ECONOMICAL "W" BEAMS REQUIRED IF THIS WERE TWO SIMPLE SPANS (ALLOWABLE BENDING $f = 24^{KSI}$)

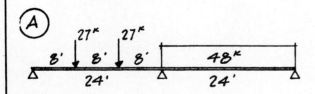

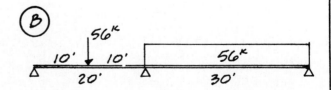

PROBLEM 17–2

CONTINUOUS BEAMS

TWO·SPAN BEAMS

ANALYZE EACH OF THE FOLLOWING USING THE DEFLECTION SOLUTION METHOD, AND DRAW THE SHEAR AND BENDING MOMENT DIAGRAMS.

(I)

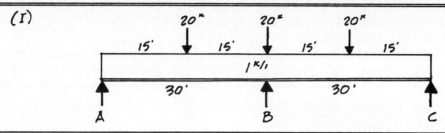

(II)

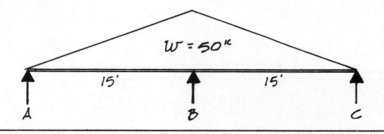

(III)

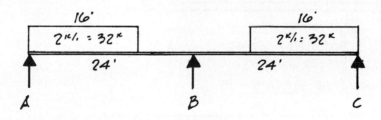

(IV)

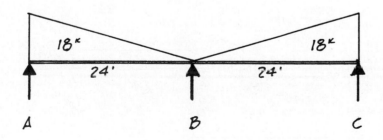

PROBLEM 17–3

CONTINUOUS BEAMS
(CONSTANT SECTION – HOMOGENEOUS MATERIAL)
USING THE THREE-MOMENT EQUATION (<u>SLOPE SOLUTION</u>): COMMON
EIΘ AT INTERIOR SUPPORT.

$$M_A L_1 + 2M_B(L_1 + L_2) + M_C L_2 = 6\left[EI\Theta_1 + EI\Theta_2\right]$$

<u>DETERMINE</u> (FOR BEAMS OF PARTS I AND II)

1. NEGATIVE BENDING MOMENT AT EACH SUPPORT
2. REACTIONS AT EACH SUPPORT.
3. SHEAR AND BENDING MOMENT DIAGRAMS (POS. & NEG. VALUES)
4. MOST ECONOMICAL STEEL "W" BEAM (ALLOWABLE BENDING f = 24 ksi)

I TWO CONTINUOUS SPANS WITH CANTILEVERS

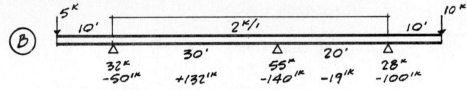

(A)

ANSWERS:

II THREE CONTINUOUS SPANS

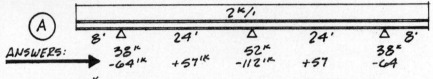

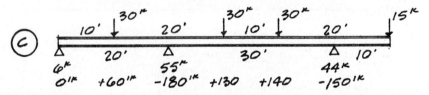

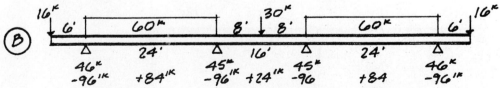

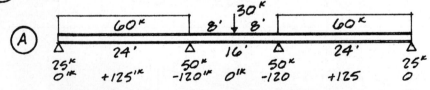

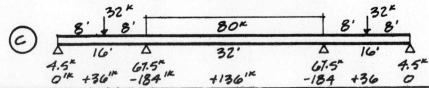

PROBLEM 17–4

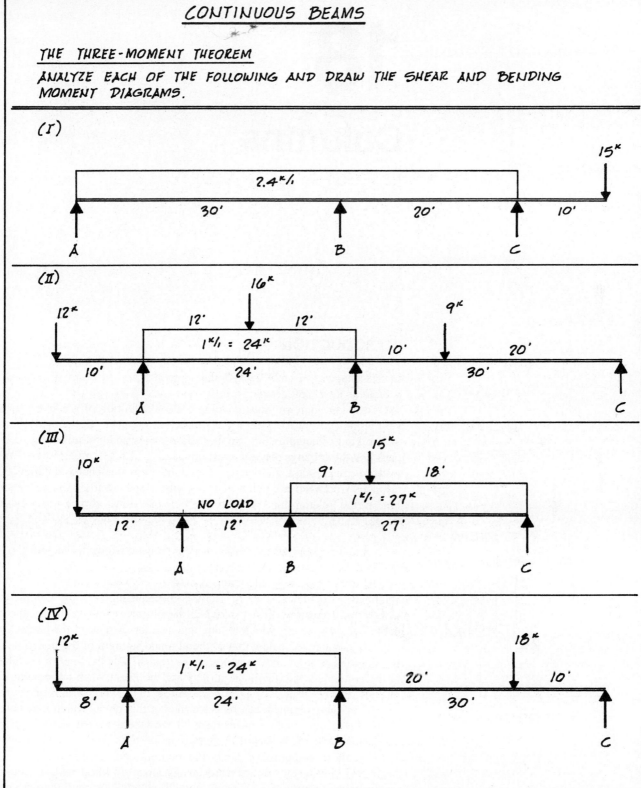

CONTINUOUS BEAMS

THE THREE-MOMENT THEOREM

ANALYZE EACH OF THE FOLLOWING AND DRAW THE SHEAR AND BENDING MOMENT DIAGRAMS.

(I)

15^k

2.4^k/,

30'   20'   10'

A   B   C

(II)

16^k

12^k

12'   12'

1^k/, = 24^k

9^k

10'   20'

10'   24'   30'

A   B   C

(III)

15^k

10^k

9'   18'

1^k/. = 27^k

NO LOAD

12'   12'   27'

A   B   C

(IV)

12^k

1^k/. = 24^k

18^k

20'   10'

8'   24'   30'

A   B   C

18
Columns

INTRODUCTION

In this chapter we shall discuss the fundamentals that form the basis for column analysis and design.

To begin with, we must develop a clear definition of a *column*. In the very broad sense, a column is a compression member. But actually, it is a unique kind of compression member. Base plates under columns, as well as piers and the footings they may rest on (see Fig. 18–1), are also transferring loads in compression, and we never think of these elements as columns. The distinction between a column and any other kind of compression member lies in the relationship between the cross-sectional dimensions and the length. It can be said that:

A column is a "long," slender member carrying a compression load.

A further, and very important, distinction between a column and a short compression member is in the failure mechanism. A short compression member will crush under a load when the ultimate strength of the material is reached. A column, which is long and slender, will bend (or buckle) at a level of loading well below that of the ultimate strength of the material. The bending of a column under a compressive load signals the limit of usefulness of the column. While the material may still be intact when the column first buckles, this nevertheless represents a failure, and a very serious one at that.

A simple experiment can be performed at this point, which may help to clarify the column failure mechanism. If we take a simple yardstick which is 3 ft long with cross-sectional dimensions of $1\frac{1}{8}$ in. by 1/8 in. and subject it to a direct compressive load, the column will buckle, as shown in Fig. 18–2. Let's say that the yardstick is made of wood with an ultimate compressive strength of 3000 p.s.i. The ultimate load according to the numbers should be

$$P = fA = (3000)(1.125)(.125) = 422\#$$

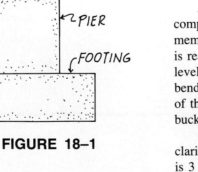

FIGURE 18–1

The experiment will reveal that the column reaches its limit of usefulness long before this load can be reached. At some small load the column will bend and deflect a distance y from its original vertical position. If the load that first produced bending remains constant, the deflection y will remain fixed and the column will support the load. If, however, the load is increased, the column will deflect further until the ultimate bending stress of the material is reached.

Experience has shown that the slightest lateral load applied to a column after it has first buckled will cause the column to fail. Therefore, the *limit of usefulness* is defined as the load that first caused the column to buckle; this load is called the *critical load, P_{CR}.*

Theoretically, at least, it would be valid to question the causes of column bending or buckling. If we had such a thing as an "ideal" column with an axial load placed *absolutely* on the centroid of the cross section, would the column bend? For the answer to this, an "ideal" column is one that conforms to the following:

1. Absolutely straight, initially
2. Absolutely homogeneous material
3. Absolutely concentric loading
4. End conditions perfectly aligned

It should be obvious that the number of "absolutes" involved would preclude the possibility of the building and installation of an ideal column. Consequently, all columns will bend under a load which is less than that indicated by the ultimate strength of the material.

General Classification of Compression Members

Earlier, the column was defined as a "long" compression member, with piers being considered as short compression members. It should be understood that the terms "long" and "short" make reference not to the length of the member alone, but rather to the length relative to cross-sectional dimensions. For example, visualize a wood column with cross-sectional dimensions of 4 in. by 4 in. and a length of, say, 15 ft. There would be no doubt that this is a "long" and slender column. Such a column would fail in bending under a reasonably small load. If, however, the same 4-in.-by-4-in. dimension was used to carry a compression load and its length was, say, 6 in., this would be thought of as a compression block rather than a column. The failure, in this case, would not be in bending, but by crushing of the material. Somewhere in between the two dramatically different lengths discussed there would be a range of lengths where the failure mechanism would not be so clear. The possibility here is that the failure would be a combination of both crushing and bending. In an attempt to classify the various possibilities in terms of length relative to cross-sectional dimensions, we shall consider the possibilities of three ranges of behavior: compression blocks or piers, columns, and long columns. It must be emphasized that the classification can only be a generalization; there is nothing absolute or definite about where one range ends and another begins. The classification and length versus cross-sectional dimensions for each range is intended only to give the reader a general sense of the proportions involved. The breakdown is as follows:

1. *Compression blocks or piers.* The failure mode of such elements is by crushing of the material. The proportions might vary from a height

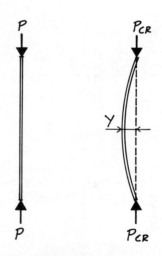

FIGURE 18–2

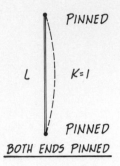

FIGURE 18-3

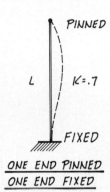

FIGURE 18-4

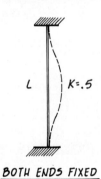

FIGURE 18-5

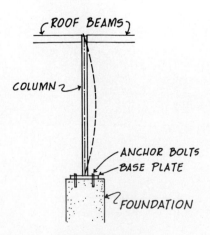

FIGURE 18-6

which is equal to the cross-sectional dimensions (a cube) to a height of about 10–15 times the least cross-sectional dimension.

2. *Columns*. An intermediate range of length, which will fail by a combination of crushing and buckling. The proportions generally vary from 10–15 times the least dimension to, possibly, 40 times the least dimension, depending on a variety of conditions.

3. *Long columns*. These will fail in bending or buckling. The proportions involved are something in the order of the length being over 40 times the least cross-sectional dimension.

Again, it must be emphasized that the preceding definitions are generalizations. No consideration is given to the material or the actual cross-sectional configuration of the column, such as whether it is solid or hollow.

In the following discussions, we will concern ourselves only with columns and long columns, but will refer to either of these conditions as *columns*, without making the distinction.

EFFECT OF END CONDITIONS ON BENDING

One of the important influences on the strength of a given column is the manner in which it is anchored at the ends. The means of anchorage, in fact, will have an effect on the length of the column, which is subject to buckling, and which for certain conditions will not be the full actual length of the column. The length of the column, which is subject to buckling, is referred to as the *effective length*. The effective length is determined by multiplying the actual length by a stiffness factor called *K*, which is based on the end conditions of the column. We will concern ourselves with the three end condition possibilities shown in Figs. 18–3, 18–4, and 18–5. The *K* factors shown indicate that a given column can carry the greatest load, because its effective length will be least, when both ends are fixed. For practical reasons, however, the most common condition is that where both ends are pinned. A typical condition of this sort is shown in Fig. 18–6. The connection to the foundation is made with anchor bolts, which do not appreciably restrain the column from rotation at this point. Consequently, this is considered a pinned connection. The roof beams, which are supported by the column, will, in turn, keep the top of the column from moving laterally, but will not keep it from rotating. Actually, the base of a column can, through elaborate detailing, be made so that it is "fixed" or moment-resisting. The cost of the detailing and fabrication, however, is such that these techniques are generally employed only on large-scale structures where wind or other lateral forces become a critical issue, and fixity at the base is required to add to the stability of the building.

Where columns are several stories tall, the intermediate floor beams will act as lateral braces, assuming the building as a whole is properly braced against lateral displacement. The lateral braces will restrain the column from displacing laterally, but will not keep the column from rotating, and the buckling mode will be as shown in Figs. 18–7 and 18–8. The length of the column under such conditions is the distance between lateral braces, and the column between these points will behave as a pinned end condition.

THE EULER FORMULA

In the mid-eighteenth century a Swiss mathematician named Leonhard Euler performed experiments to determine the buckling load for columns. From these experiments a formula was developed that yields the load under which a long column will fail by buckling. This formula is known as the Euler Formula; it served as the basis for further research and, consequently, the development of modern formulas that are used for the practical (or safe) design of columns. The derivation of the Euler Formula is a fairly complex and tedious process, which is beyond the scope or intent of this book.* The formula will be stated with appropriate definitions, and, in lieu of the derivation, we'll spend our time understanding the various parts and the limitations of the formula.

To begin with, it must be understood that the derivation of the Euler Formula was based on certain assumptions. They are as follows:

1. The column is a long column, and failure occurs in a buckling mode.
2. The unit stress in the column at the time of failure does not exceed the elastic limit of the material.
3. At initial failure by buckling, the curvature of the column is slight and, therefore, the cross section is regarded as constant throughout the length of the column.

The Euler Formula, in slightly modified form, is:

$$\frac{P_{CR}}{A} = f_{CR} = \frac{\pi^2 E}{\left(\dfrac{KL}{r}\right)^2}$$

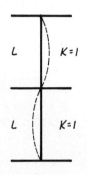

FIGURE 18-7

where P_{CR} = total buckling (or bending) load or, more precisely, the load that first produces bending, thereby defining the limit of usefulness of the column

A = the cross-sectional area of the column

f_{CR} = the *unit stress* due to P_{CR}.

E = the modulus of elasticity of the column material

KL = the effective length of the column in inches, as discussed earlier in this chapter

r = the radius of gyration, which is defined as:

$$r = \sqrt{\frac{I}{A}}$$

where I = the moment of inertia of the column

A = the cross-sectional area of the column

At this point, it must be understood that column buckling will take place about the axis with the *least* moment of inertia, which is, consequently, the axis with the least radius of gyration. However, it must also be understood that the effective length has a great deal to do with the direction in which a column will buckle. For example, if a column is pinned at the ends and unbraced in either direction, then clearly it will buckle about the axis with the least radius of gyration. If, however, the column is laterally braced in its weak axis at some point along its length, then there will be two different effective lengths involved, each relating to the two axes. In a case such as this, it is not so clear which way the column will buckle. In order to use the Euler Formula and get correct results, it must first be determined which axis

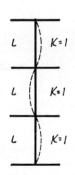

FIGURE 18-8

*For those interested in the derivation of the Euler Formula see Timoshenko, S. and Young, D. H. *Elements of the Strength of Materials*. 5th Ed. Van Nostrand Reinhold Company, New York, 1968.

is critical, where two different lengths and two different radii of gyration are involved. In order to do this, we must deal with the effective length and radius of gyration for each axis, not as isolated values, but as a ratio, which is the way these values appear in the Euler Formula. This ratio has a special name, which is

$$\frac{KL}{r} = \text{the slenderness ratio}$$

The larger the slenderness ratio, the more "slender" is the column, and when comparing two axes and lengths to determine which way buckling will occur, the largest slenderness ratio will be the critical one.

To demonstrate these ideas, let's look at several examples that will involve the application of the Euler Formula.

Example 18–1 (Fig. 18–9)

It is desired to determine the critical load (P_{CR}) for the timber column shown.

The column is pinned at the ends $(K = 1)$, and is unbraced in either direction. In this case the y-y axis is the weaker, which is easily determined by inspection. We must first determine the moment of inertia with respect to this axis.

$$I = \frac{BD^3}{12} = \frac{(8)(4)^3}{12} = 42.7 \text{ in.}^4$$

From this, we can determine the value of the radius of gyration:

$$r = \sqrt{\frac{I}{A}} = \sqrt{\frac{42.7}{32}} = 1.16 \text{ in.}$$

and

$$\frac{KL}{r} = \frac{(1)(14' \times 12''/')}{1.16} = 144.8$$

Putting this value into the Euler Formula, we can solve for P_{CR}:

$$P_{CR} = \frac{\pi^2 (1760)(32)}{(144.8)^2} = 26.5^K$$

At this load, according to the Euler Formula, the column will bend slightly in the weak direction. The last thing we must do to complete the problem is to determine if this load produces a stress within the elastic limit of the material. The elastic limit data is given in Fig. 18–9.

$$f_{CR} = \frac{P_{CR}}{A} = \frac{26.5^K}{32} = .83 \text{ k.s.i.} < 3 \text{ k.s.i. O.K.}$$

Therefore, the Euler Formula is valid, and the value determined for P_{CR} is true.

Example 18–2 (Fig. 18–10)

In this case, it is desired to determine the critical load that can be carried by the steel tube column shown in Fig. 18–10. The column is pinned at the ends and is braced in the weak direction, as shown. The column is not braced against buckling in the strong direction. Unlike the previous example, there is a question here regarding the direction in which the column will buckle. Will it initially buckle in the weak direction (about the y-y axis) where the effective length is 8 ft, or will it buckle in the strong direction (about the x-x axis) where the effective length is 16 ft? In order to determine the answer to this question, we must compare the slenderness ratios for both axes. The largest slenderness ratio will signify the axis about which buckling will take place. Therefore, the first step in this problem is to determine the slenderness ratio for each axis.

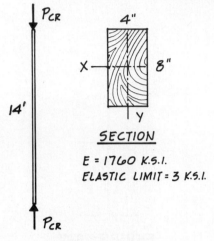

SECTION

$E = 1760$ K.S.I.
ELASTIC LIMIT = 3 K.S.I.

FIGURE 18–9

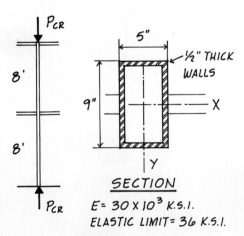

SECTION

$E = 30 \times 10^3$ K.S.I.
ELASTIC LIMIT = 36 K.S.I.

FIGURE 18–10

x-x axis. Determine I_x:

$$I_x = \frac{(5)(9)^3}{12} - \frac{(4)(8)^3}{12} = 133 \text{ in.}^4$$

$$\therefore r_x = \sqrt{\frac{133}{13}} = 3.2 \text{ in.}$$

and

$$\frac{KL_x}{r_x} = \frac{(1)(16' \times 12''/')}{3.2} = 60$$

y-y axis. Determine I_y:

$$I_y = \frac{(9)(5)^3}{12} - \frac{(8)(4)^3}{12} = 51 \text{ in.}^4$$

$$r_y = \sqrt{\frac{51}{13}} = 1.98 \text{ in.}$$

and

$$\frac{KL_y}{r_y} = \frac{(1)(8' \times 12''/')}{1.98} = 48.5$$

These computations indicate that the column is more slender in the strong direction, which is due to the fact that it is braced at the mid-height in the weak direction. Now that we have determined this, we can apply the Euler Formula, using the data given in Fig. 18–10.

$$P_{CR} = \frac{\pi^2 (30 \times 10^3)(13 \text{ in.}^2)}{(60)^2} = 1069^K$$

Check for elastic limit:

$$f_{CR} = \frac{P_{CR}}{A} = \frac{1069^K}{13 \text{ in.}^2} = 82.3 \text{ k.s.i.}$$

which is well beyond the elastic limit; therefore, the Euler Formula is invalid. The material will be destroyed before it can reach the stress given by the Euler Formula. The correct answer to this problem, therefore, is dictated by the value of stress at the elastic limit and

$$P_{CR} = f_{CR}A = (36 \text{ k.s.i.})(13 \text{ in.}^2)$$

$$= 468^K$$

LIMITATIONS OF THE EULER EQUATION

In the previous example, we computed the Euler Formula, and the check for the elastic limit showed that the formula was not valid. Actually, we could have come to this conclusion in another manner. The validity or invalidity of the Euler Formula can be determined at the outset by solving for the limiting value of the slenderness ratio. Given that we know the modulus of elasticity and the elastic limit of the material, then

$$\frac{P_{CR}}{A} = f_{CR} = \frac{\pi^2 E}{(KL/r)^2} \text{ and}$$

$$36 \text{ k.s.i.} = \frac{\pi^2 (30 \times 10^3 \text{ k.s.i.})}{(KL/r)^2}$$

$$\therefore \frac{KL}{r} = 90.7$$

This tells us that the Euler Formula can be used to determine the critical load only if $KL/r \geqq 90.7$. For slenderness ratios below this value, the unit stress in the column will exceed the elastic limit. This is what we discovered in this example where the larger slenderness ratio, based on the *x-x* axis, was used.

We can graphically illustrate the limitation of the Euler Formula. If the values of critical column stress (f_{CR}) are plotted as ordinates against corresponding values of the slenderness ratio (KL/r), a curve known as the Euler Curve is obtained, and is shown in Fig. 18–11. This shows us that

1. As f_{CR} increases, KL/r approaches zero.
2. As KL/r increases, f_{CR} approaches zero, which tells us that for certain values of the slenderness ratio, the Euler Formula will yield critical stresses beyond the elastic limit. In Fig. 18–11, for example, if point *A* represents the elastic limit for a given material, then only the portion *C–D* would be valid. The portion of the curve shown dotted represents values that would be given by the Euler Formula, but would not be valid.

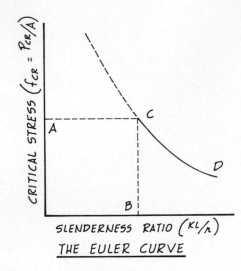

THE EULER CURVE

FIGURE 18–11

CONCLUSION

The Euler Formula represents the fundamental theory of the behavior of long columns. It should be recognized that as we get to problems of practical column design, we cannot design columns that will be working at critical stress levels. There must be factors of safety involved that will maintain the level of stress well below that which represents the critical stress. The equations used for practical design problems are known as working formulas. There are many variations in these working formulas, depending on the material being used. These considerations are beyond the level of principles and, consequently, are not considered here.

PROBLEM 18-1

COLUMNS - THE "EULER" EQUATION

(I) DETERMINE THE CRITICAL "SLENDERNESS RATIO" FOR EACH OF THE FOLLOWING CROSS-SECTIONS AND LENGTHS, BASED ON THE WEAK AXIS. LENGTH OF ALL SECTIONS = 10'.

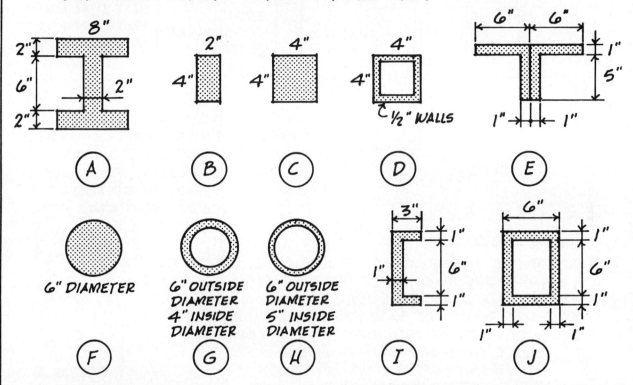

(II) DETERMINE THE BUCKLING LOAD (P_{CR}), BASED ON THE "EULER EQUATION", FOR AN 8' LONG WOOD 2"x4" (ACTUAL DIMENSIONS) WITH PINNED ENDS. E = 1.5 x 10³ K.S.I.
IF THE ELASTIC LIMIT = 6000 P.S.I., WHAT IS THE MINIMUM LENGTH FOR WHICH THE "EULER EQUATION" IS VALID?

(III) DETERMINE THE MAXIMUM LOAD FOR AN 8"x8" (FULL DIMENSION) TIMBER COLUMN, 8' LONG AND PINNED AT THE ENDS. THE ELASTIC LIMIT OF THE MATERIAL IS 4000 P.S.I.. E = 1.7 x 10³ K.S.I.

(IV) A STEEL TUBULAR COLUMN, 12' LONG, HAS A 3.5" OUTSIDE DIAMETER AND A 3" INSIDE DIAMETER. DETERMINE THE SAFE LOAD USING THE EULER EQUATION AND A FACTOR OF SAFETY OF 3. THE ENDS ARE PINNED.
ELASTIC LIMIT = 30 K.S.I.   E = 30 x 10³ K.S.I.

PROBLEM 18-2

WOOD COLUMNS

(I) 2 × 6 STUD WALL: ALLOWABLE f = 1500 P.S.I., E = 1.76 × 10³ K.S.I.

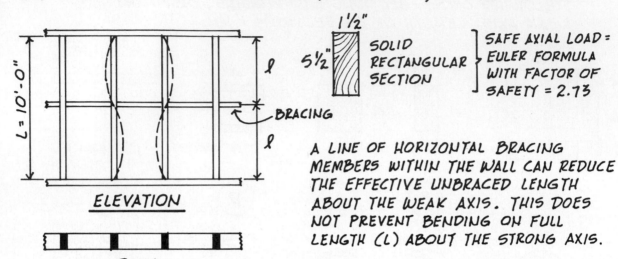

SOLID RECTANGULAR SECTION

SAFE AXIAL LOAD = EULER FORMULA WITH FACTOR OF SAFETY = 2.73

BRACING

ELEVATION

PLAN

A LINE OF HORIZONTAL BRACING MEMBERS WITHIN THE WALL CAN REDUCE THE EFFECTIVE UNBRACED LENGTH ABOUT THE WEAK AXIS. THIS DOES NOT PREVENT BENDING ON FULL LENGTH (L) ABOUT THE STRONG AXIS.

(A) DETERMINE THE GREATEST PERMISSIBLE UNBRACED LENGTH FOR A 2 × 6 FOR BUCKLING ABOUT THE WEAK AXIS, AT ℓ/r = 170.

(B) IN THE 2 × 6 STUD WALL, USING HORIZONTAL BRACING OF THE WEAK AXIS, DETERMINE THE SAFE AXIAL LOAD ON EACH STUD:

    WITH 1 LINE OF BRACING: ℓ = 5'-0" (WEAK AXIS)
    WITH 2 LINES OF BRACING: ℓ = 3'-4"   "   "
    WITH 3 LINES OF BRACING: ℓ = 2'-6"   "   "

    WITH SHEATHING NAILED EVERY 12" TO FACE OF STUDS

} L = 10'-0" STRONG AXIS

(II) GIVEN: 4 BOARDS, 1"× 6" (ACTUAL DIMENSION) - LENGTH VARIABLE, GLUED AND NAILED (WITHOUT CUTTING) AS A COLUMN SECTION.

(A) DETERMINE THE ARRANGEMENT FOR THE <u>STRONGEST SECTION</u>.

(B) DETERMINE THE ALLOWABLE AXIAL LOADS FOR:
    L = 10'-0": L = 20'-8": L = 30'-0"
THE COLUMN <u>IS NOT</u> BRACED IN ANY DIRECTION.

E = 1,760,000 P.S.I
ALLOW. f = 1800 P.S.I.
} $f = \dfrac{3.60 E}{(\ell/r)^2}$ ●   (ℓ/r LIMIT = 170)

● <u>NOTE</u>: THIS IS THE BASIC TIMBER COLUMN FORMULA FOR ANY SHAPE CROSS-SECTION, BASED ON THE EULER FORMULA WITH A FACTOR OF SAFETY = 2.73

PROBLEM 18-3

COLUMNS

(I)

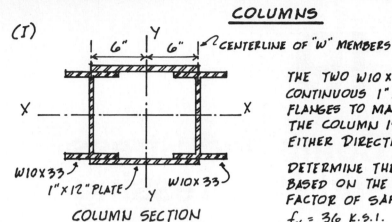

CENTERLINE OF "W" MEMBERS

W10×33

1"×12" PLATE   W10×33

COLUMN SECTION

THE TWO W10×33's ARE JOINED BY A CONTINUOUS 1"×12" PLATE WELDED TO THE FLANGES TO MAKE A SINGLE COLUMN SECTION. THE COLUMN IS 30' LONG, UNBRACED IN EITHER DIRECTION AND PINNED AT THE ENDS.

DETERMINE THE MAXIMUM SAFE AXIAL LOAD BASED ON THE EULER FORMULA, WITH A FACTOR OF SAFETY = 3 APPLIED.

$f_y = 36$ K.S.I. $E = 29,000$ K.S.I.

(II)

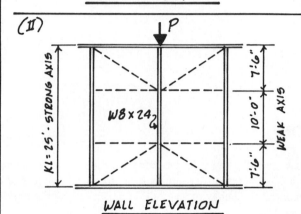

WALL ELEVATION

<u>25' STEEL WALL CONSTRUCTION:</u> $f_y = 50$ K.S.I.

(A) W8×24 COLUMN SECTIONS ARE USED AS MAIN MEMBERS WITH BRACING OF WEAK AXIS, AS SHOWN. DETERMINE THE SAFE AXIAL LOAD (P KIPS) BASED ON THE EULER FORMULA WITH A FACTOR OF SAFETY = 3 APPLIED.

(B) IF BRACING IS OMITTED, DETERMINE THE SAFE AXIAL LOAD.

(III)

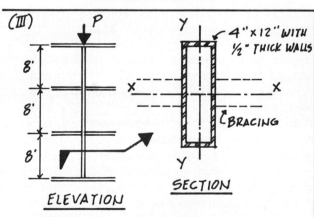

ELEVATION   SECTION

4"×12" WITH ½" THICK WALLS

BRACING

A STEEL TUBULAR COLUMN (AS SHOWN) WITH PINNED ENDS.

THE Y-Y AXIS IS BRACED, AS SHOWN.

A-36 STEEL: $f_y = 36$ K.S.I.

DETERMINE THE MAXIMUM SAFE LOAD (P KIPS) THAT CAN BE APPLIED BASED ON THE EULER FORMULA WITH A FACTOR OF SAFETY = 3 APPLIED.

(IV)

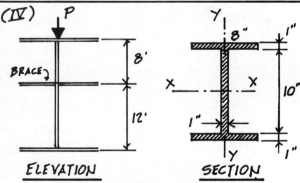

ELEVATION   SECTION

BRACE

THE TWO STORY COLUMN SHOWN IS PINNED AT THE ENDS. USING THE EULER FORMULA WITH A FACTOR OF SAFETY = 3, WHAT IS THE MAXIMUM LOAD (P KIPS) THAT MAY BE APPLIED.

THE COLUMN IS BRACED IN THE WEAK DIRECTION AND UNBRACED IN THE STRONG DIRECTION.

$E = 15 \times 10^3$ K.S.I.   ELASTIC LIMIT = 12 K.S.I.

19

Combined Stress and Prestressing

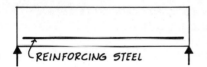

FIGURE 19-1

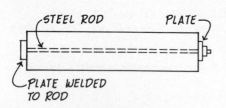

FIGURE 19-2

INTRODUCTION

In the broadest sense, prestressing may be defined as a *technique* whereby permanent stresses are deliberately introduced to a structural element or assembly in order to control the behavior that will be produced by the building loads.

Most commonly, the prestressing technique is applied to concrete for the purpose of eliminating tensile stresses that would be produced by the building loads. However, it should be recognized that prestressing, as a technique, can be applied to any material in order to control its behavior. We shall look at examples of this in a subsequent section of this chapter. First, however, let's discuss briefly the most common of prestressed materials—that is, prestressed concrete.

To begin with, it is essential that the difference between reinforced concrete and prestressed concrete be understood. It is assumed that the reader knows that concrete is a brittle material and cannot take tensile stresses. Therefore, in a reinforced concrete beam subjected to load, as shown in Fig. 19-1, reinforcing steel is placed in the bottom of the beam in order to take the tensile stress. In a prestressed concrete beam, permanent compressive stresses are introduced to the section before it is subjected to building loads. These compressive stresses are of such magnitude that they negate the tensile stresses produced by the load. Therefore, the *entire* cross section of a prestressed concrete member is effective because it is in a state of compression, which concrete is capable of resisting.

There are two ways to prestress concrete members. They are *pretensioning* and *post-tensioning*. Both of these methods make use of steel cables or rods which are placed in a state of tension in order to introduce compressive stresses to the concrete member. The way this works may be best explained by the very simple picture shown in Fig. 19-2. The steel rod in the concrete is in a protective sheath so that a bond will not be formed between it and the concrete. On the right-hand side of the figure a plate is slipped over the extended end of the rod (which is threaded), and a nut is then threaded onto the rod. Now, when the nut is turned, it will bear against the plate, forcing it inward, thereby producing tension in the rod and compression in the concrete. While the equipment and the hardware involved in prestressing is somwhat more complex, this represents, schematically, the manner in which concrete is prestressed.

The difference between pre-tensioning and post-tensioning has to do with the state of the concrete (liquid or solid) at the time the cable or rod is tensioned. In the post-tensioning process the cable is placed in position in the form, within a protective sheath to keep a bond from forming with the concrete. The concrete is then poured and allowed to set and attain some degree of compressive strength. The cable is then tensioned, thereby introducing compressive stresses in the concrete.

In the pre-tensioning process, the cable is placed in the form and tensioned, thereby stretching the cable. It is held in this stressed state and the concrete is poured in the form. After the concrete has attained a certain degree of strength, the cable is released. Because of the strain produced by the tensioning, the cable would like to snap back (much like a piece of elastic string) to its original length. In doing so, it pushes the plate at the end of the form inward against the hardened concrete, thereby introducing compressive stresses to the concrete.

The pre-tensioning method is mostly used for precast concrete members because it is somewhat less time-consuming and is therefore more suitable for the mass production process. Post-tensioning is mostly used on the job site, where members are cast in place. This is so because the cables are anchored in parts of the structure that would be pulled inward if the cables were tensioned before the beams were in place and hardened. This is illustrated in Figs. 19–3 and 19–4.

It was mentioned earlier that the prestressing technique can be applied to any structural material for the purpose of controlling its behavior. This control may be the result of concern with several possible aspects of behavior. In other words, controlling stresses within a cross section is not necessarily the only reason for prestressing. Oftentimes the prestressing technique is used for control of deflection, or the control of shrinkage cracks, where an absolutely sound and tightly closed surface is required.

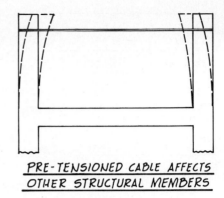

PRE-TENSIONED CABLE AFFECTS OTHER STRUCTURAL MEMBERS

FIGURE 19–3

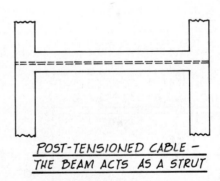

POST-TENSIONED CABLE – THE BEAM ACTS AS A STRUT

FIGURE 19–4

COMBINED STRESS—THE GENERAL CASE

When a structural member is prestressed there are several kinds of stresses introduced simultaneously. Specifically, these are axial stresses created by the direct force introduced by the prestressing operation, bending stresses produced by the eccentricity of the applied force (prestressing is normally applied eccentrically to the neutral axis of the member), and bending stresses produced by the building loads. This idea may best be explained by using the process of superposition. Figure 19–5 shows a typical prestressed situation for a simply supported beam. What's happening internally can be determined by the superposition of the pictures shown in Fig. 19–6. To begin with, consider that the axial force is applied on the axis of the member. This condition produces a stress of $f = P/A$ throughout the cross section. In reality, however, the force is applied with some eccentricity (e) from the axis. This will cause bending, as shown in Fig. 19–6b, due to the moment (Pe) producing tension in the top fibers of the beam and compression in the bottom fibers of the beam. These stresses may be measured by the bending stress equation, $f = Mc/I$. In addition to these stresses, there

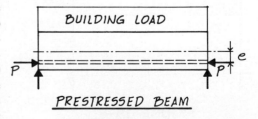

PRESTRESSED BEAM

FIGURE 19–5

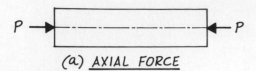

(a) _AXIAL FORCE_

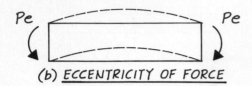

(b) _ECCENTRICITY OF FORCE_

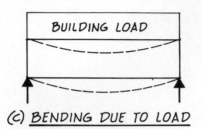

(c) _BENDING DUE TO LOAD_

FIGURE 19–6

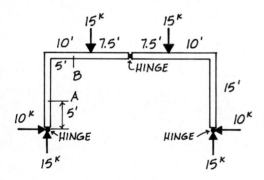

FIGURE 19–7

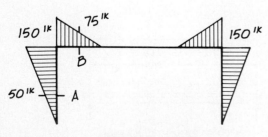

FIGURE 19–8

are bending stresses produced by the building load, as shown in Fig. 19–6c. These stresses are also measured by the bending stress equation, $f = Mc/I$, where M is the bending moment produced by the load. The character of the bending stresses produced by the load is the reverse of that produced by the eccentricity of the applied prestress force. The algebraic addition of these expressions yields the net stresses within the cross section, and this is known as the *Interaction Equation:*

$$f = \frac{P}{A} \pm \frac{M_t c}{I} \mp \frac{M_g c}{I}$$

where P = the applied axial force

A = cross-sectional area

M_t = moment produced by the eccentricity of the applied force

M_g = moment produced by the load

c = distance from the neutral axis of the section to the fiber being investigated

I = moment of inertia of the section

It should be noted that P/A is a compressive stress and is given a positive sign. This is the sign convention that will be used throughout this discussion. The bending stresses in the interaction equation produce both compressive (positive) and tensile (negative) stresses. The appropriate sign must be selected depending on the fiber being investigated.

The problem of combined stress occurs in many cases where prestressing is not involved. Therefore, before continuing with any detailed discussion of computational procedures used in the prestressing technique, we'll look at the general case of combined stress. To this end, several examples will be presented in which axial stress occurs, as well as stress due to bending moment.

Example 19–1 (Figs. 19–7 through 19–10)

It is desired to determine the net stresses at points A and B in the three-hinged frame shown in Fig. 19–7. The vertical and horizontal reactions have been determined based on the lessons of Chapter 9. The frame is made of a W24 × 76 steel wide flange with the following properties:

$$S = \frac{I}{c} = 176 \text{ in.}^3$$

$$A = 22.4 \text{ in.}^2$$

The variation in bending moment is shown in Fig. 19–8. This diagram is drawn based on the convention presented in Chapter 9. That is, the diagram is sketched using the frame shape as a base line, with the diagram drawn on the side of the frame where tension is produced by the bending moment.

In order to determine the net stresses due to axial load and bending at point A, consider the free body diagram of Fig. 19–9. From this it can be determined that

$$P = 15^K$$

$$M = 50'^K$$

The net stress on the outside fiber (the left side of the free body diagram) is determined by superimposing the compressive stress produced by the axial load with the tensile stress produced by bending. The mathematical expression is

$$f = \frac{P}{A} - \frac{M}{S}$$

Substituting the appropriate numerical values,

$$f = \frac{15^K}{22.4 \text{ in.}^2} - \frac{(50'^K)(12''/')}{176 \text{ in.}^3}$$

$$= .67 - 3.41$$

$$f \text{ (net)} = 2.74 \text{ k.s.i. tens.}$$

The net stress on the inside fiber is found by superimposing the compressive stress due to the axial load and the compressive stress due to bending.

$$\therefore f = \frac{P}{A} + \frac{M}{S}$$

and

$$f \text{ (net)} = .67 + 3.41 = 4.08 \text{ k.s.i. comp.}$$

It should be noted at this time that the 10^K horizontal force shown in the free body diagram produces a shearing stress within the section which acts in a direction perpendicular to the stresses produced by axial load and bending. Therefore, this stress is not part of the process of superposition.

In order to determine the net stress at point B due to axial stress and bending, consider the free body diagram of Fig. 19–10. The axial load and bending moment acting at this point are

$$P = 10^K$$

$$M = 75'^K$$

The net stress on the top side, where bending is producing tension, is

$$f = \frac{10^K}{22.4 \text{ in.}^2} - \frac{(75'^K)(12''/')}{176 \text{ in.}^3}$$

$$= .45 - 5.11$$

which yields a net stress of

$$f \text{ (net)} = 4.66 \text{ k.s.i. tens.}$$

On the bottom side, the axial stress and the bending stress are both compressive:

$$\therefore f \text{ (net)} = .45 + 5.11 = 5.56 \text{ k.s.i. comp.}$$

By this process, the net stress due to axial load and bending may be determined at any location.

Example 19–2 (Figs. 19–11 through 19–12)

The configuration shown in Fig. 19–11 is made of a W12 × 40 steel wide flange with an allowable stress of 24 k.s.i. The structure is fixed at the base. The necessary properties for the given section are

$$S = \frac{I}{c} = 51.9 \text{ in.}^3$$

$$A = 11.8 \text{ in.}^2$$

In this example we will determine the net stresses at a point 2 ft from the top, the maximum load P that may be placed on the left side based on an allowable stress of 24 k.s.i., and the net stresses at the fixed base. In order to determine the net stress at 2 ft from the top, we will use the free body diagram shown in Fig. 19–12. The 12^K vertical force is shown at the top of the vertical member, but because it is eccentric to the vertical member, it produces a bending moment of $48'^K$. Therefore, the free body diagram shows the 12^K load on the axis of the vertical member and a superimposed $48'^K$ moment to account for the eccentricity. The moment produces tension on the left side, at the point in question, and compression on the right side. In order to determine the net stress in the extreme fiber on the left side,

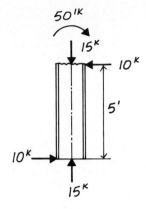

FIGURE 19–9

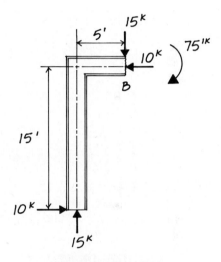

FIGURE 19–10

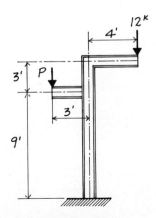

FIGURE 19–11

we use the interaction equation with

$$P = 12^K, \quad M = 48'^K$$

$$\therefore f = \frac{12^K}{11.8 \text{ in.}^2} - \frac{(48'^K)(12''/')}{51.9 \text{ in.}^3}$$

$$= 10.1 \text{ k.s.i. tens.}$$

On the right-hand side:

$$f = \frac{12^K}{11.8 \text{ in.}^2} + \frac{(48'^K)(12''/')}{51.9 \text{ in.}^3}$$

$$= 12.1 \text{ k.s.i. comp.}$$

In order to determine the maximum load P that may be placed on the left-hand cantilever, we must recognize two possibilities that could govern this maximum value. First, the load P produces a moment of $P(3')$ at the supported end of the cantilever. We can determine this value using the given allowable stress for bending and the bending stress equation alone, since there is no axial load involved at this point.

$$\therefore M = fS, \quad P(3')(12''/') = (24 \text{ k.s.i.})(51.9 \text{ in.}^3)$$

and

$$P = 34.6^K$$

Now we must check the stress at the fixed base, using the total axial load and the net effect of the moments at the base due to the 12^K force and the 34.6^K force. In order to do this, the interaction equation must be used with the following values:

Total $P = 12^K + 34.6^K = 46.6^K$

M (due to 12^K force) $= 48'^K$ producing tension on the left side and compression on the right side.

M (due to 34.6^K force) $= 103.8'^K$ producing compression on the left side and tension on the right side. Using this data:

Stress at the base (left side):

$$f = \frac{46.6^K}{11.8 \text{ in.}^2} - \frac{(48'^K)(12''/')}{51.9 \text{ in.}^3}$$

$$+ \frac{(103.8'^K)(12''/')}{51.9 \text{ in.}^3}$$

$$\therefore f \text{ (left)} = 16.9 \text{ k.s.i. comp.}$$

Stress at the base (right side):

$$f = \frac{46.6^K}{11.8 \text{ in.}^2} + \frac{(48'^K)(12''/')}{51.9 \text{ in.}^3}$$

$$- \frac{(103.8'^K)(12''/')}{51.9 \text{ in.}^3}$$

$$\therefore f \text{ (right)} = 9 \text{ k.s.i. tens.}$$

Both of these stresses (which are actually constant for the lower 9 ft on the vertical member) are below the given allowable stress and, therefore, $P = 34.6^K$ is the maximum point load that may be placed on the cantilever. It should be noted that column action, which could have an influence on this, has not been considered in order not to confuse the issue being demonstrated.

Now that we have dealt with the general case of combined stress we have the necessary principles for dealing with the technique of prestressing.

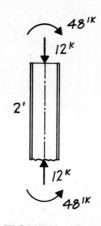

FIGURE 19–12

Before dealing with the most common application of prestressing, that is, prestressed concrete, we shall look at a special application of the technique.

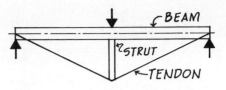

INVERTED KING-POST TRUSS

FIGURE 19–13

INVERTED KING-POST AND MULTIPLE-POST TRUSSES

An inverted king-post truss and a multiple-post truss are shown in Figs. 19–13 and 19–14. The king-post has one "post," generally in the center of the span, as shown in Fig. 19–13. The so-called "post" in these structures will, from this point on, be called the *strut*. Normally, a configuration of this type with two struts is called a "queen-post" truss, but since any number of struts may be used, depending on various conditions, the author chooses to use the phrase "multiple-post" trusses where two or more struts are involved.

King-post or multiple-post trusses of the prestressed variety (such trusses may be made that are not prestressed) are most commonly made from timber (either sawn or laminated) and sometimes steel beams. The struts are generally of the same material as the beam, and the prestressing technique is applied by means of a "tendon," which is normally a steel cable containing one or more turnbuckles, which are used for tightening the cable. In the forthcoming example problems, we shall see how the combination of the strut and tendon is used to prestress the beam. Hopefully, through the example problems, the student will also start to appreciate the advantages of prestressing and its potential to control various sorts of behavior in structural members. We'll now go through several examples.

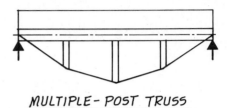

MULTIPLE-POST TRUSS

FIGURE 19–14

Example 19–3 (Figs. 19–15 through 19–20)

In this example, we will investigate several aspects of the simple king-post truss shown in Fig. 19–15. The beam is a steel wide flange, and the properties necessary for the investigation are shown in the figure. To begin, it may be of interest to know what beam size would be required if the prestressing technique, in the form of the king-post, were not being used. Consider, then, that we have a simply supported beam as shown in Fig. 19–16. The given allowable stress in bending is 24 k.s.i.; from this we can determine the required section modulus based on the maximum moment.

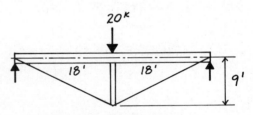

W12 × 26 STEEL BEAM, f = 24 K.S.I.
S = 33.4 IN.³ , A = 7.65 IN.²

FIGURE 19–15

$$\therefore M = \frac{PL}{4} = \frac{(20^K)(36')}{4} = 180'^K$$

and,

$$S = \frac{M}{f} = \frac{(180'^K)(12''/')}{24 \text{ k.s.i.}} = 90 \text{ in.}^3$$

According to the information on Data Sheets D–35 and D–36, this would require a W21 × 50 wide flange steel beam. But we are using a W12 × 26 to carry the same load over the same span which, without the application of the prestressing technique, would be seriously overstressed as well as deflecting considerably. We shall now see how the prestressing technique allows us to control the behavior of a structural member. Referring to Fig. 19–15, we have a W12 × 26 with a strut at the midspan and a tendon anchored at the centroid of the beam at the ends. Now, when the tendon is tightened by means, say, of a turnbuckle placed somewhere in its length, it wants to become straight. This means that it will push upward on the strut. We can represent this action by means of the free body diagram shown in Fig. 19–17. The steel beam is shown with all of the forces due to loading and those produced by the tightening of the tendon. The force in the center is due to the strut pushing upward, and the horizontal forces at the ends are due to the tendon pulling

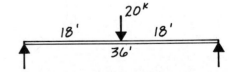

FIGURE 19–16

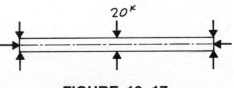

FIGURE 19–17

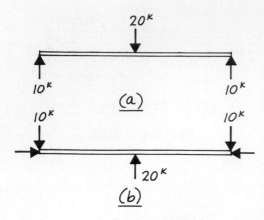

FIGURE 19–18

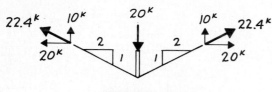

FIGURE 19–19

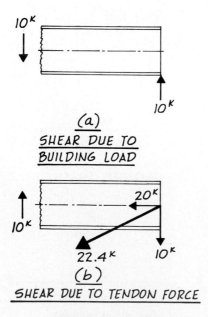

(a)

SHEAR DUE TO
BUILDING LOAD

(b)

SHEAR DUE TO TENDON FORCE

FIGURE 19–20

inward. In this case, as we shall see, we can completely control the net moment in the section as well as the deflection and shear. To begin, let's determine the force necessary in the tendon to produce a net $M = 0$ in the section. In essence, what we want to do is tighten the tendon to the extent that it will push up on the strut (which will produce a compressive force in the strut) with a force of 20^K. This will directly negate the 20^K downward force acting on the beam. We can think of this as two conditions superimposed, as shown in Fig. 19–18. Figure 19–18(a) shows the beam without the effect of prestressing, and Fig. 19–18(b) shows the effect of the prestressing necessary for $M = 0$. To determine the force in the tendon required to produce this condition, consider the free body of Fig. 19–19, where the tendon joins the bottom of the strut. From this picture it can easily be determined (because of the symmetry) that we want a force in the tendon that will have vertical components of 10^K on each side of the strut. We know the slope of the tendon (1:2), and from this we determine that the force in the tendon must be 22.4^K. This, of course, also produces a horizontal component of 20^K. Referring again to Fig. 19–18, we see that the moment due to the downward load is exactly negated by the upward force from the strut. In this case, the net bending moment at any point along the span is zero.

Let's now determine what the net deflection is at the midspan. Still referring to Fig. 19–18, we see that the deflection produced by the downward load will be exactly negated by the upward force of the strut. For this case, therefore, the net deflection at any point along the span is zero.

The net shear within the section can be determined by superimposing the free bodies of Fig. 19–20. In Fig. 19–20(a) we see that the end reaction due to the load produces a shear of 10^K in the section. Figure 19–20(b), which shows the result of the force in the tendon, indicates a shear on the section that is equal and opposite to that produced by the load. Therefore, the net shear at any point within the span is equal to zero.

In this example we have seen that through the use of prestressing, we are able to cancel the net bending moment, deflection, and shear along the entire span to zero. Such a condition is referred to as a *load-balanced* condition. There are specific requirements necessary to produce such a condition. Specifically, the strut must provide a force equal in magnitude to the load on the beam, and the load must be of the same character—that is, a point load acting in the same location as the building load. A more general statement, and perhaps more useful, for conditions required to produce load balancing is that *the shape of the tendon must be the same as the shape of the moment diagram that would be produced by the load.*

While in this example, there are no net stresses due to bending, since $M = 0$ throughout the span, there is a stress in the section which is produced by the horizontal component of the force in the tendon. This is shown in Fig. 19–20(b). Since this force acts at the centroid of the section, a uniform unit stress occurs, which is

$$f = \frac{P}{A} = \frac{20^K}{7.65 \text{ in.}^2} = 2.6 \text{ k.s.i.}$$

This unit stress is constant throughout the span.

Example 19–4 (Figs. 19–21 through 19–26)

Figure 19–21 shows a case where the tendon shape does not match the shape of the bending moment diagram produced by the load. Consequently, we cannot produce a load-balanced condition. However, we can negate the net moment to zero at certain locations. For this example, let's say that we wish to cancel the bending moment to zero at the midspan. In order to determine the forces necessary in the tendon (F_1 and F_2) to produce this condition, consider the free bodies of Fig. 19–22. Figure 19–22a shows the effect of the load, and Fig. 19–22b shows the effect of the struts pushing upward (which we will denote as P_S) due to the force applied to the tendon. The easiest way to solve this problem is to equate the moment at the midspan for the free body of Fig. 19–22a to that of Fig. 19–22b. Referring to Data Sheet D–25, we see that the general expressions for maximum bending moments in the two conditions are:

Point load at the midspan

$$M = \frac{PL}{4}$$

Equal point loads at the third-points

$$M = \frac{PL}{3} = \frac{P_S L}{3}$$

Equating these expressions and solving for P_S,

$$\frac{PL}{4} = \frac{P_S L}{3}, \quad P_S = \frac{3P}{4} = 15^K$$

To determine the forces required in the tendon, we'll use the free body diagram of Fig. 19–23, which is taken at the bottom of one of the struts. We can see, from this free body, that in order to produce a 15^K force in the strut, the vertical component in the tendon (the sloping segment) must be 15^K. Knowing the slope of the tendon (3:4), we can determine that the horizontal component is 20^K and the force in the tendon (F_1) is 25^K. The force in the horizontal tendon must be 20^K in order to satisfy $\Sigma F_x = 0$.

Let's now determine the net deflection at the midspan for this condition. In order to do this, we will again use the two conditions of Fig. 19–22 and superimpose the results. Referring to Data Sheet D–25, we see that the general expressions for deflection at the midspan are:

Point load at the midspan:

$$\Delta = \frac{PL^3}{48EI}$$

Equal point loads at the third-points:

$$\Delta = \frac{23PL^3}{648EI}$$

The algebraic summation of these will yield the net deflection:

$$\Delta = \frac{(20^K)(36' \times 12''/')^3}{48(30 \times 10^3 \text{ k.s.i.})(204 \text{ in.}^4)}$$

$$- \frac{23(15^K)(36' \times 12''/')^3}{648(30 \times 10^3 \text{ k.s.i.})(204 \text{ in.}^4)}$$

$$\Delta_{net} = -1.52'' \text{ or } 1.52'' \text{ upward}$$

Another approach to this problem would be to determine the strut force and the forces in the tendon necessary to cancel to zero the deflection at the midspan. In this case we will have a net bending moment at the midspan. In order to determine the strut force necessary for $\Delta = 0$, we will again use Fig. 19–22 and simply equate the deflection expressions, solving for P_S:

$$\frac{20^K(L^3)}{48} = \frac{23(P_S)(L^3)}{648}$$

and

$$P_S = 11.74^K$$

To determine the forces in the tendons, we'll use the free body diagram shown in Fig. 19–24. From this we can determine that

$$F_1 = 19.57^K \quad \text{and} \quad F_2 = 15.65^K$$

We can now determine the net bending moment at the midspan by using the general expressions for maximum bending moment, for the two conditions of Fig. 19–22:

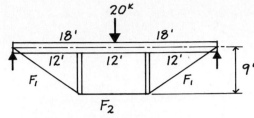

W 12 X 26 STEEL BEAM
$S = 33.4 \text{ in.}^3$, $I = 204 \text{ in.}^4$
$A = 7.65 \text{ in.}^2$, $E = 30 \times 10^3 \text{ K.S.I.}$

FIGURE 19–21

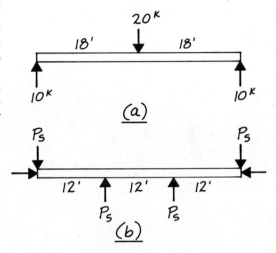

FIGURE 19–22

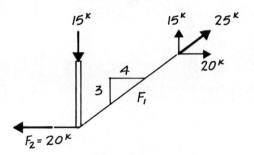

FIGURE 19–23

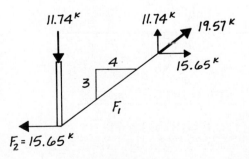

FIGURE 19–24

Positive moment:

$$M = \frac{PL}{4} = \frac{(20)(36)}{4}$$

$$= 180'^{K}$$

Negative bending:

$$M = \frac{PL}{3} = \frac{(11.7)(36)}{3}$$

$$= 140.4'^{K}$$

From this we see that there will be a net positive bending moment at the midspan of $39.6'^{K}$. For this condition, where $\Delta = 0$, we can now determine the net shear and bending moment at any point along the span by drawing the net shear and bending moment diagrams. The net loading diagram is shown in Fig. 19–25, which includes the effects of the applied load and tendon force. From the net loading diagram, the net shear and bending moment diagrams are constructed, and these are shown in Fig. 19–26. From this we see that the maximum moment in the span is the positive moment of $39.6'^{K}$ at the midspan. We'll now determine the net stresses at this point for the top and bottom fibers of the beam, using the interaction equation.

Top fibers: (compression due to bending)

$$f_T = \frac{P}{A} + \frac{M}{S} = \frac{15.7^{K}}{7.65 \text{ in.}^2} + \frac{(39.6'^{K})(12''/')}{33.4 \text{ in.}^3}$$

$$= 16.3 \text{ k.s.i. comp.}$$

Bottom fibers: (tension due to bending)

$$f_B = \frac{P}{A} - \frac{M}{S} = \frac{15.7^{K}}{7.65 \text{ in.}^2} - \frac{(39.6'^{K})(12''/')}{33.4 \text{ in.}^3}$$

$$= 12.2 \text{ k.s.i. tens.}$$

Now we'll determine the net stresses at the struts, where there is a negative moment of $20.4'^{K}$.

Top fibers: (tension due to bending)

$$f_T = \frac{15.7^{K}}{7.65 \text{ in.}^2} - \frac{(20.4'^{K})(12''/')}{33.4 \text{ in.}^3}$$

$$= 5.3 \text{ k.s.i. tens.}$$

Bottom fibers: (compression due to bending)

$$f_B = \frac{15.7^{K}}{7.65 \text{ in.}^2} + \frac{(20.4'^{K})(12''/')}{33.4 \text{ in.}^3}$$

$$= 9.4 \text{ k.s.i. comp.}$$

At the ends, where $M = 0$, there is uniform compressive stress due to the horizontal component of the force in the tendon, which is anchored on the centroid of the section:

Stress at ends:

$$f = \frac{P}{A} = \frac{15.7^{K}}{7.65 \text{ in.}^2} = 2.1 \text{ k.s.i. comp.}$$

In the previous example problems, we have seen how the prestressing technique may be used to control various kinds of behavior in structural members. While the king-post or multiple-post truss does not represent the most common application of the prestressing technique, there are occasions where this can be applied quite effectively. Perhaps one of the most notable applications of this sort occurs in rehabilitation or remodeling work, where

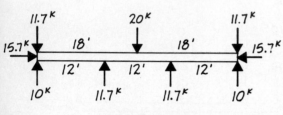

FIGURE 19–25

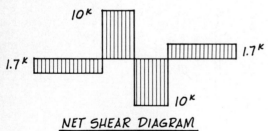

NET SHEAR DIAGRAM

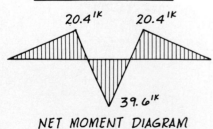

NET MOMENT DIAGRAM

FIGURE 19–26

it is desired to save the basic frame of an existing structure. In such cases, the new function of the building may require that the floor members carry a greater load than originally designed for. The application of prestressing can be used here to control stresses or deflections, or both.

A potential problem that should be noted at this time is that of column action due to the axial forces applied by prestressing. For example, in the load-balanced condition we cancelled to zero all stresses due to bending. However, there are stresses present due to the axial force (the horizontal component of the tendon force). In essence, what has been created here is a column that happens to be oriented horizontally, and the appropriate analysis would have to be made. In many cases the floor system might provide continuous bracing against column buckling, and this should be taken into consideration. Since this chapter deals only with the fundamentals of prestressing, the analyses are not taken to the consideration of column action.

INTRODUCTION TO PRESTRESSED CONCRETE

When the prestressing technique is applied to a concrete member, it is done with the intention of producing permanent compressive stresses greater than the tensile stresses produced by the building loads. Because of this, we consider a prestressed concrete beam as an elastic and homogeneous material in which the entire cross section is effective. It should be recognized, based on the lessons of Chapter 13, that this is very different from a reinforced concrete beam, which is not considered as a homogeneous section. In addition, a reinforced concrete beam is analyzed and designed based on the idea that it is a "cracked section," where the concrete on the tension side has cracked open and only the steel will carry stress. Because of this, we do not consider concrete on the tension side as part of the effective section. Therefore, the analysis of a prestressed concrete beam is quite a different matter from that of a reinforced concrete beam, as we shall see in the example problems.

Before going to the example problems, there is a consideration that must be discussed which is unique to prestressed concrete because of its inability to resist tensile stresses. That is, the various loading stages to which a concrete beam is subjected must be carefully analyzed to ensure that the applied prestressing force is not too great for the time period when the beam is not subjected to the building loads. For example, a pre-cast prestressed concrete beam will go through the initial stage of its life in the precasting yard. At this time, the prestressing force is applied and the only load available to interact with this force is the weight of the beam. Since the prestressing force is designed to interact with a higher level of load than this, overstressing can occur. We shall see the consequences of this in the example problems. Specifically, the loading stages that would be considered by the designer of a prestressed member are as follows:

BEAM WEIGHT ONLY: As mentioned previously, this occurs during the time period when the least load is present to interact with the prestressing force. Consequently, if the prestressing force is determined based on the anticipated full loading, then tensile stresses may occur on the top side of the beam.

SUSTAINED LOADING: This is an intermediate stage of loading which is the dead load of the system that the beam is supporting. This includes the

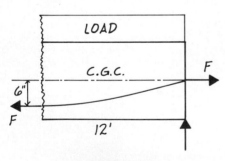

$I = 7136$ IN.^{4}
ALL. $f_c = 1.8$ K.S.I.
ALL. TENS. $= 0$
$A = 168$ IN.^{2}

SECTION

W (TOTAL) $= 24^K$

$\subset$ C.G.C.

$e_f = 6"$

PARABOLIC TENDON

LOADING DIAGRAM

FIGURE 19–27

beam weight as well as the weight of floor slabs or any dead load being supported.

FULL DESIGN LOAD: This is the total load to which the beam may be subjected based on the sustained load and the anticipated live load, which is based on the function of the building.

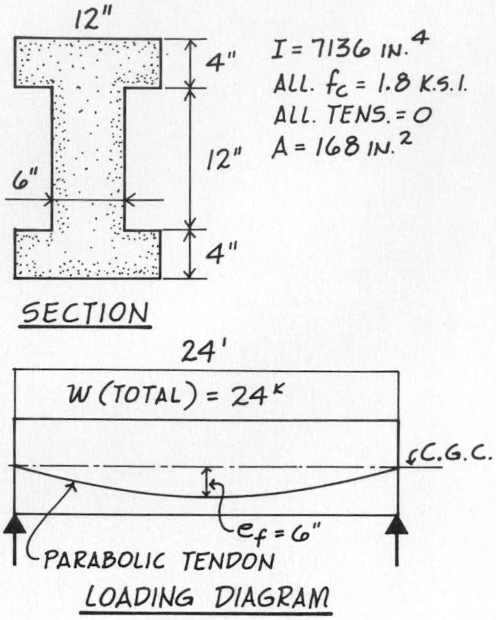

Let's now look at a few example problems in order to clarify the preceding discussion and also to make some other points that are unique to prestressed concrete.

Example 19–5 (Figs. 19–27 through 19–28)

To begin, two new notations unique to prestressed concrete must be defined. They are:

e_f = The eccentricity of the force in the tendon, relative to the C.G.C.

C.G.C. = Center of gravity of concrete, which is the centroid of the section

In this example we will deal with a parabolically shaped tendon that matches the shape of the bending moment diagram produced by the uniform load. This shape of tendon can be achieved easily in prestressed concrete when the post-tensioning approach is used. The cable, which is placed in a protective sheath, is placed in the form and allowed to develop the desired sag. Concrete is then poured and allowed to set before the cable is tensioned. The hardened concrete, of course, does not allow the cable to change its shape.

For the loading condition and section of Fig. 19–27, let's determine the horizontal force necessary in the tendon to produce a net moment of zero at the midspan. To do this we'll take a free body at the midspan, as shown in Fig. 19–28. We know that the general expression for the moment produced by the load is $WL/8$. Therefore, to solve for the force F required in the tendon:

$$\frac{WL}{8} = Fe_f$$

and

$$\frac{(24^K)(24')}{8} = F(.5'), \quad F = 144^K$$

Because the shape of the tendon is matched to the shape of the moment diagram produced by the load, this force will produce a net moment, shear, and deflection of zero throughout the span. There will, however, be a uniform compressive stress present produced by the force applied to the C.G.C. at the ends. This stress will be constant throughout the span; its value is

$$f = \frac{144^K}{168 \text{ in.}^2} = .86 \text{ k.s.i.} < 1.8 \text{ k.s.i. allowable}$$

So far, it seems that a good job was done in the design of this prestressed member. However, let's consider what happens when this member has the force applied, but it has not yet been subjected to the total load. Under the beam weight only, the force we have determined may cause problems. Therefore, let's consider the moment produced by the beam weight interacting with the prestress force of 144^K. Based on concrete weighing 150 #/ft^3, the total weight of the beam $= 4.2^K$. Therefore, the moment produced by the beam weight is

$$M_g = \frac{WL}{8} = \frac{(4.2^K)(24')}{8} = 12.6'^K$$

where M_g = moment produced by the load and

$$M_t = F(e_f) = (144^K)(.5') = 72'^K$$

LOAD

C.G.C.

F

$6"$

F

$12'$

FIGURE 19–28

where M_t = moment produced by the tendon.

These two moments taken at the midspan are opposite in character, with the net effect being

$$M = 72'^K - 12.6'^K = 59.4'^K \text{ (negative } M)$$

The stress in the top fiber for this condition is

$$f_T = \frac{144}{168} - \frac{(59.4 \times 12)(10'')}{7136} = .14 \text{ k.s.i. tens.}$$

and the stress in the bottom fibers is

$$f_B = \frac{144}{168} + \frac{(59.4 \times 12)(10'')}{7136} = 1.86 \text{ k.s.i. comp.}$$

We can see from these computations that, for this initial loading stage, tensile stresses will be present, as well as a slight overstress on the compression side. At this point a revision should be made in the prestressing force so that the stresses will comply with the allowables stated for this problem. We will not go through the revisions here, since the point of showing the consequences of considering the various loading stages has been made. It should be mentioned, however, that the ACI (American Concrete Institute) Building Code does allow a small amount of tensile stress to be present at the stage where the beam weight only is present. This is so because this is considered a short-term loading stage, and it is appropriate that some allowances be made under this condition.

We'll look at another example problem that will bring out another issue in prestressed concrete.

Example 19–6 (Figs. 19–29 through 19–32)

In this problem we have a bent tendon, as shown in the loading diagram of Fig. 19–29. This means that we are dealing with a non-load-balanced condition. However, we can cancel the moment to zero at the midspan only. Let's determine the horizontal force required in the tendon to do this, and then we'll discuss some other issues of this problem. In order to determine the horizontal force required for $M = 0$ at the midspan, we'll use the free body diagram of Fig. 19–30 and equate the moment produced by the load to that produced by the tendon.

$$\therefore \frac{WL}{8} = Fe_f$$

and

$$\frac{(32^K)(30')}{8} = F(.75'), \quad F = 160^K$$

Since this is not a load-balanced condition, there will be net bending at points along the span other than the midspan. In order to determine what's happening under these conditions, the net loading, shear, and moment diagrams will be drawn. These are shown in Fig. 19–31. An important point to be made here, which shows up on the net loading diagram, is that the bent tendon produces the same effect as a point load at the midspan. This is analogous to the king-post truss with a strut at the midspan. In order to determine the equivalent point load at the midspan, we will equate the moment produced by the tendon to the general moment expression for a point load applied at the midspan.

$$\therefore \frac{PL}{4} = Fe_f$$

and

$$\frac{P(30')}{4} = (160^K)(.75'), \quad P = 16^K$$

The net moment diagram shows that there is a positive moment of $30'^K$ at the

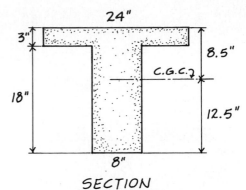

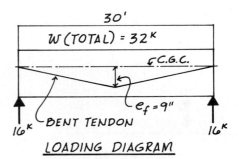

SECTION

$I = 9234 \text{ IN.}^4, \quad A = 216 \text{ IN.}^2$
$\text{ALL. } f_c = 2.25 \text{ K.S.I., ALL. TENS.} = 0$

FIGURE 19–29

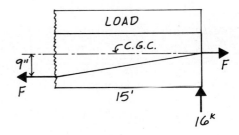

FIGURE 19–30

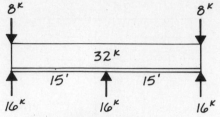

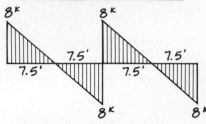

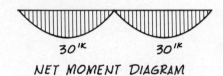

FIGURE 19–31

quarter-points of the span. We'll now evaluate the stresses at these locations.

Stress in the top fibers: $(F = 160^K)$

$$f_T = \frac{160^K}{216 \text{ in.}^2} + \frac{(30'^K \times 12''/')(8.5'')}{9234 \text{ in.}^4}$$

$$= 1.07 \text{ k.s.i. comp.}$$

Stress in the bottom fibers:

$$f_B = \frac{160}{216} - \frac{(30 \times 12)(12.5)}{9234} = .25 \text{ k.s.i. comp.}$$

Both of these stresses are compressive and are below the given allowable. Now let's investigate what happens when the beam weight only is acting. The total beam weight in this case has been determined to be 6.75^K, and the net loading, shear, and moment diagrams are shown in Fig. 19–32. The maximum moment for this condition is a negative moment of $94.7'^K$

Stress in the top fibers:

$$f_T = \frac{160^K}{216 \text{ in.}^2} - \frac{(94.7'^K \times 12''/')(8.5'')}{9234 \text{ in.}^4}$$

$$= .31 \text{ k.s.i. tens.}$$

Stress in the bottom fibers:

$$f_B = \frac{160}{216} + \frac{(94.7 \times 12)(12.5)}{9234} = 2.28 \text{ k.s.i. comp.}$$

We see here that a rather high tensile stress will occur for this loading condition and that the compressive stress is slightly beyond the allowable. Some revision would have to be made to the intensity of the prestress force in order to comply with the stated allowables.

Example 19–7 (Figs. 19–33 through 19–34)

In this example, we have a straight tendon, which is anchored below the C.G.C. This is a very common arrangement when pretensioning is being used. We'll now determine the force necessary in the tendon to produce a net moment of zero at the midspan. In order to do this, we will use the free body of Fig. 19–34 and equate the moment due to the load to that produced by the tendon.

$$\frac{WL}{8} = Fe_f$$

and

$$\frac{(24^K)(30')}{8} = F\left(\frac{1}{3}'\right), \quad F = 270^K$$

But, because the tendon has the same eccentricity at the end, a moment of $90'^K$ will be introduced here at the same time that we have balanced the midspan moment to zero. Therefore, the end section is critical and stresses must be checked at this point. With a prestress force of 270^K and a negative moment of $90'^K$:

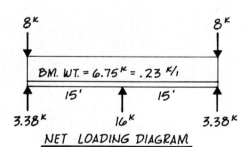

NET LOADING DIAGRAM

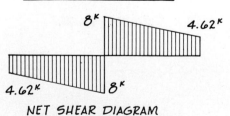

NET SHEAR DIAGRAM

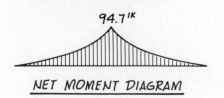

NET MOMENT DIAGRAM

FIGURE 19–32

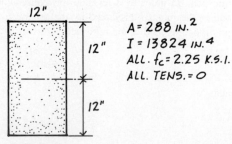

$$A = 288 \text{ in.}^2$$
$$I = 13824 \text{ in.}^4$$
$$\text{ALL. } f_c = 2.25 \text{ K.S.I.}$$
$$\text{ALL. TENS.} = 0$$

FIGURE 19–33 *SECTION*

Stress in the top fibers:

$$f_T = \frac{270^K}{288 \text{ in.}^2} - \frac{(90'^K \times 12''/')(12'')}{13824 \text{ in.}^4} = 0$$

Stress in the bottom fibers:

$$f_B = \frac{270}{288} + \frac{(90 \times 12)(12)}{13824} = 1.88 \text{ k.s.i. comp.}$$

These stresses are within the allowable limits; therefore, the design is suitable for the full loading condition. Since the procedures have been shown for checking the stresses under the action of the beam weight only, we will skip this step in this example.

CONCLUSION

It is important for the reader to understand that there are many complex issues involved in the design of prestressed concrete members which have not been presented here. What has been presented here is an overview of the subject for the purpose of making the student aware of the technique. For one who wishes to go into greater depth in these studies, a full course on the subject of prestressing is recommended.

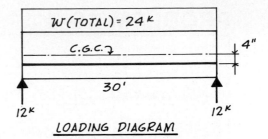

LOADING DIAGRAM

FIGURE 19–33 contd.

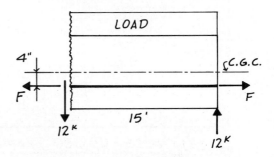

FIGURE 19–34

PROBLEM 19–1

<u>COMBINED STRESS</u>

(I) <u>CANOPY CONSTRUCTION PROJECTING FROM FACE OF BUILDING.</u>
THE SUPPORTING CABLE IS PIN-CONNECTED ON THE CENTROIDAL AXIS
OF THE STEEL BEAM. REACTIONS AT FACE OF BUILDING, AS SHOWN.

<u>FOR THE W14×34:</u>

DETERMINE THE MAGNITUDE
AND LOCATION OF THE MAX.
COMPRESSIVE AND TENSILE
UNIT STRESSES (P.S.I. OR K.S.I.)

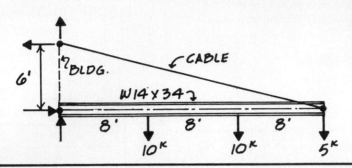

(II) <u>THREE-HINGED FRAME</u>

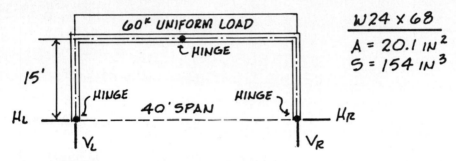

W24 × 68
A = 20.1 IN²
S = 154 IN³

(A) DETERMINE THE REACTIONS H_L, V_L, H_R AND V_R
(B) DETERMINE THE MAXIMUM TENSILE AND COMPRESSIVE UNIT
STRESSES AND THEIR LOCATIONS.

(III) <u>BEAM SUBJECTED TO BENDING AND AXIAL STRESSES.</u>

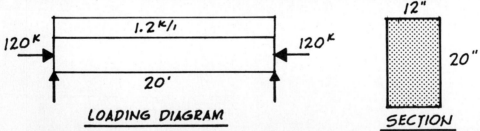

LOADING DIAGRAM   SECTION

DETERMINE THE MAXIMUM f (K.S.I.) IN TENSION AND COMPRESSION
AND DRAW THE NET STRESS DIAGRAM FOR THE FOLLOWING CASES:
(A) THE 120ᵏ AXIAL FORCE APPLIED AT THE CENTROID.
(B) THE 120ᵏ AXIAL FORCE APPLIED AT 3" BELOW THE CENTROID.

PROBLEM 19–2

PRESTRESSING

(A) INVERTED KING-POST TRUSS — STEEL "W" BEAM

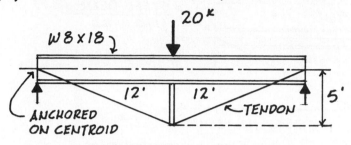

W 8 x 18

20^k

12' 12'

TENDON

5'

ANCHORED ON CENTROID

f ALL. = 24 K.S.1

f_y = 36 K.S.1.

(IGNORE WEIGHT OF THE STEEL BEAM.)

(1) WHAT TOTAL FORCE IS REQUIRED IN THE TENDON TO PRODUCE ZERO BENDING MOMENT, SHEAR AND DEFLECTION?

(2) WHAT IS THE UNIT STRESS IN THE SECTION AFTER THE APPLICATION OF THE FORCE REQUIRED IN PART (1)?

(B) MULTIPLE-POST TRUSS — STEEL "W" BEAM.

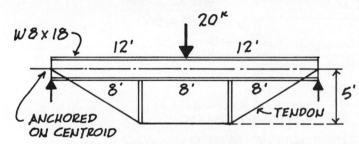

W 8 x 18

20^k

12' 12'

8' 8' 8'

ANCHORED ON CENTROID

TENDON

5'

f ALL.= 24 K.S.1.

f_y = 36 K.S.1.

(IGNORE WEIGHT OF THE STEEL BEAM.)

(1) WHAT FORCE IS REQUIRED IN THE TENDON TO PRODUCE:
  (a) ZERO BENDING MOMENT AT THE MIDSPAN — DETERMINE THE DEFLECTION AT THE MIDSPAN FOR THIS CONDITION.
  (b) ZERO DEFLECTION AT THE MIDSPAN.

(2) DRAW THE NET SHEAR AND BENDING MOMENT DIAGRAMS FOR EACH CASE.

(3) DETERMINE THE STRESSES AT THE ENDS, STRUTS AND MIDSPAN FOR EACH CASE.

(C) (1) WHICH SYSTEM (PART A OR B) PRODUCES THE MOST DESIRABLE CONDITIONS?

(2) WHAT CAN BE SAID ABOUT THE SHAPE OF THE TENDON THAT PRODUCES COMPLETE LOAD BALANCING?

(3) WHAT TENDON SHAPE IS REQUIRED TO BALANCE A UNIFORMLY DISTRIBUTED LOAD ON A SIMPLE SPAN?

PROBLEM 19–3

PRESTRESSING

(I)

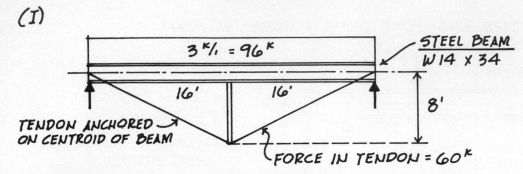

3 k/₁ = 96ᵏ — STEEL BEAM W 14 X 34

16'   16'   8'

TENDON ANCHORED ON CENTROID OF BEAM

FORCE IN TENDON = 60ᵏ

FOR THE GIVEN INVERTED KING-POST TRUSS, DETERMINE:

(A) STRESS IN THE TOP FIBERS AT THE MIDSPAN.
(B) STRESS AT THE ENDS.
(C) DEFLECTION AT THE MIDSPAN.

DRAW THE NET SHEAR AND BENDING MOMENT DIAGRAMS.

(II)

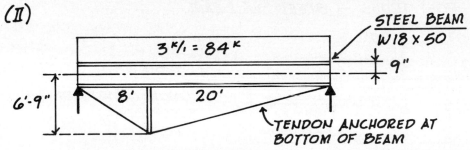

3 k/₁ = 84ᵏ — STEEL BEAM W18 X 50   9"

6'-9"   8'   20'

TENDON ANCHORED AT BOTTOM OF BEAM

THE HORIZONTAL FORCE IN THE TENDON = 100ᵏ

(A) DETERMINE STRESSES (f_T & f_B) AT THE MIDSPAN.
(B) DRAW THE NET SHEAR AND BENDING MOMENT DIAGRAMS.

(III)

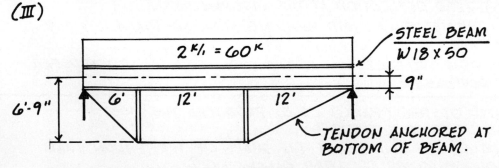

2 k/₁ = 60ᵏ — STEEL BEAM W18 X 50   9"

6'-9"   6'   12'   12'

TENDON ANCHORED AT BOTTOM OF BEAM.

(A) DETERMINE THE HORIZONTAL FORCE IN THE TENDON NEEDED TO PRODUCE A NET BENDING MOMENT = 0 AT THE MIDSPAN.
(B) DRAW THE NET SHEAR AND BENDING MOMENT DIAGRAMS.

PROBLEM 19–4

PRESTRESSED CONCRETE

(A)

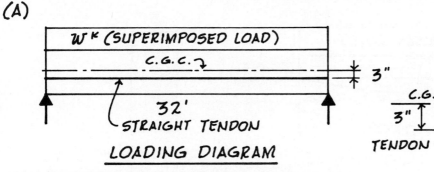

W^k (SUPERIMPOSED LOAD)

C.G.C.

3"

32'

STRAIGHT TENDON

LOADING DIAGRAM

8"

9"

C.G.C.

3"

TENDON

9"

SECTION

A = 144 IN^2
I = 3888 IN^4
ALL. f_c = 1.25 K.S.I.
ALL. TENS.= 0
E = 3.2 × 10^3 K.S.I.
f_T = STRESS IN TOP
f_B = STRESS IN BOTTOM
BM. WT. = 150 #/FT.

(1) DETERMINE THE MAXIMUM PERMISSIBLE PRESTRESS FORCE (F^k) IN THE TENDON, BASED ON THE ALL. f_c AT THE ENDS. CHECK f_T. DRAW THE STRESS DIAGRAM.

(2) DETERMINE THE MAXIMUM SUPERIMPOSED LOAD (W^k) BASED ON THE PRESTRESS FORCE DETERMINED IN PART (1). CHECK f_B.

(3) DETERMINE f_T AND f_B AT THE MIDSPAN WITH ONLY THE BEAM WEIGHT ACTING.

(4) DETERMINE THE NET DEFLECTION UNDER THE TOTAL MAXIMUM LOAD.

(B)

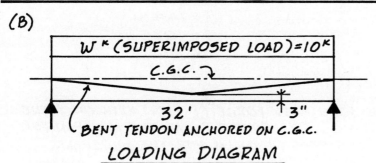

W^k (SUPERIMPOSED LOAD) = 10^k

C.G.C.

32'   3"

BENT TENDON ANCHORED ON C.G.C.

LOADING DIAGRAM

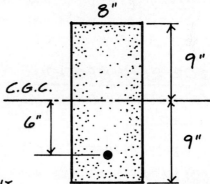

8"

9"

C.G.C.

6"

9"

MIDSPAN SECTION

SAME DATA AS PART (A)
EXCEPT:
ALL. f_c = 1.8 K.S.I.

(1) DETERMINE THE PRESTRESS FORCE (F^k) FOR M = 0 AT THE MIDSPAN (INCLUDING THE WEIGHT OF THE BEAM) – ASSUME F = HORIZONTAL COMPONENT (SLOPE IS VERY SMALL).

(2) DRAW THE NET SHEAR AND BENDING MOMENT DIAGRAMS WITH THE FULL LOAD ACTING. CHECK STRESSES AT THE CRITICAL SECTION. IS THE DESIGN O.K. OR NO GOOD?

(3) DETERMINE f_T AND f_B AT THE MIDSPAN WITH ONLY THE WEIGHT OF THE BEAM ACTING. IS THE DESIGN O.K. OR NO GOOD?

PROBLEM 19–5

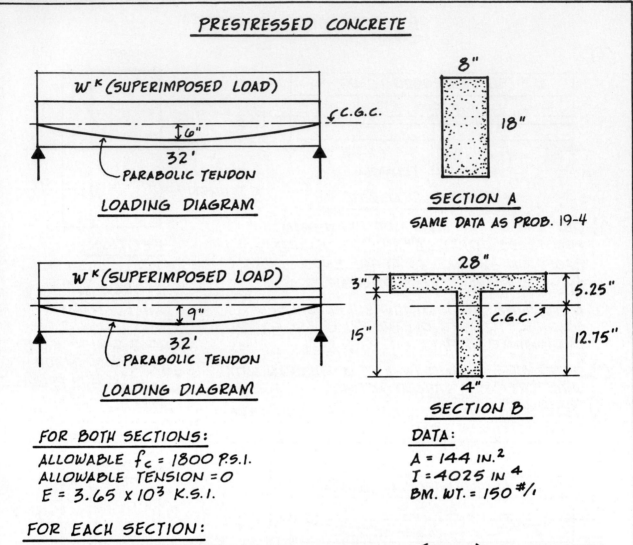

PRESTRESSED CONCRETE

w^k (SUPERIMPOSED LOAD)

C.G.C.

6"

32'

PARABOLIC TENDON

LOADING DIAGRAM

8"

18"

SECTION A
SAME DATA AS PROB. 19-4

w^k (SUPERIMPOSED LOAD)

9"

32'

PARABOLIC TENDON

LOADING DIAGRAM

28"

3"   5.25"

15"   C.G.C.

12.75"

4"

SECTION B

FOR BOTH SECTIONS:

ALLOWABLE f_c = 1800 P.S.I.
ALLOWABLE TENSION = 0
E = 3.65 × 10³ K.S.I.

DATA:

A = 144 IN.²
I = 4025 IN⁴
BM. WT. = 150 #/₁

FOR EACH SECTION:

(A) DETERMINE THE MAXIMUM PRESTRESS FORCE (F KIPS) WITH ONLY THE WEIGHT OF THE BEAM ACTING. DETERMINE f_B AT THE MIDSPAN FOR THIS CONDITION. DRAW THE STRESS DIAGRAM FOR THE MIDSPAN. DETERMINE f AT THE ENDS.

(B) DETERMINE THE SUPERIMPOSED LOAD FOR M=0 (LOAD BALANCED). DETERMINE f_T AND f_B AT THE MIDSPAN AND THE ENDS. DRAW THE STRESS DIAGRAMS.

(C) DETERMINE THE MAXIMUM SUPERIMPOSED LOAD THAT MAY BE PLACED ON THE BEAM. DETERMINE f_T AT THE MIDSPAN AND f AT THE ENDS. DRAW THE STRESS DIAGRAMS.

(D) DETERMINE THE DEFLECTION WITH ONLY THE BEAM WEIGHT ACTING.

(E) DETERMINE THE DEFLECTION WITH THE MAXIMUM TOTAL LOAD ACTING.

PROBLEM 19-6

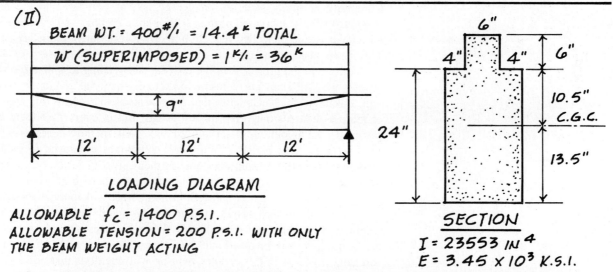

PRESTRESSED CONCRETE

(I)

w^k (SUPERIMPOSED LOAD)

C.G.C.

4"

24'

PARABOLIC TENDON ANCHORED ON C.G.C.

LOADING DIAGRAM

3"

6"

4"   4"

11.9"

14"   C.G.C.

8.1"

MIDSPAN SECTION

$I = 4182 \text{ IN}^4$
$E = 3.65 \times 10^3 \text{ K.S.I.}$

(A) DETERMINE THE MAXIMUM UNIFORMLY DISTRIBUTED LOAD (w^k) IN ADDITION TO THE BEAM WEIGHT ($200^{\#}/{}_1$), FOR A LOAD BALANCED CONDITION, IF THE HORIZONTAL FORCE IN THE TENDON = 120^k

(B) WHAT IS THE STRESS IN THE TOP FIBERS AT THE MIDSPAN WITH ONLY THE BEAM WEIGHT ACTING?

(C) DETERMINE THE NET DEFLECTION WITH ONLY THE BEAM WEIGHT ACTING.

(II)

BEAM WT. = $400^{\#}/{}_1 = 14.4^k$ TOTAL

W (SUPERIMPOSED) = $1^k/{}_1 = 36^k$

9"

12'   12'   12'

LOADING DIAGRAM

6"

4"   4"

6"

24"   10.5"   C.G.C.

13.5"

SECTION

$I = 23553 \text{ IN}^4$
$E = 3.45 \times 10^3 \text{ K.S.I.}$

ALLOWABLE $f_c = 1400$ P.S.I.
ALLOWABLE TENSION = 200 P.S.I. WITH ONLY THE BEAM WEIGHT ACTING

(A) A HORIZONTAL FORCE OF 200^k IS PLACED IN THE BENT TENDON. THE TENDON IS ANCHORED ON THE C.G.C. AT THE ENDS. BASED ON THE GIVEN ALLOWABLE STRESSES, IS THE BEAM O.K. OR NO GOOD WITH THE FULL LOAD ACTING?

(B) IS THE BEAM O.K. OR NO GOOD WITH ONLY THE BEAM WEIGHT ACTING?

(C) REGARDLESS OF THE ANSWERS TO PARTS (A) AND (B), DETERMINE THE NET DEFLECTION FOR EACH LOADING CONDITION.

20

Structural Optimization

INTRODUCTION

In Chapter 17 the concept of structural continuity was presented; it was shown that in many cases, the use of continuous beams is advantageous. Continuity normally leads to smaller beams, as compared to the requirements for simply supported beams. It should be emphasized that the primary advantage is one of weight reduction and not necessarily monetary savings. This is so because oftentimes, the cost of detailing necessary to achieve continuity may be greater than the cost saved by using less material. In this chapter we will limit our concern to that of "material economy," and we will look at several techniques that may be used to achieve this end. Some of these techniques may save money or not, but in every case there will be savings in the amount of material used and, consequently, a reduction in the dead weight of the building. While the techniques employed to reduce the amount of material needed may not, in themselves, lead to a cost savings, there are other parts of the building affected where cost savings can be achieved. Where weight is reduced, column sizes and foundation sizes may be smaller, and where weight savings are due to shallower beams, the total height of a building may be reduced. When it comes to cost, in fact, there are so many factors involved that this issue will not be considered here. Our sole concern then, in this chapter, will be with methods that will be applied in order to reduce the weight of a framing system. These methods are based on principles studied earlier in this text.

SIMPLE SPANS VERSUS CONTINUITY

To demonstrate some of the ways in which economy of material may be achieved, let's consider the five-span condition shown in Fig. 20–1.

Let's now consider the variety of alternatives available to produce this kind of framing. For the sake of demonstration, let's say that the beams are

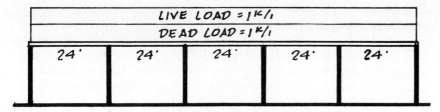

FRAMING ELEVATION

FIGURE 20-1

structural steel wide flange shapes. The first possibility we will consider, in order to make this frame, is to use five independent simply supported beams. This means that at interior supports, the beams share the support, but there is no continuity between the beams. Let's now proceed to determine the size of the steel wide flange required to satisfy the loading condition. We will use a grade of structural steel with an allowable stress in bending of 24 k.s.i. A typical beam is shown with the loading in Fig. 20–2. It should be noted that in Fig. 20–1, the live load and dead load acting on the beam have been indicated separately. The purpose for this will be demonstrated shortly. For our immediate needs, however, the total load on any one of the five typical beams is the sum of the dead load and live load. In order to size this beam, we will first determine the maximum bending moment and then the section modulus required.

$$M = \frac{WL}{8} = \frac{(48)(24)}{8} = 144'^K$$

$$S = \frac{M}{f} = \frac{144'^K \times 12''/'}{24 \text{ k.s.i.}} = 72 \text{ in.}^3$$

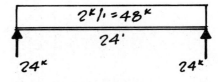

FIGURE 20-2

Referring to Data Sheets D–35 and D–36, we find that the most economical section is a W21 × 44. The member weighs 44#/' and there are five members like this, or a total of 120'. Therefore, the total weight of the beams in this framing scheme is

weight = 120' × 44#/' = 5280#

Another possibility for developing the frame shown in Fig. 20–1 is to use a continuous beam. Because of the length required (120'), this may not be practical, because structural steel sections are not normally rolled in such lengths, especially in the range of sizes that we will need to satisfy the requirements. In order to make one continuous piece, we would have to use two or three lengths to satisfy the total length of 120'. These pieces would have to be spliced with moment-resisting splices in order to produce full continuity. Such splices can be very costly and may offset any savings gained by the amount of material saved. However, our primary concern, as mentioned previously in this chapter, is with economy of material.

In Chapter 17, in the section titled "Loading Arrangements for Maximum Moments," the idea was presented that, in a continuous beam, the loading on any span will have an effect on the adjacent spans. It was also indicated that the total load on any span may vary, due to the unpredictability of the presence or absence of live loads, thereby affecting the values of the maximum negative and positive bending moments. Therefore, in order to properly design the continuous beam under consideration here, the dead and live loads must be arranged to form the critical loading patterns that will yield the maximum possible negative and positive moments. To this end, two analyses were performed, based on the rules presented in Chapter 17, and the maximum possible negative and positive moments were determined using the loading arrangements shown in Fig. 20–3. For the sake of exercise, the student may wish to perform the analyses for these situations, based on the lessons of Chapter 17. In any case, the maximum negative moment (approximately $128'^{K}$ at support E) was found to be critical. Based on the critical design moment, and considering a constant cross section, the wide flange steel beam required, using an allowable bending stress of 24 k.s.i., is a W16 × 40. The total weight for this fully continuous constant cross section is:

$$\text{Weight} = 120' \times 40\#/' = 4800\#$$

This represents a 480# savings of steel, compared to the simple-span system. Depending on how many times this five-span condition is repeated in a building, the savings in weight can be quite significant.

We'll now consider another alternative for producing the structural frame of Fig. 20–1, which will lead to still further savings of material. This alternative will be referred to as an "articulated" framing system, and we will look at this technique in some detail in the following section.

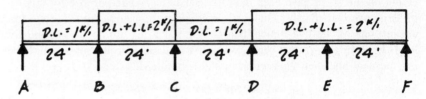

LOAD ARRANGEMENT FOR MAX. NEGATIVE MOMENT

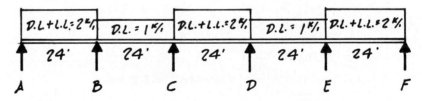

LOAD ARRANGEMENT FOR MAX. POSITIVE MOMENT

FIGURE 20–3

THE "ARTICULATED" FRAMING SYSTEM

An "articulated" framing system is one where individual segments of beams are arranged and joined together in such a way that negative moments are introduced at the supports, thereby reducing the positive moments that would exist if the beams were simply supported. While this sounds very much like a continuous system, there are certain fundamental differences and advantages, which will become clear as we proceed through the discussion and analysis. Therefore, let's now consider an "articulated" framing system to produce the structural frame of Fig. 20–1.

In order to develop negative moments at the supports, we will cantilever the beams over the supports and connect the space between the ends of the

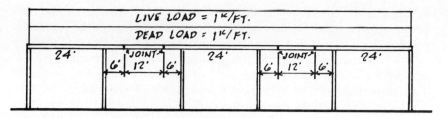

"ARTICULATED" FRAMING

FIGURE 20–4

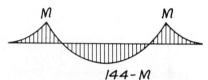

BENDING MOMENT DIAGRAM
SIMPLY SUPPORTED BEAM

BENDING MOMENT DIAGRAM
DOUBLE CANTILEVER

FIGURE 20–5

cantilevers with a short, simply supported beam, as shown in Fig. 20–4.

In this particular scheme the length of the cantilevers was made 6 ft. We will use a double cantilever for the center span and single cantilevers for the end spans. The choice of a 6-ft dimension for the cantilevers is not totally arbitrary. In a system such as this, we want to introduce negative moments in order to reduce the positive moments that we found in the simply supported beam system. For example, as simply supported beams, we found the maximum moment to be 144'^{K}. Ideally, we would like to introduce enough negative moment to reduce this value by one half. In essence, the introduction of negative moments at the supports causes a shift in the base line of the bending moment diagram. These relationships are shown in Fig. 20–5. The magnitude of the negative moment is a function of the load and the length of the cantilever. In the double-cantilever moment diagram shown in Fig. 20–5, the ideal situation would occur if we introduced a negative moment of 72'^{K}. This would reduce the positive moment to 72'^{K}; consequently, the required beam size would be much smaller than that required for a simply supported beam. Unfortunately, we cannot really achieve the ideal situation, and the reason for this should become clear as we go through the analysis of this system of framing. For our immediate purposes, let it suffice to say that in an "articulated" framing system, a reasonably good arrangement is to have the length of the cantilevers equal to about one fourth of the main span. This is the basis for the decision to have 6-ft-long cantilevers. The analysis may prove that an adjustment to this length is necessary for the greatest degree of optimization.

We will now proceed with the analysis and design of the "articulated" system shown in Fig. 20–4. To begin, we'll design the member spanning between the ends of the cantilevers. This condition is shown in Fig. 20–6.

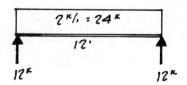

FIGURE 20–6

$$M = \frac{WL}{8} = \frac{(24)(12)}{8} = 36'^{K}$$

Using steel with an allowable bending stress = 24 k.s.i.

$$S = \frac{M}{f} = \frac{(36'^{K} \times 12''/')}{24 \text{ k.s.i.}} = 18 \text{ in.}^{3}$$

Referring to Data Sheets D–35 and D–36, we find that the most economical wide flange section is a W10 × 19. The total weight of steel required for these two members is

$$\text{Weight} = 24' \times 19\#/' = 456\#$$

We'll now analyze and design the double-cantilever beam at the center of the frame. The load being carried by the frame, shown in Fig. 20–4, is shown separately as dead load and live load. When dealing with cantilevers, the separation of dead load and live load becomes important in the determination of the maximum moment. In order to design the cantilevered beam, we must determine the maximum negative moment and the maximum positive moment, and then select a member based on the larger of the two values. The technique for doing this is different (and much simpler) than it was for the continuous beam, because we are now dealing with statically determinate members, which behave independently of each other.

Let's begin with the determination of the maximum negative moment, which is the moment due to the cantilever at the support. In order to do this, we can simply take a free body of the cantilever, as shown in Fig. 20–7. The maximum negative moment is based on the maximum load that can be acting on the cantilever, which is the sum of the dead load and the live load. The point load at the end of the cantilever is the reaction from the adjacent simple span, which is supported by the end of the cantilever. The value of the point load is based, in this case, on the simple span being fully loaded with the dead and live loads. Therefore, the maximum negative moment (which is the same at both supports for this double cantilever) is

$$M = (12^K)(6') + (12^K)(3') = 108'^K$$

It should be noted that the negative moment is the moment due to the cantilever, regardless of the loads acting on the main span.

We will now proceed to determine the maximum positive moment in the double cantilever. This will occur when the negative moment due to the cantilevers is as small as possible. The smallest conceivable load on the cantilevers (which produces the smallest negative moment) is the dead load only on the cantilevers and dead load only being supported by the adjacent simple span. The maximum positive moment will occur with full loading on the main span and dead loads only on the cantilevers. The loading pattern used to determine the maximum positive moment is shown in Fig. 20–8, along with the shear and bending moment diagrams. The maximum moment for this condition is found to be $90'^K$. It should be recognized that if the full loading were used on the cantilevers, the positive moment would be less. We have now determined that the maximum possible moments for this double-cantilever beam are

$$\text{Neg. } M = 108'^K$$

$$\text{Pos. } M = 90'^K$$

The beam must be designed for the larger of the two values. Therefore,

$$S = \frac{M}{f} = \frac{(108'^K \times 12''/')}{24 \text{ k.s.i.}} = 54 \text{ in.}^3$$

In scanning Data Sheets D–35 and D–36, we find that the most economical section is a W18 × 35. This member is 36 ft long. Therefore,

$$\text{Weight} = 36' \times 35\#/' = 1260\#$$

Two points must be made at this time about the analysis of the cantilevered beam and the load arrangements used to determine the maximum possible moments. First, it must be emphasized that the maximum possible negative moment occurs when the cantilever has as much load on it as possible. This is totally independent of whatever load is acting on the main span. The student may verify this by using the full load on the cantilevers, as shown in Fig. 20–7, and *any* load on the main span. It will be found that no matter what load is used on the main span, the cantilever moment will remain constant. Second, it must be recognized that the two maximum moments found for this member cannot exist simultaneously, since they are based on two different loading pattern possibilities.

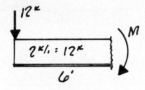

FIGURE 20–7

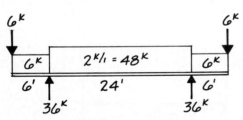

LOADING DIAGRAM

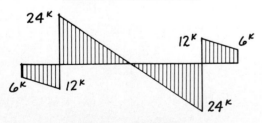

SHEAR DIAGRAM

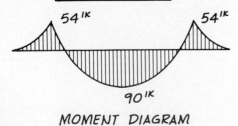

MOMENT DIAGRAM

FIGURE 20–8

It was mentioned earlier in this section that ideally, we would like to have a perfect balance (i.e., negative moment = positive moment) between the maximum moments. It was suggested that this is not really possible. The analysis of the cantilevered beam with different load arrangements for maximum negative and positive moments should serve to clarify this. If, indeed, we were dealing with permanent and predictable loadings, we could easily determine the length of the cantilevers necessary to produce a perfect balance between the negative and positive moments. In reality, however, live loads are not necessarily permanent and they are usually not predictable with any great degree of accuracy.

At this point, let's discuss further the decision to use 6-ft-long cantilevers. The analysis of the double cantilever has shown us that the maximum values for negative and positive moments are not too far apart. While it is true that these two values cannot coexist, this nevertheless represents a kind of balance between maximum moments, for the purpose of optimization. At this point we may decide to adjust the length of the cantilevers in order to reduce the negative moment slightly, which will also increase the positive moment. The best we can do, using the analysis already done as a basis for judgment, is to have these values converge to some value between the maximum negative moment of $108'^K$ and the maximum positive moment of $90'^K$. A slight shortening of the cantilever will move these values in the proper direction. For our purposes, however, the choice of a 6-ft cantilever will be accepted because the maximum negative and positive moments are reasonably close. For the sake of exercise, it is suggested that the student attempt to "fine tune" the length of the cantilever so that a better balance is achieved between the maximum values.

The final step is the analysis and design of the single cantilevers. We must again consider the load arrangement possibilities and determine the maximum negative and positive moments. The loading condition that produces the worst possible negative moment is precisely the same as the one used for the double cantilever. That is, the full load on the cantilever and full load on the adjacent simple span, which is being supported by the end of the cantilever. A free body diagram of the cantilever, with the appropriate loading, is shown in Fig. 20–9. From this picture it is determined that the maximum negative moment is $108'^K$. The maximum positive moment for the single cantilever will occur when the cantilever has the least possible load on it (dead load only) and full loading on the main span. The loading pattern that produces the maximum possible positive moment is shown in Fig. 20–10, along with the shear and bending moment diagrams. This analysis shows us that the maximum positive moment is $118'^K$, which is greater than the maximum possible negative moment of $108'^K$. Therefore, we must design this wide flange section for a moment of $118'^K$.

$$S = \frac{M}{f} = \frac{118'^K \times 12''/'}{24 \text{ k.s.i.}} = 59 \text{ in.}^3$$

Referring to Data Sheets D–35 and D–36, we find that the most economical section is a W16 × 40. There are two members like this with a total length of 60 ft. Therefore,

$$\text{Weight} = 60' \times 40\#/' = 2400\#$$

To summarize, let's compare the "articulated" system with the simple span system and the fully continuous beam. Total weights were earlier determined to be

Simple spans: Weight = 5280#

Continuous beam: Weight = 4800#

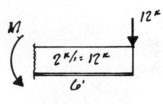

FIGURE 20–9

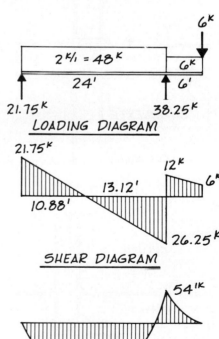

LOADING DIAGRAM

SHEAR DIAGRAM

MOMENT DIAGRAM

FIGURE 20–10

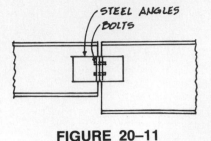

FIGURE 20–11

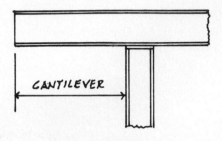

CANTILEVER

FIGURE 20–12

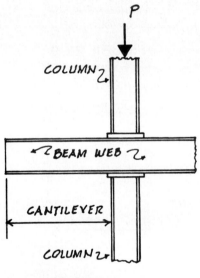

P

COLUMN

BEAM WEB

CANTILEVER

COLUMN

FIGURE 20–13

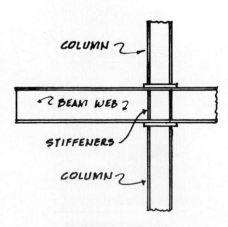

COLUMN

BEAM WEB

STIFFENERS

COLUMN

FIGURE 20–14

The weight of the structural steel required for the "articulated" system is

Two simple beams:	Weight =	456#
Two single cantilevers:	Weight =	2400#
Double cantilever:	Weight =	1260#
Total weight		= 4116#

In comparing the three alternatives, we can see that a substantial savings in weight is achieved through the use of "articulated" framing.

It should be pointed out that for the purpose of analysis, the essential difference between the "articulated' framing system and the fully continuous beam is that the point of contraflexure (zero bending moment) is known in the "articulated" system. It is the point where the connection is made between the end of the cantilever and the simple spans. The location of this point of zero bending moment is totally controlled and, as discussed earlier in this section, is based on the judgment of the designer. The connection at the end of the cantilever may be of the kind shown in Fig. 20–11. This is a "pinned" connection, which will allow relative rotation between the two members. Consequently, there will be no moment at this point. By establishing the location of the point of zero bending moment, through proper detailing, we have established a system that is statically determinate. That is, each member can be analyzed based on the equations of static equilibrium. In the fully continuous beam, the location of the point of contraflexure is not known at the outset of the analysis. Consequently, the continuous beam is statically indeterminate, and the analysis requires something other than the direct use of the equations of static equilibrium, such as the Theorem of Three Moments, which was presented in Chapter 17.

Before concluding this section, it may be useful to point out any limitations of the "articulated" framing system, especially when designing in structural steel. Specifically, the "articulated" framing system is mostly suitable and most easily applicable to a roof structure. This is so because of the cantilevers involved. In order to have cantilevers, it is necessary that the beams be continuous over the columns, or supporting walls, as shown in Fig. 20–12. If an "articulated" system were used for a floor structure, then there would have to be a column resting on top of the beam, as shown in Fig. 20–13. This kind of situation creates a problem of load transfer through the web of the beam. Since the column load is concentrated and relatively high in magnitude, and the web of a wide flange beam is relatively thin in cross section, the unit stresses produced in the web are very intense. This can easily cause buckling of the web. Consequently, the normal procedure is to provide stiffening plates for the web of the beam, as shown in Fig. 20–14. The size and number of these plates may be determined by analysis. The important point here, however, is that this kind of detailing (which requires welding) is somewhat costly and may to a large degree offset the savings produced by the use of an "articulated" framing system.

VARIABLE SECTIONS

In a structural member subjected to bending, it should by now be rather obvious that the maximum bending moment occurs only at specific locations. When designing a member of constant cross section (such as a wide flange steel beam), a constant strength is provided throughout the entire span. It is possible to reduce the total weight of a structural member by responding to strength requirements at locations of maximum moment and reducing the section where these requirements diminish. To demonstrate, we'll continue to use the "articulated" system that was analyzed in the preceding section, and we will redesign the members, using cross sections that vary in response to strength requirements.

To begin, we'll redesign the double cantilever, using the information established by the analysis in the preceding section. In that analysis we found that the maximum negative moment was $108'^K$ and the maximum positive moment was $90'^K$. The maximum moments were determined on the basis of the loading patterns shown in Figs. 20–15 and 20–16. In order to satisfy these requirements, a W18 × 35 steel wide flange beam was selected. This was the most economical section that would satisfy the critical moment of $108'^K$. It should be recognized, however, that this section, which is constant throughout the entire length, exceeds the requirements at all other locations where the moment is lower than the value of $108'^K$.

In order to reduce the amount of material required, we will now try a wide flange section that is smaller than the W18 × 35, and add steel plates to the top and bottom flanges where necessary in order to develop the required moment of inertia. The selection of the member to be used is based on judgment and an understanding of the concept of moment of inertia. Specifically, it should be understood that any drastic reduction in depth from the depth of the section originally required (the W18 × 35) may require very large plates to be added at critical moment sections in order to develop the appropriate moment of inertia. This would be counterproductive to the goal of weight reduction. The idea here is to choose a member smaller than that required by the analysis, but not drastically shallower in depth. As a beginning point for this kind of design, it is usually good to select a member of the next smaller size in depth than that indicated in the original analysis. Consequently, referring to Data Sheets D–28 through D–34, a W16 × 26 was chosen to use a basis for the development of a variable section. After the analysis of this section is completed, we may wish to reconsider the choice of the W16 × 26. We'll now proceed with the analysis, using a W16 × 26 and design plates to add to the top and bottom flanges where needed.

Using the results of the analysis that has already been performed, we know

$$\text{Max. Negative Moment} = 108'^K$$

The maximum moment-carrying capacity for the W16 × 26, based on a grade of steel with an allowable bending stress of 24 k.s.i., is (see Data Sheets D–28 through D–34 for properties of sections)

$$M = \frac{f}{c} I = \frac{(24 \text{ k.s.i.})(301 \text{ in.}^4)}{(7.85'')(12''/')} = 77'^K$$

This, of course, does not satisfy the required moment-carrying capacity of $108'^K$. Therefore, we will add plates to this section in order to increase the moment of inertia. The first step in the design of the plates is to make a judgment regarding the thickness of the plate. We must do this so that we will know what the distance is from the neutral axis to the outermost fiber of the section (the c distance). In this demonstration, the decision was made to try a .5"-thick plate. We may wish to revise this dimension later. In any case, the section with the plates added is shown in Fig. 20–17. The issue here is the determination of the width (B) of the plates.

We have already determined that the moment-carrying capacity of the W16 × 26 is $77'^K$. Therefore, the plates must furnish the additional moment-carrying capacity required.

$$M_{\text{(Plates)}} = 108'^K - 77'^K = 31'^K$$

The moment of inertia required by the section with the plates, as shown in

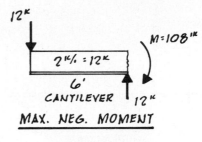

MAX. NEG. MOMENT

FIGURE 20–15

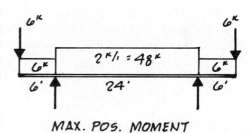

MAX. POS. MOMENT

FIGURE 20–16

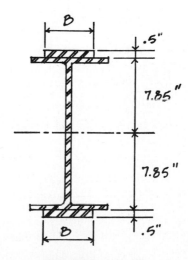

FIGURE 20–17

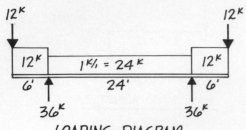

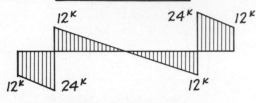

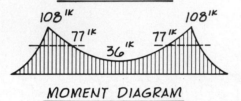

LOADING DIAGRAM

SHEAR DIAGRAM

MOMENT DIAGRAM

FIGURE 20–18

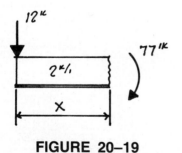

FIGURE 20–19

Fig. 20–17, in order to furnish a moment-carrying capacity of $108^{\prime K}$, is

$$I = \frac{Mc}{f} = \frac{(108^{\prime K} \times 12^{\prime\prime}/^{\prime})(8.35^{\prime\prime})}{24 \text{ k.s.i.}} = 451 \text{ in.}^4$$

Since the W16 $\times$ 26 furnishes a moment of inertia of 301 in.^{4}, then the plates must furnish the additional moment of inertia.

$$I_{\text{(Plates)}} = 451 \text{ in.}^4 - 301 \text{ in.}^4 = 150 \text{ in.}^4$$

Therefore, each plate must furnish 75 in.^{4}. Using the transfer equation for moment of inertia,

$$I_{\text{(Plate)}} = 75 \text{ in.}^4 = \frac{BD^2}{12} + A\bar{x}^2$$

Since the plates we are using have a thickness of .5″, the first part of this expression is negligible. Therefore,

$$I_{\text{(Plate)}} = 75 \text{ in.}^4 = A\bar{x}^2 = B(.5)(8.1)^2$$

and

$$B = 2.3″$$

We now know the cross-sectional dimensions of the plate. The next step in the design of this variable section is to determine the length of the plates required at the negative moment sections. It should be recognized that the plates are no longer required where the bending moment has diminished to $77^{\prime K}$, which is the moment-carrying capacity of the W16 $\times$ 26. In order to determine the cutoff points for the plates at the negative moment sections, we will use the relationships between the shear and bending moment diagrams, which are shown in Fig. 20–18. It is important to note at this time that for the purpose of determining the location of the plate cutoffs, the loading pattern used is important. It was noted earlier in this chapter that the maximum negative moment was determined based on the maximum possible load that is supported by the cantilever. For this condition, the load on the main span was unimportant, since this has no effect on the cantilever moment. When determining the plate cutoff locations, however, the loading pattern is important. We want to determine the condition that will cause the plates to be as long as necessary under the worst conditions. This condition, for the negative moment plates, occurs when the load on the main span produces the lowest rate of change possible in the moment variation. This is shown in the diagrams of Fig. 20–18. This figure shows that the critical loading, which must be used for the negative moment plate cutoffs, occurs when the loading on the cantilevers is the maximum possible and the main span supports only the dead load.

In order to determine the location of the plate cutoff to the left of the left-hand support in this double cantilever (of course, it will be the same at the right-hand support), we will use the free body shown in Fig. 20–19, and determine the distance x from the end of the cantilever to the point where the moment is $77^{\prime K}$. Therefore,

$$12x + \frac{2x^2}{2} = 77^{\prime K}$$

and, solving this quadratic,

$$x = 4.6' \text{ (or 1.4' from the support)}$$

To determine the location of the plate cutoff to the right of the left-hand support, we will use the area under the shear diagram. Referring to the shear diagram in Fig. 20–18, and starting from the point of zero shear, we are looking for an area under the shear diagram that is equal to the difference in values on the moment diagram. The relationship between the shear and bending moment diagrams that are being used is shown in Fig. 20–20. The rate of change of the shear diagram is $1^K/'$, and the area (shown shaded) must be equal to the difference in values on the moment diagram, which is $41'^K$. Therefore,

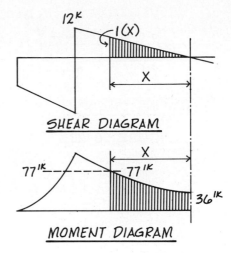

$$(1^K/')(x)\left(\frac{x}{2}\right) = 41'^K$$

and $x = 9'$ from the point of zero shear (or $3'$ from the support).

Based on this information, we must add four plates, each 2.3 in. wide and 4.4 ft long to the W 16 × 26. The total weight of the steel plates required for the negative moments in the double cantilever is:

Note: Steel weighs 3.4#/in.^{2}/lin. ft

$$\text{Wt. of plates} = (.5)(2.3)(3.4)(17.6) = 69\#$$

SHEAR DIAGRAM

MOMENT DIAGRAM

FIGURE 20–20

Since the maximum possible positive moment for the double cantilever is $90'^K$, and the moment capacity of the W16 × 26 is $77'^K$, we must also add plates at the midspan. Using the information given in Fig. 20–17, and again using .5″ thick plates, the moment of inertia required for the positive moment is

$$I = \frac{Mc}{f} = \frac{(90'^K \times 12''/')(8.35'')}{24 \text{ k.s.i.}} = 376 \text{ in.}^4$$

The W16 × 26 furnishes 301 in.^{4}. Therefore,

$$I_{\text{(Plates)}} = 376 \text{ in.}^4 - 301 \text{ in.}^4 = 75 \text{ in.}^4$$

Therefore, each plate must furnish 37.5 in.^{4}. Using the transfer equation for moment of inertia,

$$I_{\text{(Plate)}} = 37.5 \text{ in.}^4 = B(.5)(8.1)^2$$

and

$$B = 1.2''$$

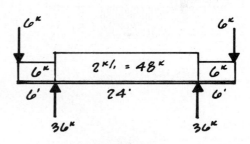

FIGURE 20–21

To determine the length of the plates required for the positive moment, we must use the loading shown in Fig. 20–21. This loading produces the shear and bending moment diagrams shown in Fig. 20–22. In order to determine the location of the cutoff points, we'll use the concept that the area of the shear diagram (shown shaded) must equal the difference in values on the bending moment diagram, then

$$(2^K/')(x)\left(\frac{x}{2}\right) = 13'^K$$

and $x = 3.6'$ (to each side of the centerline). Therefore, the length of each plate at the midspan is $7.2'$, and

$$\text{Wt. of plates} = (.5)(1.2)(3.4)(14.4) = 29\#$$

The total weight of steel required for the double cantilever is

W16 × 26	936#
Neg. Moment Plates	69#
Pos. Moment Plates	29#
Total	= 1034#

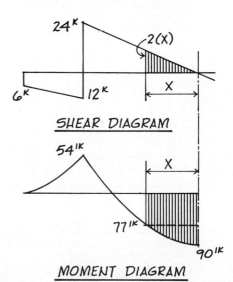

SHEAR DIAGRAM

MOMENT DIAGRAM

FIGURE 20–22

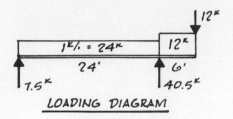

LOADING DIAGRAM

FIGURE 20–23

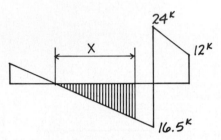

SHEAR DIAGRAM

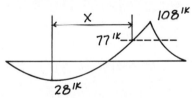

MOMENT DIAGRAM

FIGURE 20–24

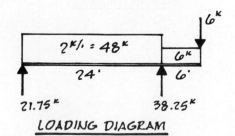

LOADING DIAGRAM

FIGURE 20–25

Comparing this to the "articulated" system where a constant section (W 18 × 35) was used for the double cantilever, we find that 226# of steel was saved by using the variable section.

We will now go through the same process and redesign the single cantilever in the "articulated" system as a variable section. As a constant wide flange section, a W16 × 40 was required for the single cantilever. We will redesign the single cantilever using a W16 × 26 and adding steel plates where necessary. To begin, we'll design the plates necessary to satisfy the maximum negative moment. The critical loading for this condition is shown in Fig. 20–23. This loading produces a critical value for the negative moment of 108$^{'K}$. This is the same maximum negative moment as in the double cantilever. Therefore, the size of the plates required for the negative moment is:

$$\text{Neg. Moment Plates:} \quad .5'' \times 2.3''$$

We must now determine the lengths of the negative moment plates. To do this, we will use the shear and bending moment diagrams shown in Fig. 20–24. To the right-hand side of the support, the plate cutoff computation is precisely the same as used for the double cantilever. This distance was found to be 1.4 ft from the support. In order to find the cutoff location on the left-hand side of the support, we will use the relationships between the shear and bending moment diagrams. Starting from the point of zero shear, we are looking for an area under the shear diagram that is equal to the difference in values on the moment diagram. In this case we are looking for the location on the moment diagram where the moment is 77$^{'K}$, which is the value that the W16 × 26 is capable of furnishing. The area under the shear diagram, starting from the point of zero shear (where the moment is a positive 28$^{'K}$) must equal 105$^{'K}$. Therefore,

$$(1'^{K}/')(x)\left(\frac{x}{2}\right) = 105'^{K}$$

and $x = 14.5$ ft, or 2 ft from the support.

The total length of the negative moment plates is 3.4 ft each, or a total of 6.8 ft for top and bottom plates.

Let's now determine the size and length of the plates required for the positive moment. In the earlier analysis of the "articulated" system, it was found that the maximum possible positive moment is 118$^{'K}$. This was based on the loading pattern shown in Fig. 20–25. Based on the moment of 118$^{'K}$,

$$I = \frac{(118'^{K} \times 12''/')(8.35'')}{24 \text{ k.s.i.}} = 493 \text{ in.}^4$$

Since the W16 × 26 furnishes a moment of inertia of 301 in.^{4}, then

$$I_{(\text{Plates})} = 493 \text{ in.}^4 - 301 \text{ in.}^4 = 192 \text{ in.}^4$$

or 96 in.^{4}/Plate. Therefore,

$$I = A\bar{x}^2 = 96 \text{ in.}^4 = (B)(.5)(8.1)^2$$

and

$$B = 3 \text{ in.}$$

We must now determine the length of the steel plates required to satisfy the positive moment. For this purpose, we will use the shear and bending moment diagrams shown in Fig. 20–26. Using the point of zero shear as the reference,

$$(2^{K}/')(x)\left(\frac{x}{2}\right) = 41'^{K}$$

and $x = 6.4$ ft, to either side of the point of zero shear. Therefore, the total length of steel plate required (top and bottom) is 25.6 ft.

Let's now summarize this discussion by computing the total weight of steel required for the single cantilever.

W16 × 26	780#
Neg. Moment Plates	27#
Pos. Moment Plates	131#
Total	= 938#

Comparing this to the design of the "articulated" system where a constant section (W16 × 40) was used, we have saved 262# for *each* single cantilever. The *total* weight saved for the "articulated" system, using a variable section for the single and double cantilevers, is 750#. In comparing the variable section "articulated" system with the originally analyzed alternative of five simple spans, which required a total weight of 5280#, we find that 1914# of steel were saved. It should be noted that perhaps a bit more weight may be saved if we choose to revise the simple span, in the "articulated" system, for which a W10 × 19 was used. However, this member is already very light and there is not much length involved. Therefore, a redesign of this member as a variable section would not be very fruitful.

CONCLUSION

While the principles for the design of a variable section were presented in the context of steel beams, they are, of course, applicable to any material. In fact, variable sections of laminated timber, such as the tapered roof beam shown in Fig. 20–27, are not uncommon. While shearing stresses are not normally critical in steel beams (and were not considered in this chapter), it should be recalled from Chapter 11 that shearing stresses in wood may be critical and must be considered in the design process. In any case, the student should be aware that the most important guide in the development of a variable section, regardless of the material being used, is the shear and bending moment diagrams. From these diagrams the required moment of inertia, or depth, may be determined at critical locations, so that allowable stresses (shear and/or bending) may be satisfied. Using these diagrams as the guide, structural members may be properly shaped, with the result being a savings in the amount of material required to do a given job.

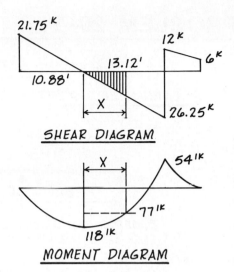

SHEAR DIAGRAM

MOMENT DIAGRAM

FIGURE 20–26

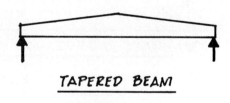

TAPERED BEAM

FIGURE 20–27

PROBLEM 20–1

ARTICULATED FRAMING

(I) LAMINATED TIMBER

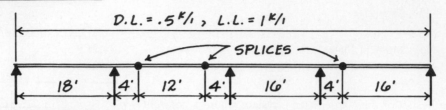

D.L.= .5^{k}/ı , L.L.= 1^{k}/ı

SPLICES

18' 4' 12' 4' 16' 4' 16'

BASED ON AN ALLOWABLE BENDING STRESS = 2 K.S.I.
AND AN ALLOWABLE SHEARING UNIT STRESS = 200 P.S.I.

● DETERMINE THE MOST ECONOMICAL (LEAST CROSS-SECTIONAL AREA)
LAMINATED TIMBER BEAM FOR EACH SEGMENT, BASED ON A 5 $\frac{1}{8}$" WIDTH,
CONSIDERING CRITICAL ARRANGEMENTS OF DEAD AND LIVE LOADS.

(II) STEEL WIDE FLANGE BEAMS

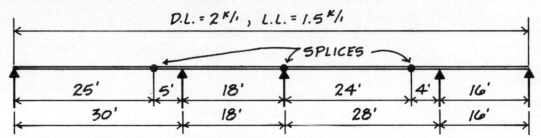

D.L.= 2^{k}/ı , L.L.= 1.5^{k}/ı

SPLICES

25' 5' 18' 24' 4' 16'
30'   18'   28'   16'

(A) DETERMINE THE MAXIMUM NEGATIVE AND POSITIVE MOMENT IN EACH
PIECE BASED ON THE CRITICAL ARRANGEMENTS OF LIVE AND DEAD LOADS.
(B) DETERMINE THE MOST ECONOMICAL STEEL WIDE FLANGE SECTIONS
BASED ON AN ALLOWABLE BENDING STRESS = 30 K.S.I.
(C) COMPARE WITH SIMPLY SUPPORTED BEAMS.

(III) STEEL WIDE FLANGE BEAMS

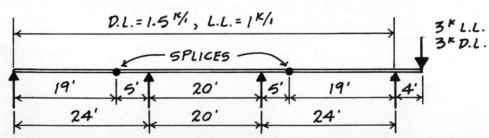

D.L.=1.5^{k}/ı , L.L.= 1^{k}/ı

3^{k} L.L.
3^{k} D.L.

SPLICES

19' 5' 20' 5' 19' 4'
24'   20'   24'

(A) DETERMINE THE MAXIMUM NEGATIVE AND POSITIVE MOMENT IN EACH
PIECE BASED ON THE CRITICAL ARRANGEMENTS OF LIVE AND DEAD LOADS.
(B) DETERMINE THE MOST ECONOMICAL WIDE FLANGE SECTIONS BASED
ON AN ALLOWABLE BENDING STRESS OF 24 K.S.I.
(C) DETERMINE THE MOST ECONOMICAL WIDE FLANGE SECTIONS BASED
ON SPLICES AT THE SUPPORTS (EXCEPT, OF COURSE, THE CANTILEVER).

PROBLEM 20-2

STEEL: VARIABLE SECTIONS

(1) <u>GIVEN</u>: 64' SIMPLE SPAN - UNIFORM LOADING = $4^{k}/_{l.f} \times 64' = 256^{k}$
ALLOWABLE BENDING $f = 24$ K.S.I.

(A)   DETERMINE THE MOST ECONOMICAL "W" SECTION NEEDED
FOR BENDING AT ALLOWABLE f.

(B)

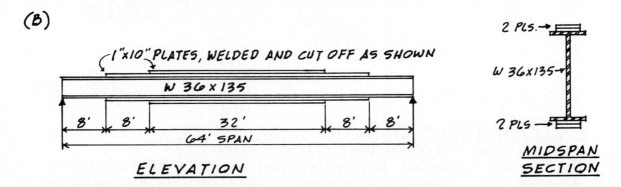

1"x10" PLATES, WELDED AND CUT OFF AS SHOWN

W 36 x 135

8'   8'   32'   8'   8'

64' SPAN

ELEVATION

2 PLS. →
W 36 x 135 →
2 PLS →

**MIDSPAN
SECTION**

● DETERMINE THE MAXIMUM BENDING STRESS f AT THE MIDSPAN
AND PLATE CUT-OFF POINTS.

---

(2) <u>LOADING</u>:

ALL. BENDING $f = 24$ K.S.I.

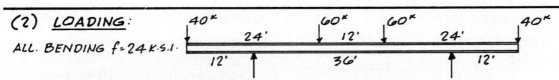

40^{k} 60^{k} 60^{k} 40^{k}

24'   12'   24'

12'   36'   12'

(A)   a. DRAW THE SHEAR AND BENDING MOMENT DIAGRAMS
     b. DETERMINE WHICH W 21 SECTION IS NEEDED FOR BENDING AT
       ALLOWABLE f.

(B)

½"x12" PLATE, WELDED AND CUT-OFF AS SHOWN

W 21 x 62

←X→←Y→   ←Y→←X→

ELEVATION

1 PL. ↰
W 21 x 62 →
1 PL. ↲

**SECTION AT
SUPPORTS**

a. DETERMINE THE MAX. BENDING f AT MIDSPAN AND SUPPORTS.
b. DETERMINE DISTANCES X' & Y' WHERE THE PLATES MAY BE CUT
    OFF WITHOUT OVER-STRESSING THE W 21 x 62 SECTION.

PROBLEM 20–3

VARIABLE SECTIONS

(1) LAMINATED TIMBER ROOF GIRDER

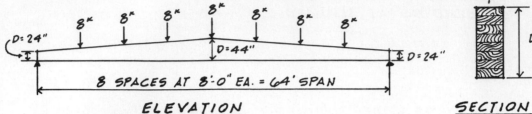

8 SPACES AT 8'-0" EA. = 64' SPAN

ELEVATION

9"

D" VARIES

SECTION

ALLOW. STRESSES
SHEAR f_v = 200 P.S.1
BENDING f = 2150 P.S.1.

NOTE: FOR A RECTANGULAR SECTION $\begin{cases} \text{MAX. } f_v = 1.5\,V/A \\ I/c = BD^2/6 \end{cases}$

DETERMINE:

(a) MAXIMUM SHEARING UNIT STRESS (ACTUAL f_v).

(b) MAXIMUM BENDING UNIT STRESS ON THE BEAM AT EACH LOAD POINT.

(c) COMPARING __ACTUAL__ UNIT STRESSES WITH __ALLOWABLE__ UNIT STRESSES IT WILL BE SEEN THAT CERTAIN SECTIONS ARE OVER-STRESSED.
__REVISE THE DESIGN OF THE GIRDER:__ RETAINING THE 9" WIDTH,
● DETERMINE THE DEPTH OF SECTION (NEAREST WHOLE INCH) NEEDED AT THE ENDS AND AT EACH LOAD POINT.

---

(2) STEEL GIRDER

ALLOWABLE BENDING f = 24 K.S.1.

45ᵏ   45ᵏ

16'   16'   16'

250#/FT. = 12ᵏ

48' SPAN

(a) DETERMINE THE MOST ECONOMICAL "W" SECTION.

(b) DETERMINE WHICH W 27 x COULD BE USED.

(UNIFORM LOADING INCLUDES ALLOW-ANCE FOR GIRDER WEIGHT)

---

(3) IF THE GIRDER IN PART (2) IS A W 27 x 84 + 2 PLATES 12" x 5/8" WELDED TO THE GIRDER AND CUT OFF AS SHOWN.

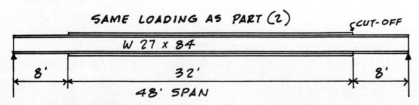

SAME LOADING AS PART (2)

CUT-OFF

W 27 x 84

8'   32'   8'

48' SPAN

● DETERMINE THE MAXIMUM BENDING f AT THE MIDSPAN AND AT THE PLATE CUT-OFFS

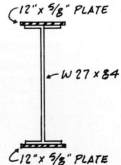

12" x 5/8" PLATE

W 27 x 84

12" x 5/8" PLATE

PROBLEM 20–4

VARIABLE SECTIONS

(1) LAMINATED TIMBER CANTILEVER BEAM

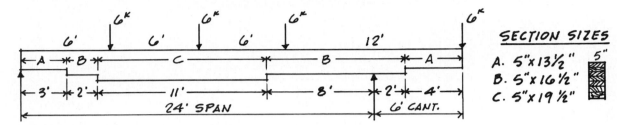

SECTION SIZES

A. 5"x13½"

B. 5"x16½"

C. 5"x19½"

BENDING MOMENT DIAGRAM

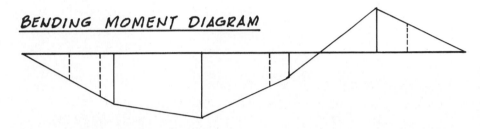

(a) DETERMINE THE MAGNITUDE OF BENDING MOMENTS AT THE LOAD AND REACTION POSITIONS AND WHERE CHANGES OF SECTIONS OCCUR.

(b) DETERMINE THE MAXIMUM BENDING UNIT STRESSES AT THE EXTREME FIBERS OF SECTIONS A, B & C.

(2) GIVEN: TWO BOARDS, EACH 12' LONG AND 1"x6" SECTION (ACTUAL DIMENSIONS) TO BE GLUED TOGETHER AND USED AS A SIMPLE BEAM FOR THE FOLLOWING LOADING. DETERMINE P.

ALLOWABLE BENDING f= 1800 P.S.I.

ALLOWABLE SHEAR fᵥ = 200 P.S.I.

(a) BEAM OF CONSTANT SECTION BOARDS MAY BE ARRANGED FOR THREE DIFFERENT SECTIONS, AS SHOWN.

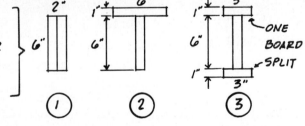

(b) VARIABLE SECTION BEAM: DETERMINE THE MAXIMUM UNIT STRESS IN EACH SECTION.

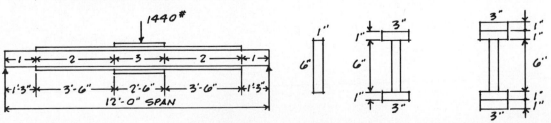

Appendix

DATA SHEET D-1

ABRIDGED METRIC (SI) CONVERSIONS

Definition of Metric Terms

Symbol	
m	meter (a measure of length)
mm	millimeter = 1/1000 of a meter
kg	kilogram = 1000 grams (a measure of mass)
N	newton (a measure of force)
kN	kilonewton (1000 newtons)
Pa	Pascal—a unit of stress = 1 N per square meter
MPa	mega Pascal = 1,000,000 Pascals
kPa	kilo Pascal = 1000 Pascals

Conversions

U.S. to SI Metric	SI Metric to U.S.
1 ft = 0.3048 m = 304.8 mm	1 m = 39.370 in.
1 ft^2 = 9.29 $\times$ 10^{-2}m^2	1 m^2 = 10.764 ft^2
1 in. = 25.40 mm	1 mm = 0.0394 in.
1 in.^{2} = 645.2 mm^2	1 mm^2 = 1.550 $\times$ 10^{-3} in.^{2}
1 in.^{3} = 1.639 $\times$ 10^{4}mm^3	1 mm^3 = 61.024 $\times$ 10^{-6} in.^{3}
1 in.^{4} = 4.162 $\times$ 10^{5}mm^4	1 mm^4 = 2.403 $\times$ 10^{-6} in.^{4}
1 lb = 0.4536 kg	1 kg = 2.2046 lb
1 lb (force) = 4.448 N	1 kN = 224.8 lb (force)
1 lb/ft^2 = 47.88 Pa	1 kPa = 0.145 p.s.i.
1 k.s.i. = 6.895 MPa	1 MPa = 145 p.s.i.
1 lb/ft = 1.488 kg/m	1 kg/m = 0.672 lb/ft

DATA SHEET D-2

CENTROIDS AND AREAS

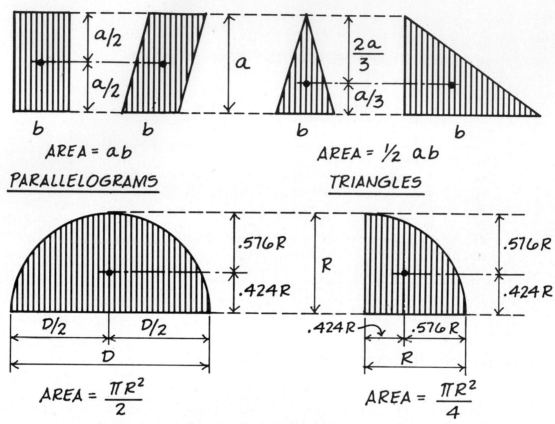

AREA = ab

PARALLELOGRAMS

AREA = ½ ab

TRIANGLES

$$AREA = \frac{\pi R^2}{2}$$

$$AREA = \frac{\pi R^2}{4}$$

CIRCULAR AREAS

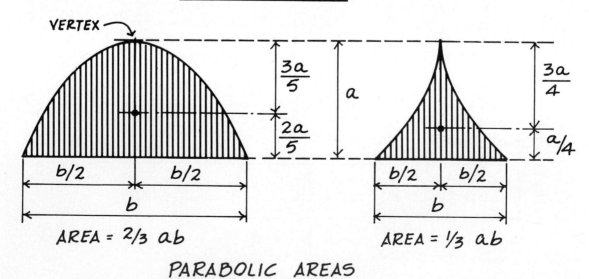

AREA = ⅔ ab

AREA = ⅓ ab

PARABOLIC AREAS

DATA SHEET D-3

WEIGHTS OF BUILDING MATERIALS

Materials	Weight Lb. per Sq. Ft.	Materials	Weight Lb. per Sq. Ft.
CEILINGS		**PARTITIONS**	
Channel suspended system	1	Clay Tile	
Lathing and plastering	See Partitions	3 in.	17
Acoustical fiber tile	1	4 in.	18
		6 in.	28
		8 in.	34
		10 in.	40
FLOORS		Gypsum Block	
Steel Deck	See Manufacturer	2 in.	9½
		3 in.	10½
Concrete-Reinforced 1 in.		4 in.	12½
Stone	12½	5 in.	14
Slag	11½	6 in.	18½
Lightweight	6 to 10	Wood Studs 2 × 4	
		12–16 in. o.c.	2
Concrete-Plain 1 in.		Steel partitions	4
Stone	12	Plaster 1 inch	
Slag	11	Cement	10
Lightweight	3 to 9	Gypsum	5
		Lathing	
Fills 1 inch		Metal	½
Gypsum	6	Gypsum Board ½ in.	2
Sand	8		
Cinders	4		
		WALLS	
Finishes		Brick	
Terrazzo 1 in.	13	4 in.	40
Ceramic or Quarry Tile ¾ in.	10	8 in.	80
Linoleum ¼ in.	1	12 in.	120
Mastic ¾ in.	9	Hollow Concrete Block (Heavy Aggregate)	
Hardwood ⅞ in.	4	4 in.	30
Softwood ¾ in.	2½	6 in.	43
		8 in.	55
		12½ in.	80
ROOFS		Hollow Concrete Block (Light Aggregate)	
Copper or tin	1	4 in.	21
		6 in.	30
3-ply ready roofing	1	8 in.	38
3-ply felt and gravel	5½	12 in.	55
5-ply felt and gravel	6	Clay tile (Load Bearing)	
		4 in.	25
Shingles		6 in.	30
Wood	2	8 in.	33
Asphalt	3	12 in.	45
Clay tile	9 to 14	Stone 4 in.	55
Slate ¼	10	Glass Block 4 in.	18
		Windows, Glass, Frame & Sash	8
Sheathing		Curtain Walls	See Manufacturer
Wood ¾ in.	3	Structural Glass 1 in.	15
Gypsum 1 in.	4	Corrugated Cement Asbestos ¼ in.	3
Insulation 1 in.			
Loose	½		
Poured in place	2		
Rigid	1½		

From the *Manual of Steel Construction*, 8th Edition. Reprinted with permission granted by the courtesy of the American Institute of Steel Construction.

DATA SHEET D-4

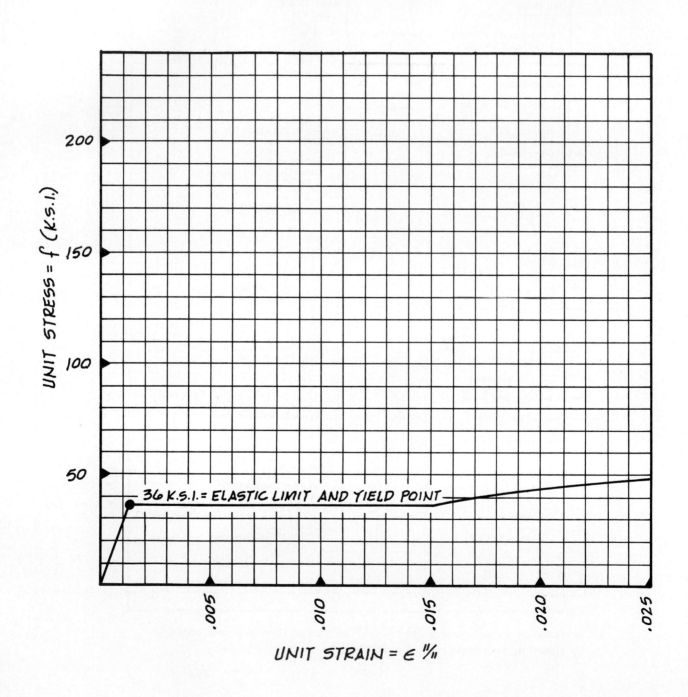

DATA SHEET D-5

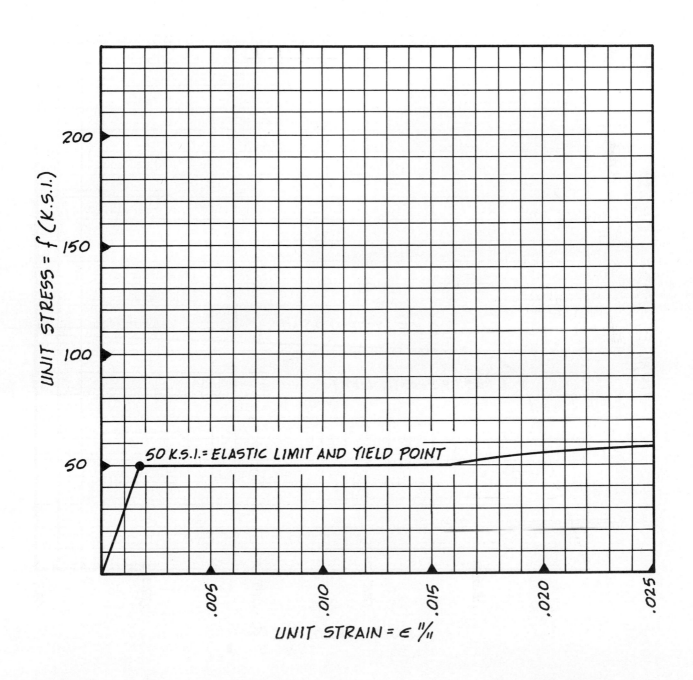

STRESS/STRAIN DIAGRAM – GRADE S-2 STEEL

MODULUS OF ELASTICITY = E = 29,000,000 P.S.I.

50 K.S.I.= ELASTIC LIMIT AND YIELD POINT

UNIT STRESS = f (K.S.I.)

UNIT STRAIN = ϵ "/"

DATA SHEET D-6

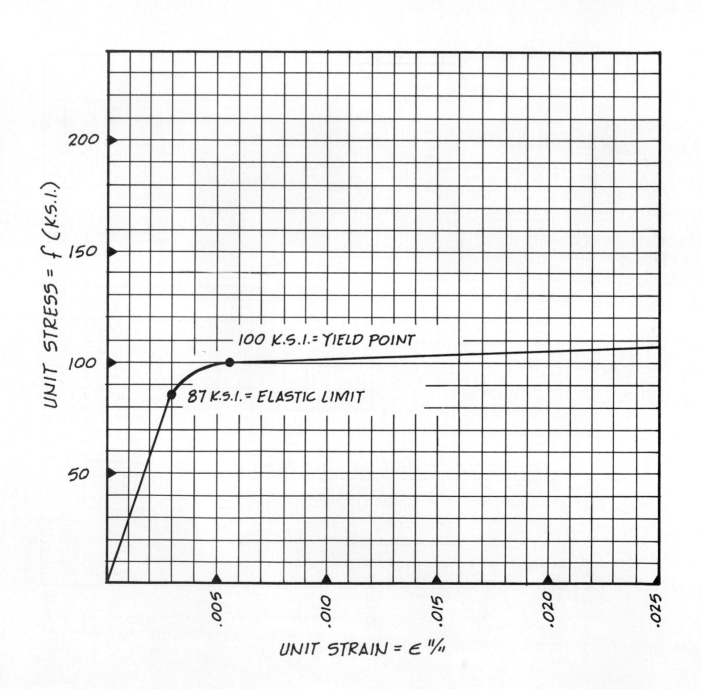

STRESS/STRAIN DIAGRAM – GRADE S-3 STEEL

MODULUS OF ELASTICITY = E = 29,000,000 P.S.I.

DATA SHEET D-7

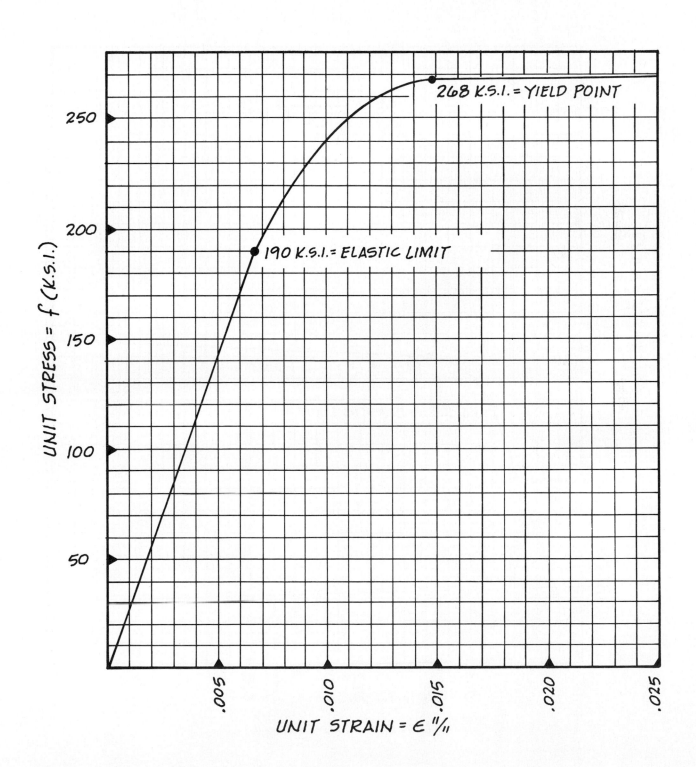

STRESS/STRAIN DIAGRAM - 7 WIRE STRAND (CABLE)

MODULUS OF ELASTICITY = E = 29,000,000 P.S.I.

268 K.S.I. = YIELD POINT

190 K.S.I. = ELASTIC LIMIT

UNIT STRESS = f (K.S.I.)

UNIT STRAIN = ε "/"

DATA SHEET D-8

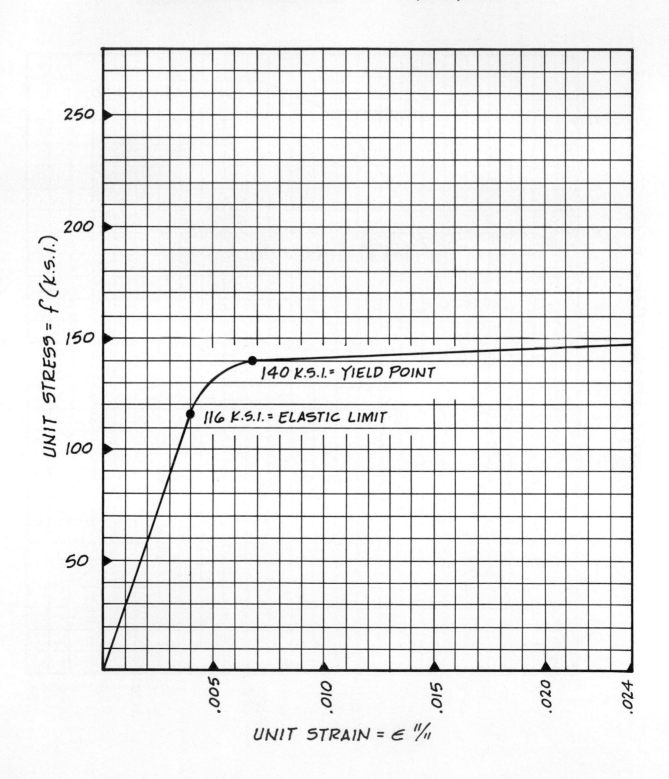

STRESS/STRAIN DIAGRAM - HIGH STRENGTH STEEL BARS

MODULUS OF ELASTICITY = E = 29,000,000 P.S.I.

140 K.S.I. = YIELD POINT

116 K.S.I. = ELASTIC LIMIT

UNIT STRESS = f (K.S.I.)

UNIT STRAIN = ε "/"

DATA SHEET D-9

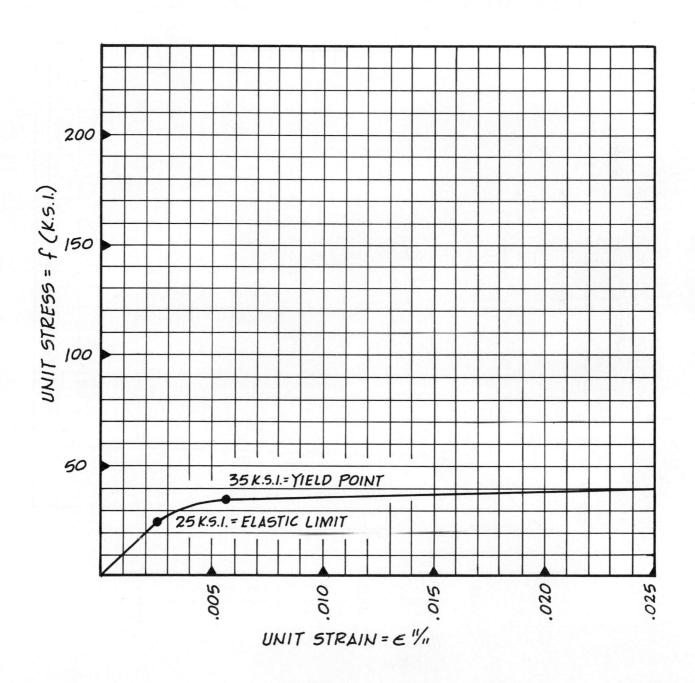

STRESS/STRAIN DIAGRAM-GRADE A-1 ALUMINUM

MODULUS OF ELASTICITY = E = 10,000,000 P.S.I.

DATA SHEET D-10

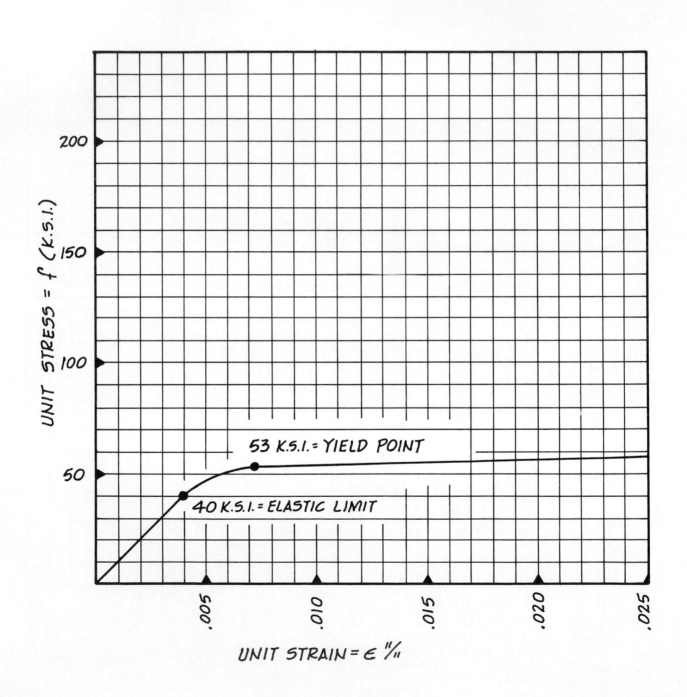

STRESS/STRAIN DIAGRAM - GRADE A-2 ALUMINUM

MODULUS OF ELASTICITY = E = 10,000,000 P.S.I.

53 K.S.I. = YIELD POINT

40 K.S.I. = ELASTIC LIMIT

UNIT STRESS = f (K.S.I.)

200

150

100

50

.005 .010 .015 .020 .025

UNIT STRAIN = ε "/"

DATA SHEET D-11

COEFFICIENTS OF EXPANSION

Material	Coefficient of Expansion In./In./ Degree F.
Structural Steel	.0000065
Aluminum	.0000128
Wrought Iron	.0000067
Copper	.0000098
Brick	.0000035–.0000050
Cement Mortar	.0000070
Concrete	.0000055–.0000070
Limestone	.0000040
Plaster	.0000090
Wood (Fir), Parallel to Grain	.0000025
Wood (Fir), Perpendicular to Grain	.0000200–.0000300
Glass	.0000045
Plexiglas	.0000450–.0000500
Styrofoam	.0000400
Polyethylene	.0001000

COEFFICIENTS OF FRICTION

Wood on Wood	.40
Metal on Metal	.20
Metal on Masonry	.35
Metal on Wood	.40
Masonry on Masonry	.65
Masonry on Wood	.40
Masonry on Moist Clay	.35

Note: The values shown are intended to provide a basis for the solution of problems. They should be regarded as approximations, since accurate values will vary depending on the condition of the surfaces, presence of moisture, etc.

DATA SHEET D-12

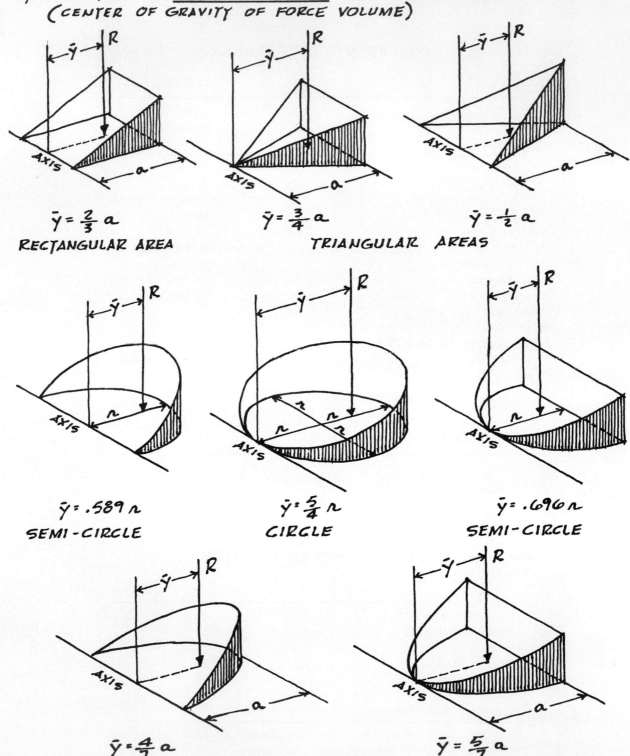

PROPORTIONAL UNIT STRESS APPLIED TO AN AREA

$\bar{y}$ = LOCATION OF <u>RESULTANT FORCE</u> FROM AXIS OF ZERO UNIT STRESS (CENTER OF GRAVITY OF FORCE VOLUME)

$\bar{y} = \frac{2}{3}a$

RECTANGULAR AREA

$\bar{y} = \frac{3}{4}a$

$\bar{y} = \frac{1}{2}a$

TRIANGULAR AREAS

$\bar{y} = .589\,r$

SEMI-CIRCLE

$\bar{y} = \frac{5}{4}r$

CIRCLE

$\bar{y} = .696\,r$

SEMI-CIRCLE

$\bar{y} = \frac{4}{7}a$

$\bar{y} = \frac{5}{7}a$

PARABOLIC AREAS

DATA SHEET D-13

MOMENT OF INERTIA OF VARIOUS CROSS SECTIONS WITH RESPECT TO AXES INDICATED (• INDICATES CENTROIDAL AXIS)

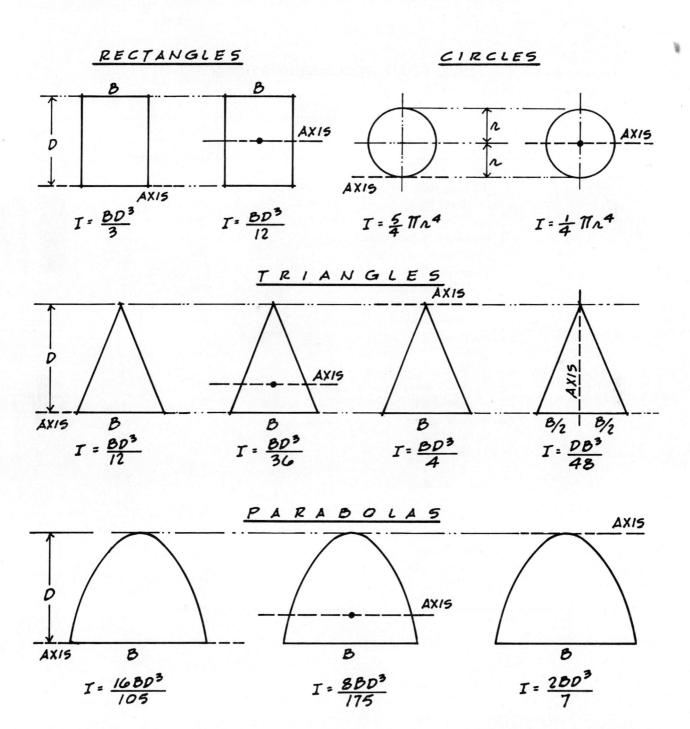

RECTANGLES

$$I = \frac{BD^3}{3}$$

$$I = \frac{BD^3}{12}$$

CIRCLES

$$I = \frac{5}{4}\pi r^4$$

$$I = \frac{1}{4}\pi r^4$$

TRIANGLES

$$I = \frac{BD^3}{12}$$

$$I = \frac{BD^3}{36}$$

$$I = \frac{BD^3}{4}$$

$$I = \frac{DB^3}{48}$$

PARABOLAS

$$I = \frac{16BD^3}{105}$$

$$I = \frac{8BD^3}{175}$$

$$I = \frac{2BD^3}{7}$$

DATA SHEET D-14

LIVE LOAD RECOMMENDATIONS

Function	Live Load (Pounds/ft^2)
Assembly	
Assembly Halls, Auditoriums, Churches, etc.	
Fixed Seats	60
Movable Seats	100
Restaurants, Gymnasiums, Grandstands, etc.	100
Theaters:	
Aisle and Lobbies	100
Balconies	60
Stage Floors	150
Business Facilities	
Offices	80
File Rooms:	
Letter Files	80
Card Files	125
Educational Facilities	
Libraries:	
Reading Rooms	60
Stacks	150
School Buildings:	
Classrooms	40
Corridors	100
Industrial Facilities	
Manufacturing:	
Light	125
Heavy	250
Laboratories	100

DATA SHEET D-15

LIVE LOAD RECOMMENDATIONS (CONTINUED)

Function	Live Load (Pounds/Ft^2)
Institutional Facilities	
Hospitals:	
Wards and Private Rooms	40
Operating Rooms	60
Corridors	80
Penal Institutions:	
Cell Blocks	40
Corridors	100
Residential Facilities	
Private Dwellings:	
First Floor	40
Upper Floors	30
Uninhabitable Attics	20
Multifamily:	
Apartments	40
Corridors	60

Note: This list of live loads is given only to provide the student with a sense of the variations for different functions and is not to be used for structural design purposes. Values may vary, depending on the building code being used in a particular location. When designing a structure, the appropriate building code should be used for live load recommendations.

DATA SHEET D-16

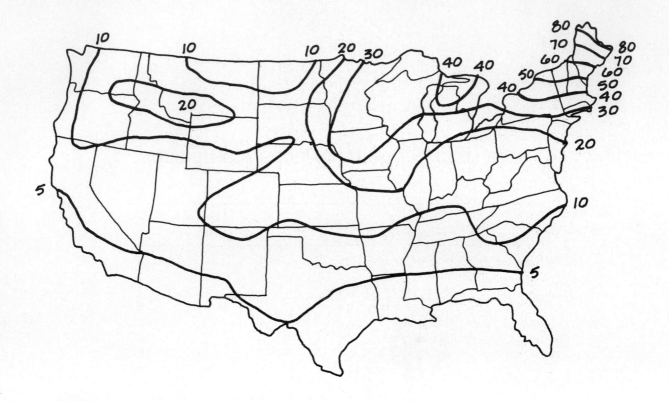

GENERALIZED SNOW LOAD MAP OF THE CONTINENTAL UNITED STATES

Notes.

   1. The values shown on this page are snow loads in ''pounds per square foot.'' Generally, snow may be thought of as weighing about .5#/ft^2 per inch of thickness for dry snow and increasing for dense, wet snow.
   2. The values shown will gradually diminish to the south of the contours.
   3. The information presented on this page is very general; it is provided to give the student a sense of the regional variations in snow loads. It is not intended for purposes of structural design. Where precise information is required, the appropriate building code should be consulted.

DATA SHEET D-17

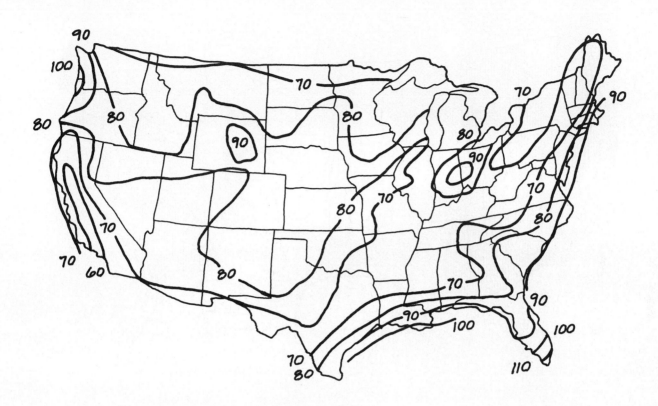

GENERALIZED WIND VELOCITY MAP OF THE CONTINENTAL UNITED STATES

Notes.

1. The values shown on this page are wind velocities in "miles per hour." Approximate static pressures produced by these velocities in terms of "pounds per square foot" acting on the surface of a building may be determined by

$$P = .003V^2$$

where P = Pressure (#/ft^2)
V = Velocity (mi/h)

2. The information presented on this page is very general; it is provided to give the student a sense of the regional variations in wind forces. It is not intended for purposes of structural design. Where precise information is required, the appropriate building code should be consulted.

DATA SHEET D-18

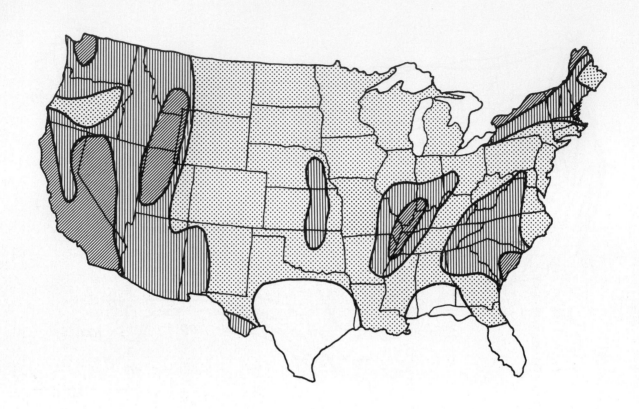

GENERALIZED SEISMIC RISK MAP OF THE UNITED STATES

Greatest Risk (Major Damage)

Moderate Risk (Moderate Damage)

Low Risk (Minor Damage)

Very Low Risk (No Damage)

Much of the states of Alaska and Hawaii is in "Greatest Risk" and "Moderate Risk" zones.

Note.

    The information furnished on this page is very general; it is provided merely to give the student a sense of the regional variations in seismic risks. It is not intended for purposes of structural design. Where precise information is required, the appropriate building code should be consulted.

DATA SHEET D-19

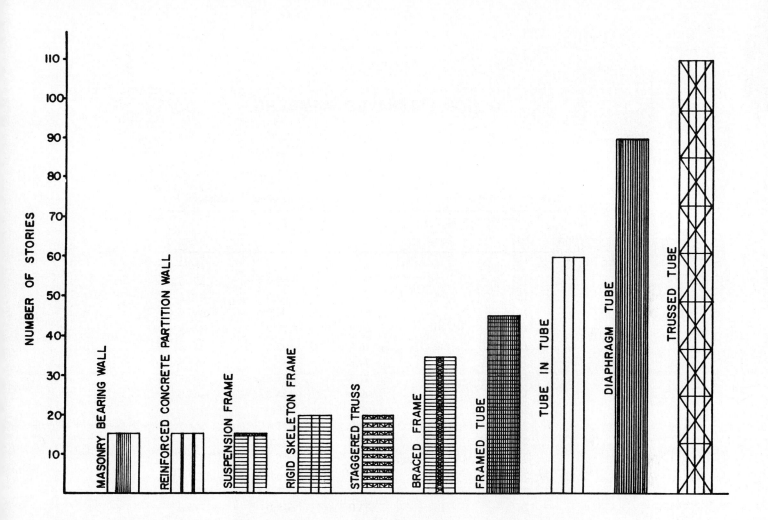

OPTIMUM HIGH-RISE STRUCTURAL SYSTEMS

From "Survey and Design of Multistory Buildings," by Samuel F. Johnson, Jr., July 1970—A Design Report for the Department of Civil and Environmental Engineering, Washington University. Reprinted with the permission of the author and the Department of Civil and Environmental Engineering, Washington University.

DATA SHEET D-20

GLUED LAMINATED MEMBERS

Unit Properties of Sections (Partial Listing), 1½ Lamination Thickness

Width		Depth	No. of Lams	A	S_x	I_x
$3\frac{1}{8}$	×	9	6	28.1	42.2	189.8
		$10\frac{1}{2}$	7	32.8	57.4	301.5
		12	8	37.5	75.0	450.0
		$13\frac{1}{2}$	9	42.2	94.9	640.7
		15	10	46.9	117.2	878.9
$5\frac{1}{8}$	×	15	10	76.9	192.2	1441.4
		$16\frac{1}{2}$	11	84.6	232.5	1918.5
		18	12	92.3	276.8	2490.8
		$19\frac{1}{2}$	13	99.9	324.8	3166.8
		21	14	107.6	376.6	3955.2
		$22\frac{1}{2}$	15	115.3	432.4	4864.7
		24	16	123.0	492.0	5904.0
		$25\frac{1}{2}$	17	130.7	555.4	7081.6
$6\frac{3}{4}$	×	21	14	141.8	496.1	5209.3
		$22\frac{1}{2}$	15	151.9	569.5	6407.2
		24	16	162.0	648.0	7776.0
		$25\frac{1}{2}$	17	172.1	731.5	9327.0
		27	18	182.3	820.1	11071.7
		$28\frac{1}{2}$	19	192.4	913.8	13021.4
		30	20	202.5	1012.5	15187.5
		$31\frac{1}{2}$	21	212.6	1116.3	17581.4
		33	22	222.8	1225.1	20214.6
		$34\frac{1}{2}$	23	232.9	1339.0	23098.3

Symbol Identification: A = Area = in.^{2}; S = Section Modulus = in.^{3}; I = Moment of Inertia = in.^{4}

DATA SHEET D-21

GLUED LAMINATED MEMBERS

Unit Properties of Sections (Partial Listing), 1½ Lamination Thickness

Width	Depth	No. of Lams	A	S_x	I_x
8¾ ×	27	18	236.3	1063.1	14352.2
	28½	19	249.4	1184.5	16879.6
	30	20	262.5	1312.5	19687.5
	31½	21	275.6	1447.0	22790.7
	33	22	288.8	1588.1	26204.1
	34½	23	301.9	1735.8	29942.2
	36	24	315.0	1890.0	34020.0
	37½	25	328.1	2050.8	38452.2
	39	26	341.3	2218.1	43253.4
	40½	27	354.4	2392.0	48438.6
	42	28	367.5	2572.5	54022.5
	43½	29	380.6	2759.5	60019.8
	45	30	393.8	2953.1	66445.3
10¾ ×	33	22	354.8	1951.1	32193.6
	34½	23	370.9	2132.5	36786.2
	36	24	387.0	2322.0	41796.0
	37½	25	403.1	2519.5	47241.2
	39	26	419.3	2725.1	53139.9
	40½	27	435.4	2938.8	59150.3
	42	28	451.5	3160.5	66370.5
	43½	29	467.6	3390.3	73738.6
	45	30	483.8	3628.1	81632.8
	46½	31	499.9	3874.0	90071.2
	48	32	516.0	4128.0	99072.0
	49½	33	532.1	4390.0	108652.3
	51	34	548.3	4660.1	118833.2
	52½	35	564.4	4938.3	129629.9
	54	36	580.5	5224.5	141061.5
	55½	37	596.6	5518.8	153146.2

Symbol Identification: A = Area = in.^{2}; S = Section Modulus = in.^{3}; I = Moment of Inertia = in.^{4}

DATA SHEET D-22

PROPERTIES OF SAWN LUMBER SECTIONS

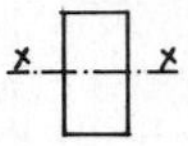

Nominal Size b × d	Actual Size b × d	Area in.²	I_x in.⁴	S_x in.³
1 × 4	3/4 × 3½	2.63	2.68	1.53
1 × 6	" × 5½	4.13	10.40	3.78
1 × 8	" × 7¼	5.44	23.82	6.57
1 × 10	" × 9¼	6.94	49.47	10.70
1 × 12	" × 11¼	8.44	88.99	15.83
2 × 4	1½ × 3½	5.25	5.36	3.06
2 × 6	" × 5½	8.25	20.80	7.56
2 × 8	" × 7¼	10.88	47.64	13.14
2 × 10	" × 9¼	13.88	98.93	21.39
2 × 12	" × 11¼	16.88	177.98	31.64
3 × 4	2½ × 3½	8.75	8.93	5.10
3 × 6	" × 5½	13.75	34.66	12.60
3 × 8	" × 7¼	18.13	79.39	21.90
3 × 10	" × 9¼	23.13	164.89	35.65
3 × 12	" × 11¼	28.13	296.63	52.73
4 × 4	3½ × 3½	12.25	12.50	7.15
4 × 6	" × 5½	19.25	48.53	17.65
4 × 8	" × 7¼	25.38	111.15	30.66
4 × 10	" × 9¼	32.38	230.84	49.91
4 × 12	" × 11¼	39.38	415.28	73.83
6 × 6	5½ × 5½	30.25	76.26	27.73
6 × 8	" × 7½	41.25	193.36	51.56
6 × 10	" × 9½	52.25	392.96	82.73
6 × 12	" × 11½	63.25	697.07	121.23
6 × 14	" × 13½	74.25	1127.67	167.06
6 × 16	" × 15½	85.25	1706.78	220.23

DATA SHEET D-23

REINFORCED CONCRETE DATA

Values of E and n for Various $f'c$

$f'c$	n	E_c*
2500	10	2900
3000	9	3200
3500	8.5	3450
4000	8	3650
4500	7.5	3900
5000	7	4100
5500	6.7	4300
6000	6.4	4500
6500	6.2	4670
7000	6	4850

*All values of E_c are to be $\times\ 10^3$

REINFORCING BAR DATA

Bar No.	Area	Dia (in.)
3	0.11	0.375
4	0.20	0.500
5	0.31	0.625
6	0.44	0.750
7	0.60	0.875
8	0.79	1.000
9	1.00	1.128
10	1.27	1.270
11	1.56	1.410
14	2.25	1.693
18	4.00	2.257

DATA SHEET D-24

FRACTIONAL AREAS OF ENCLOSURE RECTANGLES

CURVES TANGENT TO
HORIZONTAL AT VERTEX •

<u>NOTE</u> REVERSE POSSIBILITIES

FRACTION OF RECTANGULAR
AREAS SHOWN → ◯

CENTROID OF FRACTIONAL AREA
LOCATED BY HORIZONTAL
DIMENSIONS

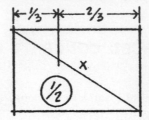

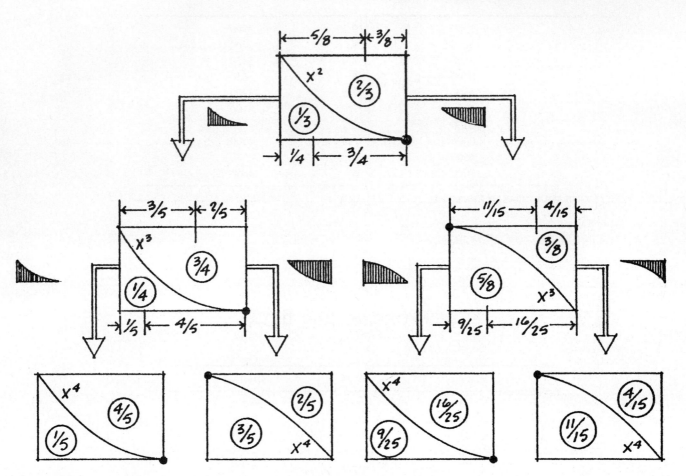

DATA SHEET D-25

MAXIMUM VALUES: SLOPE, DEFLECTION, AND BENDING MOMENT

NOTE: VALUES OF SLOPE AND DEFLECTION TO BE DIVIDED BY "EI"

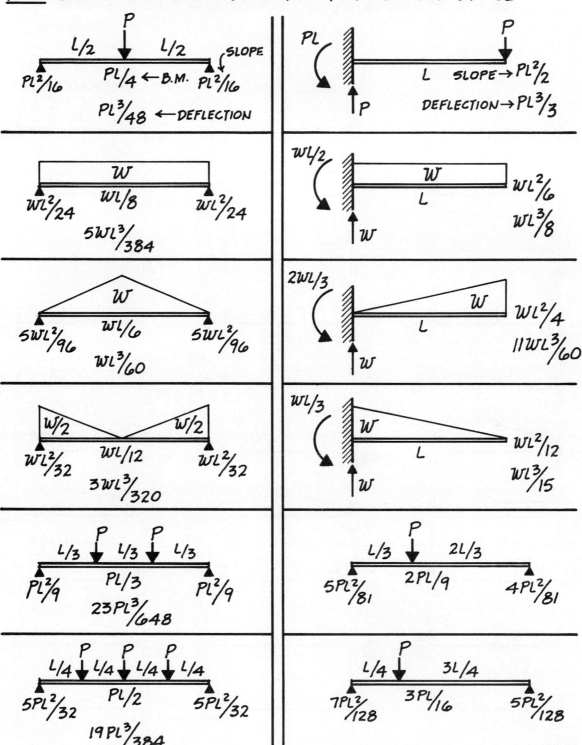

DATA SHEET D-26

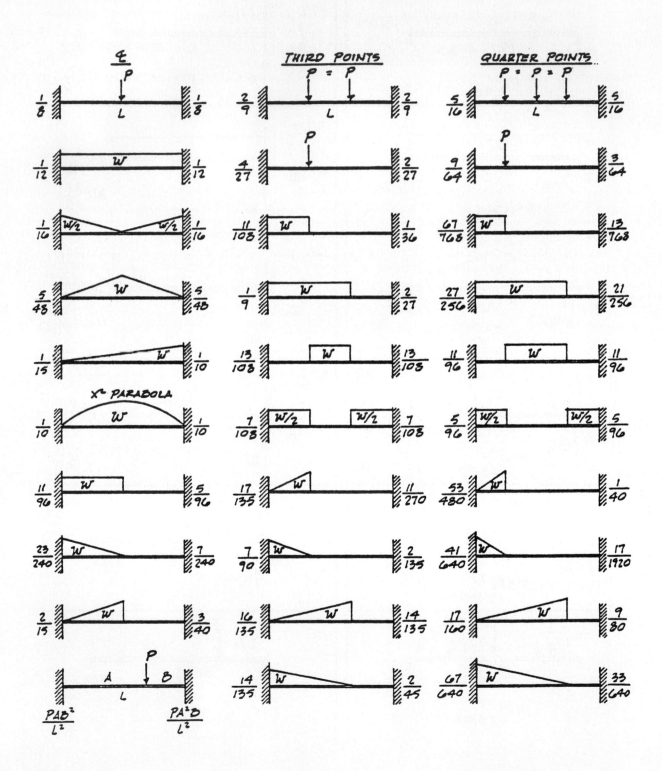

FIXED END MOMENT COEFFICIENTS

FIXED END MOMENT = CWL OR CPL: C = COEFFICIENT, W OR P = LOAD, L = SPAN

DATA SHEET D-27

NOTES FOR DATA SHEETS D-28 THROUGH D-36

1. The information given on Data Sheets D–28 through D–36 is taken from "Bethlehem Structural Shapes," Catalog 3277B, 1980 edition, and is reprinted here with the permission of Bethlehem Steel Corporation.

2. The data furnished on the following pages (Data Sheets D–28 through D–36) is provided as a convenience for the student using this textbook in order to facilitate the learning process. From time to time, slight changes are made to this information. Therefore, these pages are not intended for purposes of actual structural design. When doing the design for a building, current data should be consulted.

3. A few words are necessary regarding the nomenclature used for steel sections on the following data sheets. Steel sections are designated according to:
   (a) shape of the section
   (b) nominal depth (inches) of the section
   (c) weight (lbs/lin ft) of the section.
   For example, consider the following designation:

$$W \ 36 \times 300$$

   (a) The symbol "W" indicates that this is a wide flange shape. There are many other shapes in steel, but the wide flange is most commonly used for beams in architectural structures and this is what we will use in these studies.
   (b) The number "36" indicates the nominal depth of the section in inches. Nominal depth means that this section is made in the 36-in. rollers at the mill. The actual depth will usually be slightly different from the nominal depth.
   (c) The number "300" indicates the weight of the section in terms of pounds per linear foot. Since steel is priced by weight, this number in the designation is important for cost considerations.

DATA SHEET D-28

WIDE FLANGE SHAPES

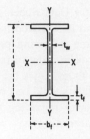

Theoretical Dimensions and Properties for **Designing**

Section Number	Weight per Foot	Area of Section	Depth of Section	Flange		Web Thickness	Axis X-X			Axis Y-Y			
				Width	Thickness		I_x	S_x	r_x	I_y	S_y	r_y	r_T
		A	d	b_f	t_f	t_w							
	lb	in.²	in.	in.	in.	in.	in.⁴	in.³	in.	in.⁴	in.³	in.	in.
W36 x 300	300	88.3	36.74	16.655	1.680	0.945	20300	1110	15.2	1300	156	3.83	4.39
280	280	82.4	36.52	16.595	1.570	0.885	18900	1030	15.1	1200	144	3.81	4.37
260	260	76.5	36.26	16.550	1.440	0.840	17300	953	15.0	1090	132	3.78	4.34
245	245	72.1	36.08	16.510	1.350	0.800	16100	895	15.0	1010	123	3.75	4.32
230	230	67.6	35.90	16.470	1.260	0.760	15000	837	14.9	940	114	3.73	4.30
W36 x 210	210	61.8	36.69	12.180	1.360	0.830	13200	719	14.6	411	67.5	2.58	3.09
194	194	57.0	36.49	12.115	1.260	0.765	12100	664	14.6	375	61.9	2.56	3.07
182	182	53.6	36.33	12.075	1.180	0.725	11300	623	14.5	347	57.6	2.55	3.05
170	170	50.0	36.17	12.030	1.100	0.680	10500	580	14.5	320	53.2	2.53	3.04
160	160	47.0	36.01	12.000	1.020	0.650	9750	542	14.4	295	49.1	2.50	3.02
150	150	44.2	35.85	11.975	0.940	0.625	9040	504	14.3	270	45.1	2.47	2.99
135	135	39.7	35.55	11.950	0.790	0.600	7800	439	14.0	225	37.7	2.38	2.93
W33 x 241	241	70.9	34.18	15.860	1.400	0.830	14200	829	14.1	932	118	3.63	4.17
221	221	65.0	33.93	15.805	1.275	0.775	12800	757	14.1	840	106	3.59	4.15
201	201	59.1	33.68	15.745	1.150	0.715	11500	684	14.0	749	95.2	3.56	4.12
W33 x 152	152	44.7	33.49	11.565	1.055	0.635	8160	487	13.5	273	47.2	2.47	2.94
141	141	41.6	33.30	11.535	0.960	0.605	7450	448	13.4	246	42.7	2.43	2.92
130	130	38.3	33.09	11.510	0.855	0.580	6710	406	13.2	218	37.9	2.39	2.88
118	118	34.7	32.86	11.480	0.740	0.550	5900	359	13.0	187	32.6	2.32	2.84
W30 x 211	211	62.0	30.94	15.105	1.315	0.775	10300	663	12.9	757	100	3.49	3.99
191	191	56.1	30.68	15.040	1.185	0.710	9170	598	12.8	673	89.5	3.46	3.97
173	173	50.8	30.44	14.985	1.065	0.655	8200	539	12.7	598	79.8	3.43	3.94
W30 x 132	132	38.9	30.31	10.545	1.000	0.615	5770	380	12.2	196	37.2	2.25	2.68
124	124	36.5	30.17	10.515	0.930	0.585	5360	355	12.1	181	34.4	2.23	2.66
116	116	34.2	30.01	10.495	0.850	0.565	4930	329	12.0	164	31.3	2.19	2.64
108	108	31.7	29.83	10.475	0.760	0.545	4470	299	11.9	146	27.9	2.15	2.61
99	99	29.1	29.65	10.450	0.670	0.520	3990	269	11.7	128	24.5	2.10	2.57

All shapes on these pages have parallel-faced flanges.

DATA SHEET D-29

WIDE FLANGE SHAPES

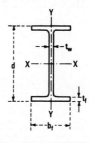

Theoretical Dimensions and Properties for **Designing**

Section Number	Weight per Foot	Area of Section	Depth of Section	Flange		Web Thick-ness	Axis X-X			Axis Y-Y			r_T
				Width	Thick-ness								
		A	d	b_f	t_f	t_w	I_x	S_x	r_x	I_y	S_y	r_y	
	lb	in.²	in.	in.	in.	in.	in.⁴	in.³	in.	in.⁴	in.³	in.	in.
W27 x 178	178	52.3	27.81	14.085	1.190	0.725	6990	502	11.6	555	78.8	3.26	3.72
	161	47.4	27.59	14.020	1.080	0.660	6280	455	11.5	497	70.9	3.24	3.70
	146	42.9	27.38	13.965	0.975	0.605	5630	411	11.4	443	63.5	3.21	3.68
W27 x 114	114	33.5	27.29	10.070	0.930	0.570	4090	299	11.0	159	31.5	2.18	2.58
	102	30.0	27.09	10.015	0.830	0.515	3620	267	11.0	139	27.8	2.15	2.56
	94	27.7	26.92	9.990	0.745	0.490	3270	243	10.9	124	24.8	2.12	2.53
	84	24.8	26.71	9.960	0.640	0.460	2850	213	10.7	106	21.2	2.07	2.49
W24 x 162	162	47.7	25.00	12.955	1.220	0.705	5170	414	10.4	443	68.4	3.05	3.45
	146	43.0	24.74	12.900	1.090	0.650	4580	371	10.3	391	60.5	3.01	3.43
	131	38.5	24.48	12.855	0.960	0.605	4020	329	10.2	340	53.0	2.97	3.40
	117	34.4	24.26	12.800	0.850	0.550	3540	291	10.1	297	46.5	2.94	3.37
	104	30.6	24.06	12.750	0.750	0.500	3100	258	10.1	259	40.7	2.91	3.35
W24 x 94	94	27.7	24.31	9.065	0.875	0.515	2700	222	9.87	109	24.0	1.98	2.33
	84	24.7	24.10	9.020	0.770	0.470	2370	196	9.79	94.4	20.9	1.95	2.31
	76	22.4	23.92	8.990	0.680	0.440	2100	176	9.69	82.5	18.4	1.92	2.29
	68	20.1	23.73	8.965	0.585	0.415	1830	154	9.55	70.4	15.7	1.87	2.26
W24 x 62	62	18.2	23.74	7.040	0.590	0.430	1550	131	9.23	34.5	9.80	1.38	1.71
	55	16.2	23.57	7.005	0.505	0.395	1350	114	9.11	29.1	8.30	1.34	1.68
W21 x 147	147	43.2	22.06	12.510	1.150	0.720	3630	329	9.17	376	60.1	2.95	3.34
	132	38.8	21.83	12.440	1.035	0.650	3220	295	9.12	333	53.5	2.93	3.31
	122	35.9	21.68	12.390	0.960	0.600	2960	273	9.09	305	49.2	2.92	3.30
	111	32.7	21.51	12.340	0.875	0.550	2670	249	9.05	274	44.5	2.90	3.28
	101	29.8	21.36	12.290	0.800	0.500	2420	227	9.02	248	40.3	2.89	3.27
W21 x 93	93	27.3	21.62	8.420	0.930	0.580	2070	192	8.70	92.9	22.1	1.84	2.17
	83	24.3	21.43	8.355	0.835	0.515	1830	171	8.67	81.4	19.5	1.83	2.15
	73	21.5	21.24	8.295	0.740	0.455	1600	151	8.64	70.6	17.0	1.81	2.13
	68	20.0	21.13	8.270	0.685	0.430	1480	140	8.60	64.7	15.7	1.80	2.12
	62	18.3	20.99	8.240	0.615	0.400	1330	127	8.54	57.5	13.9	1.77	2.10
W21 x 57	57	16.7	21.06	6.555	0.650	0.405	1170	111	8.36	30.6	9.35	1.35	1.64
	50	14.7	20.83	6.530	0.535	0.380	984	94.5	8.18	24.9	7.64	1.30	1.60
	44	13.0	20.66	6.500	0.450	0.350	843	81.6	8.06	20.7	6.36	1.26	1.57

All shapes on these pages have parallel-faced flanges.

DATA SHEET D-30

WIDE FLANGE SHAPES

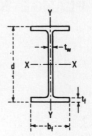

Theoretical Dimensions and Properties for **Designing**

Section Number	Weight per Foot	Area of Section	Depth of Section	Flange		Web Thickness	Axis X-X			Axis Y-Y			
				Width	Thickness		I_x	S_x	r_x	I_y	S_y	r_y	r_T
		A	d	b_f	t_f	t_w							
	lb	in.²	in.	in.	in.	in.	in.⁴	in.³	in.	in.⁴	in.³	in.	in.
W18 x 119	119	35.1	18.97	11.265	1.060	0.655	2190	231	7.90	253	44.9	2.69	3.02
	106	31.1	18.73	11.200	0.940	0.590	1910	204	7.84	220	39.4	2.66	3.00
	97	28.5	18.59	11.145	0.870	0.535	1750	188	7.82	201	36.1	2.65	2.99
	86	25.3	18.39	11.090	0.770	0.480	1530	166	7.77	175	31.6	2.63	2.97
	76	22.3	18.21	11.035	0.680	0.425	1330	146	7.73	152	27.6	2.61	2.95
W18 x 71	71	20.8	18.47	7.635	0.810	0.495	1170	127	7.50	60.3	15.8	1.70	1.98
	65	19.1	18.35	7.590	0.750	0.450	1070	117	7.49	54.8	14.4	1.69	1.97
	60	17.6	18.24	7.555	0.695	0.415	984	108	7.47	50.1	13.3	1.69	1.96
	55	16.2	18.11	7.530	0.630	0.390	890	98.3	7.41	44.9	11.9	1.67	1.95
	50	14.7	17.99	7.495	0.570	0.355	800	88.9	7.38	40.1	10.7	1.65	1.94
W18 x 46	46	13.5	18.06	6.060	0.605	0.360	712	78.8	7.25	22.5	7.43	1.29	1.54
	40	11.8	17.90	6.015	0.525	0.315	612	68.4	7.21	19.1	6.35	1.27	1.52
	35	10.3	17.70	6.000	0.425	0.300	510	57.6	7.04	15.3	5.12	1.22	1.49
W16 x 100	100	29.4	16.97	10.425	0.985	0.585	1490	175	7.10	186	35.7	2.52	2.81
	89	26.2	16.75	10.365	0.875	0.525	1300	155	7.05	163	31.4	2.49	2.79
	77	22.6	16.52	10.295	0.760	0.455	1110	134	7.00	138	26.9	2.47	2.77
	67	19.7	16.33	10.235	0.665	0.395	954	117	6.96	119	23.2	2.46	2.75
W16 x 57	57	16.8	16.43	7.120	0.715	0.430	758	92.2	6.72	43.1	12.1	1.60	1.86
	50	14.7	16.26	7.070	0.630	0.380	659	81.0	6.68	37.2	10.5	1.59	1.84
	45	13.3	16.13	7.035	0.565	0.345	586	72.7	6.65	32.8	9.34	1.57	1.83
	40	11.8	16.01	6.995	0.505	0.305	518	64.7	6.63	28.9	8.25	1.57	1.82
	36	10.6	15.86	6.985	0.430	0.295	448	56.5	6.51	24.5	7.00	1.52	1.79
W16 x 31	31	9.12	15.88	5.525	0.440	0.275	375	47.2	6.41	12.4	4.49	1.17	1.39
	26	7.68	15.69	5.500	0.345	0.250	301	38.4	6.26	9.59	3.49	1.12	1.36

All shapes on these pages have parallel-faced flanges.

DATA SHEET D-31

WIDE FLANGE SHAPES

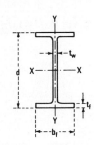

Theoretical Dimensions and Properties for **Designing**

Section Number	Weight per Foot	Area of Section	Depth of Section	Flange		Web Thick-ness	Axis X-X			Axis Y-Y			
				Width	Thick-ness		I_x	S_x	r_x	I_y	S_y	r_y	r_T
		A	d	b_f	t_f	t_w							
	lb	in.²	in.	in.	in.	in.	in.⁴	in.³	in.	in.⁴	in.³	in.	in.
W14 x	730*	215	22.42	17.890	4.910	3.070	14300	1280	8.17	4720	527	4.69	4.99
	665*	196	21.64	17.650	4.520	2.830	12400	1150	7.98	4170	472	4.62	4.92
	605*	178	20.92	17.415	4.160	2.595	10800	1040	7.80	3680	423	4.55	4.85
	550*	162	20.24	17.200	3.820	2.380	9430	931	7.63	3250	378	4.49	4.79
	500*	147	19.60	17.010	3.500	2.190	8210	838	7.48	2880	339	4.43	4.73
	455*	134	19.02	16.835	3.210	2.015	7190	756	7.33	2560	304	4.38	4.68
W14 x	426	125	18.67	16.695	3.035	1.875	6600	707	7.26	2360	283	4.34	4.64
	398	117	18.29	16.590	2.845	1.770	6000	656	7.16	2170	262	4.31	4.61
	370	109	17.92	16.475	2.660	1.655	5440	607	7.07	1990	241	4.27	4.57
	342	101	17.54	16.360	2.470	1.540	4900	559	6.98	1810	221	4.24	4.54
	311	91.4	17.12	16.230	2.260	1.410	4330	506	6.88	1610	199	4.20	4.50
	283	83.3	16.74	16.110	2.070	1.290	3840	459	6.79	1440	179	4.17	4.46
	257	75.6	16.38	15.995	1.890	1.175	3400	415	6.71	1290	161	4.13	4.43
	233	68.5	16.04	15.890	1.720	1.070	3010	375	6.63	1150	145	4.10	4.40
	211	62.0	15.72	15.800	1.560	0.980	2660	338	6.55	1030	130	4.07	4.37
	193	56.8	15.48	15.710	1.440	0.890	2400	310	6.50	931	119	4.05	4.35
	176	51.8	15.22	15.650	1.310	0.830	2140	281	6.43	838	107	4.02	4.32
	159	46.7	14.98	15.565	1.190	0.745	1900	254	6.38	748	96.2	4.00	4.30
	145	42.7	14.78	15.500	1.090	0.680	1710	232	6.33	677	87.3	3.98	4.28
W14 x	132	38.8	14.66	14.725	1.030	0.645	1530	209	6.28	548	74.5	3.76	4.05
	120	35.3	14.48	14.670	0.940	0.590	1380	190	6.24	495	67.5	3.74	4.04
	109	32.0	14.32	14.605	0.860	0.525	1240	173	6.22	447	61.2	3.73	4.02
	99	29.1	14.16	14.565	0.780	0.485	1110	157	6.17	402	55.2	3.71	4.00
	90	26.5	14.02	14.520	0.710	0.440	999	143	6.14	362	49.9	3.70	3.99
W14 x	82	24.1	14.31	10.130	0.855	0.510	882	123	6.05	148	29.3	2.48	2.74
	74	21.8	14.17	10.070	0.785	0.450	796	112	6.04	134	26.6	2.48	2.72
	68	20.0	14.04	10.035	0.720	0.415	723	103	6.01	121	24.2	2.46	2.71
	61	17.9	13.89	9.995	0.645	0.375	640	92.2	5.98	107	21.5	2.45	2.70
W14 x	53	15.6	13.92	8.060	0.660	0.370	541	77.8	5.89	57.7	14.3	1.92	2.15
	48	14.1	13.79	8.030	0.595	0.340	485	70.3	5.85	51.4	12.8	1.91	2.13
	43	12.6	13.66	7.995	0.530	0.305	428	62.7	5.82	45.2	11.3	1.89	2.12

*These shapes have a 1°-00′ (1.75%) flange slope. Flange thicknesses shown are average thicknesses. Properties shown are for a parallel flange section.

All other shapes on these pages have parallel-faced flanges.

DATA SHEET D-32

WIDE FLANGE SHAPES

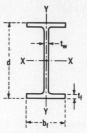

Theoretical Dimensions and Properties for **Designing**

Section Number	Weight per Foot	Area of Section	Depth of Section	Flange		Web Thickness	Axis X-X			Axis Y-Y			
				Width	Thickness		I_x	S_x	r_x	I_y	S_y	r_y	r_T
		A	d	b_f	t_f	t_w							
	lb	in.²	in.	in.	in.	in.	in.⁴	in.³	in.	in.⁴	in.³	in.	in.
W14 x 38		11.2	14.10	6.770	0.515	0.310	385	54.6	5.88	26.7	7.88	1.55	1.77
34		10.0	13.98	6.745	0.455	0.285	340	48.6	5.83	23.3	6.91	1.53	1.76
30		8.85	13.84	6.730	0.385	0.270	291	42.0	5.73	19.6	5.82	1.49	1.74
W14 x 26		7.69	13.91	5.025	0.420	0.255	245	35.3	5.65	8.91	3.54	1.08	1.28
22		6.49	13.74	5.000	0.335	0.230	199	29.0	5.54	7.00	2.80	1.04	1.25
W12 x 190		55.8	14.38	12.670	1.735	1.060	1890	263	5.82	589	93.0	3.25	3.50
170		50.0	14.03	12.570	1.560	0.960	1650	235	5.74	517	82.3	3.22	3.47
152		44.7	13.71	12.480	1.400	0.870	1430	209	5.66	454	72.8	3.19	3.44
136		39.9	13.41	12.400	1.250	0.790	1240	186	5.58	398	64.2	3.16	3.41
120		35.3	13.12	12.320	1.105	0.710	1070	163	5.51	345	56.0	3.13	3.38
106		31.2	12.89	12.220	0.990	0.610	933	145	5.47	301	49.3	3.11	3.36
96		28.2	12.71	12.160	0.900	0.550	833	131	5.44	270	44.4	3.09	3.34
87		25.6	12.53	12.125	0.810	0.515	740	118	5.38	241	39.7	3.07	3.32
79		23.2	12.38	12.080	0.735	0.470	662	107	5.34	216	35.8	3.05	3.31
72		21.1	12.25	12.040	0.670	0.430	597	97.4	5.31	195	32.4	3.04	3.29
65		19.1	12.12	12.000	0.605	0.390	533	87.9	5.28	174	29.1	3.02	3.28
W12 x 58		17.0	12.19	10.010	0.640	0.360	475	78.0	5.28	107	21.4	2.51	2.72
53		15.6	12.06	9.995	0.575	0.345	425	70.6	5.23	95.8	19.2	2.48	2.71
W12 x 50		14.7	12.19	8.080	0.640	0.370	394	64.7	5.18	56.3	13.9	1.96	2.17
45		13.2	12.06	8.045	0.575	0.335	350	58.1	5.15	50.0	12.4	1.94	2.15
40		11.8	11.94	8.005	0.515	0.295	310	51.9	5.13	44.1	11.0	1.93	2.14
W12 x 35		10.3	12.50	6.560	0.520	0.300	285	45.6	5.25	24.5	7.47	1.54	1.74
30		8.79	12.34	6.520	0.440	0.260	238	38.6	5.21	20.3	6.24	1.52	1.73
26		7.65	12.22	6.490	0.380	0.230	204	33.4	5.17	17.3	5.34	1.51	1.72
W12 x 22		6.48	12.31	4.030	0.425	0.260	156	25.4	4.91	4.66	2.31	0.848	1.02
19		5.57	12.16	4.005	0.350	0.235	130	21.3	4.82	3.76	1.88	0.822	0.997
16		4.71	11.99	3.990	0.265	0.220	103	17.1	4.67	2.82	1.41	0.773	0.963
14		4.16	11.91	3.970	0.225	0.200	88.6	14.9	4.62	2.36	1.19	0.753	0.946

All shapes on these pages have parallel-faced flanges.

DATA SHEET D-33

WIDE FLANGE SHAPES

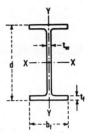

Theoretical Dimensions and Properties for **Designing**

Section Number	Weight per Foot	Area of Section	Depth of Section	Flange		Web Thickness	Axis X-X			Axis Y-Y			
				Width	Thickness		I_x	S_x	r_x	I_y	S_y	r_y	r_T
		A	d	b_f	t_f	t_w							
	lb	in.²	in.	in.	in.	in.	in.⁴	in.³	in.	in.⁴	in.³	in.	in.
W10 x 112		32.9	11.36	10.415	1.250	0.755	716	126	4.66	236	45.3	2.68	2.88
100		29.4	11.10	10.340	1.120	0.680	623	112	4.60	207	40.0	2.65	2.85
88		25.9	10.84	10.265	0.990	0.605	534	98.5	4.54	179	34.8	2.63	2.83
77		22.6	10.60	10.190	0.870	0.530	455	85.9	4.49	154	30.1	2.60	2.80
68		20.0	10.40	10.130	0.770	0.470	394	75.7	4.44	134	26.4	2.59	2.79
60		17.6	10.22	10.080	0.680	0.420	341	66.7	4.39	116	23.0	2.57	2.77
54		15.8	10.09	10.030	0.615	0.370	303	60.0	4.37	103	20.6	2.56	2.75
49		14.4	9.98	10.000	0.560	0.340	272	54.6	4.35	93.4	18.7	2.54	2.74
W10 x 45		13.3	10.10	8.020	0.620	0.350	248	49.1	4.33	53.4	13.3	2.01	2.18
39		11.5	9.92	7.985	0.530	0.315	209	42.1	4.27	45.0	11.3	1.98	2.16
33		9.71	9.73	7.960	0.435	0.290	170	35.0	4.19	36.6	9.20	1.94	2.14
W10 x 30		8.84	10.47	5.810	0.510	0.300	170	32.4	4.38	16.7	5.75	1.37	1.55
26		7.61	10.33	5.770	0.440	0.260	144	27.9	4.35	14.1	4.89	1.36	1.54
22		6.49	10.17	5.750	0.360	0.240	118	23.2	4.27	11.4	3.97	1.33	1.51
W10 x 19		5.62	10.24	4.020	0.395	0.250	96.3	18.8	4.14	4.29	2.14	0.874	1.03
17		4.99	10.11	4.010	0.330	0.240	81.9	16.2	4.05	3.56	1.78	0.845	1.01
15		4.41	9.99	4.000	0.270	0.230	68.9	13.8	3.95	2.89	1.45	0.810	0.987
12		3.54	9.87	3.960	0.210	0.190	53.8	10.9	3.90	2.18	1.10	0.785	0.965
W8 x 67		19.7	9.00	8.280	0.935	0.570	272	60.4	3.72	88.6	21.4	2.12	2.28
58		17.1	8.75	8.220	0.810	0.510	228	52.0	3.65	75.1	18.3	2.10	2.26
48		14.1	8.50	8.110	0.685	0.400	184	43.3	3.61	60.9	15.0	2.08	2.23
40		11.7	8.25	8.070	0.560	0.360	146	35.5	3.53	49.1	12.2	2.04	2.21
35		10.3	8.12	8.020	0.495	0.310	127	31.2	3.51	42.6	10.6	2.03	2.20
31		9.13	8.00	7.995	0.435	0.285	110	27.5	3.47	37.1	9.27	2.02	2.18
W8 x 28		8.25	8.06	6.535	0.465	0.285	98.0	24.3	3.45	21.7	6.63	1.62	1.77
24		7.08	7.93	6.495	0.400	0.245	82.8	20.9	3.42	18.3	5.63	1.61	1.76
W8 x 21		6.16	8.28	5.270	0.400	0.250	75.3	18.2	3.49	9.77	3.71	1.26	1.41
18		5.26	8.14	5.250	0.330	0.230	61.9	15.2	3.43	7.97	3.04	1.23	1.39
W8 x 15		4.44	8.11	4.015	0.315	0.245	48.0	11.8	3.29	3.41	1.70	0.876	1.03
13		3.84	7.99	4.000	0.255	0.230	39.6	9.91	3.21	2.73	1.37	0.843	1.01
10		2.96	7.89	3.940	0.205	0.170	30.8	7.81	3.22	2.09	1.06	0.841	0.994

All shapes on these pages have parallel-faced flanges.

DATA SHEET D-34

WIDE FLANGE SHAPES

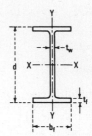

Theoretical Dimensions and Properties for **Designing**

Section Number	Weight per Foot	Area of Section A	Depth of Section d	Flange Width b_f	Flange Thickness t_f	Web Thickness t_w	Axis X-X I_x	Axis X-X S_x	Axis X-X r_x	Axis Y-Y I_y	Axis Y-Y S_y	Axis Y-Y r_y	r_T
	lb	in.²	in.	in.	in.	in.	in.⁴	in.³	in.	in.⁴	in.³	in.	in.
W6 x 25	25	7.34	6.38	6.080	0.455	0.320	53.4	16.7	2.70	17.1	5.61	1.52	1.66
20	20	5.87	6.20	6.020	0.365	0.260	41.4	13.4	2.66	13.3	4.41	1.50	1.64
15	15	4.43	5.99	5.990	0.260	0.230	29.1	9.72	2.56	9.32	3.11	1.45	1.61
W6 x 16	16	4.74	6.28	4.030	0.405	0.260	32.1	10.2	2.60	4.43	2.20	0.967	1.08
12	12	3.55	6.03	4.000	0.280	0.230	22.1	7.31	2.49	2.99	1.50	0.918	1.05
9	9	2.68	5.90	3.940	0.215	0.170	16.4	5.56	2.47	2.20	1.11	0.905	1.03
W5 x 19	19	5.54	5.15	5.030	0.430	0.270	26.2	10.2	2.17	9.13	3.63	1.28	1.38
16	16	4.68	5.01	5.000	0.360	0.240	21.3	8.51	2.13	7.51	3.00	1.27	1.37
†W4 x 13	13	3.83	4.16	4.060	0.345	0.280	11.3	5.46	1.72	3.86	1.90	1.00	1.10

MISCELLANEOUS SHAPES

Theoretical Dimensions and Properties for **Designing**

Section Number	Weight per Foot	Area of Section A	Depth of Section d	Flange Width b_f	Flange Thickness t_f	Web Thickness t_w	Axis X-X I_x	Axis X-X S_x	Axis X-X r_x	Axis Y-Y I_y	Axis Y-Y S_y	Axis Y-Y r_y	r_T
	lb	in.²	in.	in.	in.	in.	in.⁴	in.³	in.	in.⁴	in.³	in.	in.
†M5 x 18.9	18.9	5.55	5.00	5.003	0.416	0.316	24.1	9.63	2.08	7.86	3.14	1.19	1.32

†W4 x 13 and M5 x 18.9 have flange slopes of 2.0 and 7.4 pct respectively. Flange thickness shown for these sections are average thicknesses. Properties are the same as if flanges were parallel.

All other shapes on these pages have parallel-faced flanges.

DATA SHEET D-35

ELASTIC SECTION MODULUS

These tables are in accordance with the AISC Specification for the Design, Fabrication & Erection of Structural Steel for Buildings, Supplement No. 3 (1974).

Sections shown in **bold face** are "Weight Economy Sections."

S_x in.³	Shape	F'_y ksi	S_x in.³	Shape	F'_y ksi	S_x in.³	Shape	F'_y ksi
1110	**W36x300**	**	439	W36x135	**	213	W27x84	**
			415	W14x257	**	209	W14x132	**
1030	**W36x280**	**	414	W24x162	**	209	W12x152	**
			411	W27x146	**	204	W18x106	**
953	**W36x260**	**						
			406	**W33x130**	**	**196**	**W24x84**	**
895	**W36x245**	**	380	W30x132	**	192	W21x93	**
			375	W14x233	**	190	W14x120	**
837	**W36x230**	**	371	W24x146	**	188	W18x97	**
829	W33x241	**				186	W12x136	**
			359	**W33x118**	**			
757	**W33x221**	**	355	W30x124	**	**176**	**W24x76**	**
			338	W14x211	**	175	W16x100	**
719	**W36x210**	**				173	W14x109	58.6
707	W14x426	**	**329**	**W30x116**	**	171	W21x83	**
			329	W24x131	**	166	W18x86	**
684	**W33x201**	**	329	W21x147	**	163	W12x120	**
			310	W14x193	**	157	W14x99	48.5
664	**W36x194**	**				155	W16x89	**
663	W30x211	**	**299**	**W30x108**	**			
656	W14x398	**	299	W27x114	**	**154**	**W24x68**	**
			295	W21x132	**	151	W21x73	**
623	**W36x182**	**	291	W24x117	**	146	W18x76	**
607	W14x370	**	281	W14x176	**	145	W12x106	**
598	W30x191	**	273	W21x122	**	143	W14x90	40.4
580	**W36x170**	**						
559	W14x342	**	**269**	**W30x99**	**	**140**	**W21x68**	**
			267	W27x102	**	134	W16x77	**
542	**W36x160**	**	263	W12x190	**			
539	W30x173	**	258	W24x104	58.5	**131**	**W24x62**	**
506	W14x311	**	254	W14x159	**	131	W12x96	**
			249	W21x111	**			
504	**W36x150**	**				**127**	**W21x62**	**
502	W27x178	**	**243**	**W27x94**	**	127	W18x71	**
487	W33x152	**	235	W12x170	**	126	W10x112	**
459	W14x283	**	232	W14x145	**	123	W14x82	**
455	W27x161	**	231	W18x119	**	118	W12x87	**
			227	W21x101	**	117	W18x65	**
448	**W33x141**	**				117	W16x67	**
			222	**W24x94**	**			

**Theoretical maximum yield stress exceeds 60 ksi.

DATA SHEET D-36

ELASTIC SECTION MODULUS

These tables are in accordance with the AISC Specification for the
Design, Fabrication & Erection of Structural Steel for Buildings,
Supplement No. 3 (1974).

Sections shown in **bold face** are "Weight Economy Sections."

S_x in.³	Shape	F'_y ksi	S_x in.³	Shape	F'_y ksi	S_x in.³	Shape	F'_y ksi
114	**W24x55**	**	**57.6**	**W18x35**	**	**21.3**	**W12x19**	**
112	W14x74	**	56.5	W16x36	**	20.9	W8x24	**
112	W10x100	**	54.6	W14x38	**			
111	W21x57	**	54.6	W10x49	53.0	**18.8**	**W10x19**	**
108	W18x60	**	52.0	W8x58	**	18.2	W8x21	**
107	W12x79	**	51.9	W12x40	**			
103	W14x68	**	49.1	W10x45	**	**17.1**	**W12x16**	**
98.5	W10x88	**				16.7	W6x25	**
			48.6	**W14x34**	**	16.2	W10x17	**
98.3	**W18x55**	**				15.2	W8x18	**
97.4	W12x72	52.3	**47.2**	**W16x31**	**			
			45.6	W12x35	**	**14.9**	**W12x14**	54.3
94.5	**W21x50**	**	43.3	W8x48	**	13.8	W10x15	**
92.2	W16x57	**	42.1	W10x39	**	13.4	W6x20	**
92.2	W14x61	**				11.8	W8x15	**
			42.0	**W14x30**	55.3			
88.9	**W18x50**	**				**10.9**	**W10x12**	47.5
87.9	W12x65	43.0	**38.6**	**W12x30**	**	10.2	W6x16	**
85.9	W10x77	**				10.2	W5x19	**
			38.4	**W16x26**	**	9.91	W8x13	**
81.6	**W21x44**	**	35.5	W8x40	**	9.72	W6x15	31.8
81.0	W16x50	**				9.63	M5x18.9	**
78.8	W18x46	**	**35.3**	**W14x26**	**	8.51	W5x16	**
78.0	W12x58	**	35.0	W10x33	50.5			
77.8	W14x53	**				**7.81**	**W8x10**	45.8
75.7	W10x68	**	**33.4**	**W12x26**	57.9	7.31	W6x12	**
72.7	W16x45	**	32.4	W10x30	**			
70.6	W12x53	55.9	31.2	W8x35	**	**5.56**	**W6x9**	50.3
70.3	W14x48	**				5.46	W4x13	**
			29.0	**W14x22**	**			
68.4	**W18x40**	**	27.9	W10x26	**			
66.7	W10x60	**	27.5	W8x31	50.0			
64.7	**W16x40**	**	**25.4**	**W12x22**	**			
64.7	W12x50	**	24.3	W8x28	**			
62.7	W14x43	**						
60.4	W8x67	**	**23.2**	**W10x22**	**			
60.0	W10x54	**						
58.1	W12x45	**						

**Theoretical maximum yield stress exceeds 60 ksi.

DATA SHEET D-37

RECTANGULAR SECTION NOMOGRAPH

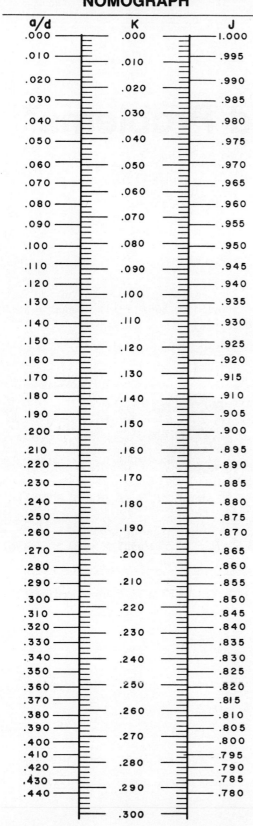

a/d	K	J

LIMITING VALUES FOR RATIO OF a/d

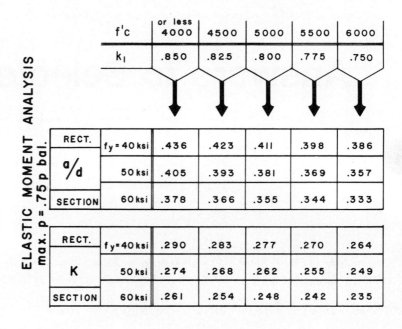

ELASTIC MOMENT ANALYSIS — max. p = .75 p bal.

f'_c	or less 4000	4500	5000	5500	6000
k_1	.850	.825	.800	.775	.750

RECT. a/d SECTION		4000	4500	5000	5500	6000
	f_y = 40 ksi	.436	.423	.411	.398	.386
	50 ksi	.405	.393	.381	.369	.357
	60 ksi	.378	.366	.355	.344	.333

RECT. K SECTION		4000	4500	5000	5500	6000
	f_y = 40 ksi	.290	.283	.277	.270	.264
	50 ksi	.274	.268	.262	.255	.249
	60 ksi	.261	.254	.248	.242	.235

BASIS FOR VALUES ON THE NOMOGRAPH

For a rectangular section:

$$\overline{M} = .85f'_c abjd$$

$$jd = d - a/2, \quad a = 2d(l - j)$$

$$\therefore \overline{M} = .85f'_c b(2d)(l - j)(jd)$$

$$\overline{M} = 1.70f'_c bj(l - j)d^2$$

$$K = 1.70j(l - j)$$

Therefore,

$$\overline{M} = Kf'_c bd^2 \rightarrow K = \frac{\overline{M}}{f'_c bd^2}$$

$$\overline{M} = A_s f_y jd \rightarrow A_s = \frac{\overline{M}}{f_y jd}$$

Answers to Selected Problems*

Problem 2–1: (1) $H = 866\#$, $V = 500\#$; $H = 707\#$, $V = 707\#$; $H = 500\#$, $V = 866\#$; (3) Horizontal $= 1000\#$, Vertical $= 500\#$; (5) (a) $R = 6000\#$, (b) $R = 600\#$, (c) $R = 1918\#$, (d) $R = 6124\#$.

Problem 2–2: (B) $R = 18.66^K$; (D) $R = 500\#$; (F) $R = 1902\#$.

Problem 2–3: (I) (A) $R = 0$; (B) $118\#$; (C) $223\#$; (III) $x = 19\#$, $y = 28\#$.

✗**Problem 3–1:** (B) $R_1 = 4.5^K$, $R_2 = 3.5^K$; (D) $R_1 = 6.67^K$, $R_2 = 15.33^K$; (F) $R_1 - R_2 = 6^K$; (H) $R_1 = 5.6^K$, $R_2 = 3.4^K$; (K) $R_1 = 6.38^K$, $R_2 = 8.62^K$; (M) $R_1 = 11.6^K$, $R_2 = 9.4^K$; (O) $R_1 = 11.5^K$, $R_2 = 3.5^K$.

Problem 3–2: (A) $R_1 = 17.3^K$, $R_2 = 36.7^K$; (C) $R_1 = 14^K$, $R_2 = 24^K$; (E) $R_1 = 3.3^K$, $R_2 = 14^K$, $R_3 = 12.7^K$.

Problem 3–3: (B) $R_1 = 2^K$ (down), $R_2 = 24^K$; (D) $R_1 = 25.9^K$, $R_2 = 3.9^K$ (down), $R_3 = 4^K$ (right); (F) $R_1 = .5^K$ (down), $R_2 = 28.5^K$.

Problem 3–4: (B) $R = 3^K$, $V = 2.2^K$, $H = 3^K$ (left); (D) $R_H = R_V = 13.5^K$, $H = 13.5^K$ (right), $V = 31.5^K$; (F) $R_H = 36^K$ (left), $R_V = 20.8^K$, $H = 36^K$ (right), $V = 9.2^K$.

Problem 3–5: (A) 14^k (down), $6.29'$ left of A; (C) $141.4\#$ at $45°$, $.8'$ left of A; (E) 15.2^K at $76.7°$ from horizontal, $15.1'$ right of "A"; (G) 42.6^K at $50.1°$ from horizontal, $14.7'$ left of "A".

Problem 4–1: (B) $\bar{x} = 7.29'$, $\bar{y} = 3.51'$, $\bar{z} = 5'$; (D) $\bar{x} = 5'$, $\bar{y} = 2.63'$, $\bar{z} = 2.55'$.

Problem 4–2: (II) $P = 5.67^K$; (IV) $P = 90.6\#/'$.

Problem 4–3: (A) $\bar{x} = 6.78''$; (C) $\bar{x} = 2.84''$; (E) $\bar{x} = 5.20''$.

Problem 4–4: (A) $\bar{x} = 5.75''$; (C) $\bar{x} = 9.6''$; (E) $\bar{x} = 3.65''$.

*Note: Answers that are essentially graphic are not provided.

Problem 4–5: (I) Steel pipe: $f = 9302$ p.s.i.; Wood post: $f = 653$ p.s.i.; plane x: $f = 9302$ p.s.i.; plane y: $f = 333$ p.s.i.; plane z: $f = 3000$/p.s.f.; (III) Height $= 4800'$ and $240'$.

Problem 5–4: (B) 11.2^K Comp.; $Cl = 9^K$; (D) 30^K Tension.

Problem 6–1: (1) *Truss #2:* $W_1 = 9^K$ C, $W_2 = 5^K$ T, $W_3 = 12^K$ C, $W_4 = 15^K$ T, $W_5 = 24^K$ C, $W_6 = 40^K$ T, $W_7 = 60^K$ C, $W_8 = 35^K$ T, $W_9 = 12^K$ C, $W_{10} = 15^K$ C, $B_1 = 0$, $B_2 = 8^K$ C, $B_3 = 40^K$ C, $B_4 = 40^K$ C, $T_1 = 4^K$ C, $T_2 = 4^K$ C, $T_3 = 8^K$ T, $T_4 = 12^K$ T, $T_5 = 12^K$ T; (2) *Truss #2:* $a = 26.8^K$ T, $z = 24^K$ C, $c = 25.5^K$ T, $x = 42^K$ C, $y = 24^K$ T, $v = 72^K$ C, $e = 42.4^K$ T, $g = 101.9^K$ F, $f = 108^K$ C, $j = 5^K$ T, $u = 112^K$ C, $t = 72^K$ T, $s = 120^K$ C, $1 = 10^K$ T, $r = 76^K$ T, $o = 84^K$ T, $n = 15^K$ T, $p = 132^K$ C.

Problem 6–2: (I) $a = 16^K$ T, $b = 20^K$ C, $c = 12^K$ C, $d = 16^K$ T, $e = 40^K$ T, $f = 48^K$ C, $g = 36^K$ C, $h = 48^K$ T, $j = 60^K$ T, $k = 96^K$ C; (III) $A = 134.2^K$ C, $B = 36.1^K$ T, $C = 100^K$ T.

Problem 6–3: (II) $A = 96^K$ C, $B = 26^K$ C, $C = 96^K$ T; (IV) $a = 0$, $b = 0$, $c = 4.2^K$ C, $d = 3^K$ T, $e = 3^K$ T, $f = 3^K$ C, $g = 12.7^K$ C, $h = 12.7^K$ T, $j = 6^K$ C.

Problem 7–2: (1) $999'6''$; (3) (a) $p = 12.85^K$, $F = 2.70^K$; (b) $P = 21.3^K$, $F = 7.67^K$.

Problem 7–3: (1) (a) (A) $f = 11.6$ k.s.i.; (B) $f = 4.35$ k.s.i.; (b) $.60''$; (2) (a) 103#; (b) $95.4°$F.

Problem 7–4: (A) 1175.1 in.^{4}; (C) 4065.7 in.^{4}; (E) 371.0 in.^{4}.

Problem 7–5: (A) 314.2 in.^{4}; (C) 327.0 in.^{4}; (E) 465.7 in.^{4}.

Problem 7–6: (B) $120'^K$; (D) $36'^K$; (F) $87.8'^K$; (H) $154.2'^K$.

Problem 7–7: (I) $b = 12.5''$; (II) $W = 71^K$; (III) $P = 12^K$.

Problem 7–8: (I) *Sec. B:* $I = 464$ in.^{4}, $W = 7.73^K$; *Sec. D:* $I = 256$ in.^{4}, $W = 5.33^K$; (II) $P = 1.6^K$; (III) (A) $I_1 = 226.8$ in.^{4}, $I_2 = 302.4$ in.^{4}; (C) $I_1 = 302.4$ in.^{4}, $I_2 = 403.2$ in.^{4}.

Problem 9–1: (A) $V_L = V_R = 10^K$, $H_L = H_R = 6^K$; (C) $V_L = 1.6^K$, $V_R = 6.8^K$, $H_L = 5.9^K$ (left), $H_R = 4.5^K$ (left).

Problem 9–2: (A) $V_L = 7^K$, $V_R = 13^K$, $H_L = 1^K$, $H_R = 7^K$; (C) $V_L = 5.4^K$, $V_R = 34.6^K$, $H_R = 23.1^K$, $H_L = 11.9^K$ (left).

Problem 9–3: (II) $A = 4^K$, $B = 22^K$, $C = 20^K$, $D = 20^K$; (IV) $A = 2.9^K$, $B = 22.1^K$, $C = 7.9^K$, $D = 2.9^K$.

Problem 9–4: (1) $V_L = V_R = 12^K$, $H_L = H_R = 16^K$; (3) $V_L = 16^K$, $V_R = 8^K$, $H_L = H_R = 16^K$; (5) $V_L = 14^K$, $V_R = 10^K$, $H_L = H_R = 16^K$; (7) $V_L = 14^K$, $V_R = 10^K$, $H_L = H_R = 12^K$; (Bridge) *Part I:* $B = 100^K$, $V = 200^K$, $R = 141.4^K$, $x = 20'$, $y = 35'$, $z = 45'$.

Problem 9–5: (A) (1) $z = 56.6^K$, $T = 80^K$, $V_L = 40^K$, $V_R = 40^K$, $h_1 = 20'$, $h_2 = 35'$, $h_3 = 45'$; (C) $V = 72^K$, $H = 72^K$, $h_1 = 20'$, $h_2 = 30'$, $h_3 = 36.7'$.

Problem 9–6: (B) $A_H = 40^K$, $A_V = 45^K$, $B_H = 40^K$, $B_V = 30^K$, $x = 11.25'$, $y = 22.5'$; (D) $A_H = B_H = 40.7^K$, $A_V = 49^K$, $B_V = 61^K$, $h_1 = 40.8'$, $h_2 = 45.2'$.

Problem 9–7: (1) $V_L = 16^K$, $V_R = 8^K$, $H_L = H_R = 16^K$; (3) $V_L = 12^K$, $V_R = 12^K$, $H_L = H_R = 16^K$; (5) $V_L = 14^K$, $V_R = 10^K$, $H_L = H_R = 15^K$, $x = 5.6'$, $y = 9.6'$.

Problem 9–8: (A) $A_V = 30^K$, $B_V = 45^K$, $A_H = B_H = 25^K$, $x = 12'$, $y = 18'$; (C) $V_L = 100^K$, $V_R = 40^K$, $H_L = H_R = 80^K$.

Problem 9–9: (A) $V_L = 80^K$, $V_R = 60^K$, $H = 100^K$, $h_1 = 16'$, $h_2 = 22'$, $h_3 = 20'$, $h_4 = 12'$; (C) $V_L = V_R = 400^K$, $H = 500^K$, $h_1 = 6.6'$, $h_2 = 10.8'$, $h_3 = 13.3'$, $h_4 = 14.6'$.

Problem 10–1: *Note: Only max. f_v given:* (B) 36.5 p.s.i.; (D) 63.7 p.s.i.; (F) 75.6 p.s.i.

Problem 10–2: (I) (A) $f = .73$ k.s.i.; (B) $f_v = 86$ p.s.i.; (III) $L = 18.7'$.

Problem 10–3: (I) O.K.; (III) (a) $W = 3188\#$, (b) $W = 3042\#$.

Problem 10–4: (A) Bending $f = 15.1$ k.s.i., Shear $f_v = 1.9$ k.s.i.; (C) Bending $f = 6.37$ k.s.i., Shear $f_v = 1.18$ k.s.i.

Problem 10–5: (I) (A) $f = 2.02$ k.s.i. (at cantilever); (III) Total $= 197$ conn.

Problem 11–1: (1) (a) $W = 548\#$, $W = 890\#$, $W = 1170\#$, $W = 1520\#$; (b) N.G. (Bending); (c) $2'' \times 12''$; (3) (a) $D = 24''$; (b) O.K.

Problem 11–2: (1) 2×12's overstressed, others O.K.

Problem 11–3: (2) B1 $= 5\frac{1}{8}'' \times 15''$, B2 $= 5\frac{1}{8}'' \times 15''$, B3 $= 5\frac{1}{8}'' \times 24''$, B4 $= 5\frac{1}{8}'' \times 16\frac{1}{2}''$, B5 $= 5\frac{1}{8}'' \times 25\frac{1}{2}''$, B6 $= 5\frac{1}{8}'' \times 24''$.

Problem 12–1: (A) *Section 1:* Timber $f = 1.78$ k.s.i.: Alum. $f = 9.66$ k.s.i.; *Section 2:* Timber $f = 2.71$ k.s.i.: Alum. $f = 8.76$ k.s.i.; *Section 3:* Timber $f = 1.16$ k.s.i.: Alum. $f = 6.26$ k.s.i.; (C) (1) *Section 1:* $W = 18.8^K$; *Section 2:* $W = 12.4^K$; *Section 3:* $W = 29^K$; (2) $W = 9.4^K$.

Problem 12–2: (I) $W = 11.7^K$; (III) (A): $P_1 = 14.3^K$: $P_2 = 48^K$; (B) 15.7'.

Problem 12–3: (II) $W = 13.9^K$; (III) Wood $f = 1.09$ k.s.i., Steel $f = 14.6$ k.s.i., Aluminum $f = 8.75$ k.s.i..

Problem 12–4: (I) (1) Cantilevers: left side, $L = 9.5'$; right side, $L = 8.1'$; (2) $.37'' \times 11.4''$; (III) $P = 23.5^K$.

Problem 12–5: (I) $.27'' \times 10.4''$; (II) $L = 4.8'$; $.13''$ thick.

Problem 13–1: (A) (1) $x = 6.7''$; (2) $I = 4268$ in.^{4}; (3) $78.7'^K$, steel; (4) $f_c = 1.48$ k.s.i.; (C) (1) $x = 6.27''$; (2) $I = 2837.5$ in.^{4}; (3) $50.4'^K$, steel; (4) $f_c = 1.40$ k.s.i.; (E) (1) $x = 8.2''$; (2) $I = 4080$ in.^{4}; (3) $74.6'^K$, concrete; (4) $f_s = 19$ k.s.i..

Problem 13–2: (I) (A) $x = 7.5''$, $A_s = 4$ in.^{2}, $M = 104'^K$; (B) $x = 7.5''$, $A_s = 4.82$ in.^{2}, $M = 129.7'^K$, (C) $x = 7.5''$, $A_s = 3.4$ in.^{2}, $M = 84.3'^K$; (D) $x = 9.2''$, $A_s = 2.88$ in.^{2}, $M = 88.3'^K$; (E) $x = 8''$, $A_s = 2.8$ in.^{2}, $M = 73.1'^K$; (F) $x = 5.4''$, $A_s = 4.1$ in.^{2}, $M = 80'^K$; (III) (1) $x = 7.2''$, $d = 20.2''$; (2) $M = 96.3'^K$.

Problem 13–3: (II) (1) $x = 8.5''$; (2) $I = 13058$ in.^{4}; (3) $M = 230.4'^K$; (4) $W = 26.1^K$; (5) overreinforced.

Problem 13–4: (I) $d = 24''$, $A_s = 4.73$ in.^{2}; (III) $A_s = .62$ in.^{2}.

Problem 14–1: (1) (a) $f = 22.4$ k.s.i.; (b) $f = 16.6$ k.s.i.; (2) (a) $W = 82.3^K$; (b) $L = 36.6'$.

Problem 14–2: (I) (1) $M = 186'^K$; (2) $W = 62^K$; (3) W21 $\times$ 50; (III) (1) W36 $\times$ 182; (2) $f_c = 1.06$ k.s.i., $f_s = 24$ k.s.i.

Problem 14–3: (1) B1 = W18 $\times$ 50, B2 = W24 $\times$ 84, B3 = W27 $\times$ 84, B4 = W21 $\times$ 62, B5 = W21 $\times$ 44, B6 = W24 $\times$ 55, B7 = W12 $\times$ 16, B8 = W12 $\times$ 19, B9 = W14 $\times$ 22.

Problem 15–1: (A) (1) $76.6'^K$; (2) $167.2'^K$; (3) $227.9'^K$; (4) $227.9'^K$; (C) $136.5'^K$; (E) $296.7'^K$.

Problem 15–2: (B) (1) $132.9'^K$; (2) $216.3'^K$; (3) $350.7'^K$; (4) $402'^K$; (D) S.L.L. $= 62.4$ #/ft^2.

Problem 15–3: (A) $A_s = 6.66$ in.^{2}; (C) $A_s = 6.8$ in.^{2}.

Problem 16–1: (B) (2) $\Delta = 1.48''$; (D) (2) $\Delta = .60''$.

Problem 16–2: (I) (A) $EI\Delta = 6520.5$ k.ft^3; (B) $EI\Delta = 8618.7$ k.ft^3; (C) $EI\Delta = 17WL^3/1536$.

Problem 16–3: (A) F.E.M. $= 48'^K$; (C) F.E.M. $= 5PL/27$; (E) F.E.M. (left) $= 181.3'^K$, F.E.M. (right) $= 127.3'^K$; (G) F.E.M. $= 132'^K$; (J) F.E.M. $= 36'^K$ (Positive).

Problem 17–1: (I) (A) (1) $R_1 = 8^K$, $R_2 = 92^K$, $R_3 = 8^K$; (3) W21 $\times$ 62; (4) W24 $\times$ 68; (B) $R_1 = 14^K$, $R_2 = 68^K$, $R_3 = 14^K$; (3) W21 $\times$ 44; (4) W18 $\times$ 50.

Problem 17–2: (II) $A = 5^K$, $B = 40^K$, $C = 5^K$; (IV) $A = 9.9^K$, $B = 16.2^K$, $C = 9.9^K$.

Problem 17–3: See problem sheet for answers.

Problem 17–4: (I) $A = 30^K$, $B = 67.5^K$, $C = 37.5^K$; (III) $A = 12^K$, $B = 25.1^K$, $C = 14.9^K$.

Problem 18–1: (II) $P_{CR} = 4.3^K$, $L = 2.4'$; (IV) $P = 16.1^K$.

Problem 18–2: (I) (A) $6.25'$; (B) 2.72^K, 6.13^K, 9.17^K, 9.17^K.

Problem 18–3: (II) (A) 87.8^K; (B) 19.5^K; (IV) 205.1^K.

Problem 19–1: (I) f (Comp.) $= 25.75$ k.s.i., f (Tens.) $= 13.75$ k.s.i.; (II) (A) $H_L = H_R = 20^K$, $V_L = V_R = 30^K$; (B) f (Comp.) $= 24.87$ k.s.i., f (Tens.) $= 22.38$ k.s.i..

Problem 19–2: (A) (1) 26^K; (2) 4.56 k.s.i.; (B) (1) (a) 28.2^K; (b) 22^k.

Problem 19–3: (I) (A) $f = 5.83$ k.s.i. (Tens.); (B) $f = 5.37$ k.s.i. (Comp.); (C) $\Delta = .75''$ (down); (III) (A) 33.3^K.

Problem 19–4: (A) (1) 90^K; (2) 6.4^k; (3) $f_T = 533$ p.s.i. (Comp.), $f_B = 717$ p.s.i. (Comp.); (4) $\Delta = .26''$ (down).

Problem 19–5: *Section A:* (A) $F = 76.8^K$, $f_B = 1.07$ k.s.i. (Comp.), f (ends) $= .533$ k.s.i. (Comp.); (B) $W = 4.8^K$, $f = .533$ k.s.i.; (C) $W = 9.6^K$, $f_T = 1.07$ k.s.i. (Comp.); (D) $\Delta = .25''$ (up); (E) $\Delta = .25''$ (down).

Problem 19–6: (I) (A) 8.5^K; (B) $f_T = 176$ p.s.i. (Tens.); (C) $\Delta = .17''$ (up).

Problem 20–1: (I) Sections (left to right): $5\frac{1}{8}'' \times 24''$, $5\frac{1}{8}'' \times 15''$, $5\frac{1}{8}'' \times 27''$, $5\frac{1}{8}'' \times 19\frac{1}{2}''$; (III) (B) Sections (left to right): W18 $\times$ 35, W21 $\times$ 44, W18 $\times$ 35.

Problem 20–2: (2) (A) (b) W21 $\times$ 111; (B) (a) Midspan: $f = 22.7$ k.s.i., Supports: $f = 23.3$ k.s.i.; (b) $x = 5.65'$, $y = 3.78'$.

Problem 20–3: (2) (a) W33 $\times$ 130; (b) W27 $\times$ 146; (3) Midspan: $f = 23.5$ k.s.i., Cut-off: $f = 22.5$ k.s.i..

Problem 20–4: (1) (a) From right to left: $M_1 = 24'^K$, $M_2 = 36'^K$, $M_3 = 27'^K$, $M_4 = 36'^K$, $M_5 = 54'^K$, $M_6 = 45'^K$, $M_7 = 37.5'^K$, $M_8 = 22.5'^K$; (b) Sec. A: $f = 1895$ p.s.i.; Sec. B: $f = 2040$ p.s.i.; Sec. C: $f = 2040$ p.s.i..

Index

360